Agriculture Utilization of Urban and Industrial By-Products

Related Society Publications

Sewage Sludge: Land Utilization and the Environment

Utilization, Treatment and Disposal of Waste on Land

For more information on these titles, please contact the SSSA Headquarters Office; Attn: Marketing; 677 South Segoe Road; Madison, WI 53711-1086. Phone: (608) 273-8080 ext. 322. Fax: (608) 273-2021.

Agriculture Utilization of Urban and Industrial By-Products

Proceedings of a symposium sponsored by Divisions S-6 and S-7 of the Soil Science Society of America and A-5 of the American Society of Agronomy in Cincinnati, Ohio, 7–12 Nov. 1993.

Editors
D.L. Karlen, R.J. Wright, and W.O. Kemper

Organizing Committee
D.L. Karlan and W.D. Kemper

Editorial Committee
D.L. Karlen, R.J. Wright, and W.D. Kemper

Editor-in-Chief ASA
Gary A. Peterson

Editor-in-Chief CSSA
P. S. Baenziger

Editor-in-Chief SSSA
Robert J. Luxmoore

Managing Editor
J.M. Bartels

ASA Special Publication Number 58

American Society of Agronomy, Inc.
Crop Science Society of America, Inc.
Soil Science Society of America, Inc.
Madison, Wisconsin, USA

1995

Cover Design: D.L. Karlan

American Society of Agronomy, Inc.
Crop Science Society of America, Inc.
Soil Science Society of America, Inc.
677 South Segoe Road, Madison, WI 53711 USA

Library of Congress Cataloging-in-Publication Data

Agricultural utilization of urban and industrial by-products : proceed-
ings of a symposium sponsored by Divisions S-6 and S-7 of the Soil
Science Society of America and A-5 of the American Society of
Agronomy in Cincinnati, Ohio, 7–12 November 1993 / editors, D.L.
Karlen, R.J. Wright, and W.D. Kemper ; organizing committee, D.L.
Karlen and W.D. Kemper ; managing editor, J.M. Bartels . . . [et al.].
        p.   cm. — (ASA special publication ; no. 58)
    Includes bibliographical references.
    ISBN 0-89118-123-7
    1. Sewage sludge as fertilizer—Congresses.   2. Factory and trade
waste as fertilizer—Congresses.   I. Karlen, D. L. (Douglas L.)
II. Wright, R. J.   III. Kemper, William Doral, 1928-   .
IV. Soil Science Society of America. Division S-6.   V. Soil Science
Society of America. Division S-7.   VI. American Society of Agrono-
my. Division A-5.   VII. Series.
S1.A453   no. 58
[S657]
630 s—dc20   94-44326
[362.72′8]   CIP

Printed in the United States of America

CONTENTS

FOREWORD

Technological advances in the twentieth century have mechanized and improved the efficiency of food and fiber production. As a result, people have been freed from the burden of producing these basic necessities and have congregated into cities, to form a more sophisticated society that has developed many goods to make life less burdensome and to enhance communication, movement and leisure activities.

The concentration of people into cities and the manufacturing of todays goods and development of new services have not only concentrated the production of wastes, but have created many new wastes from various energy production and manufacturing processes. Recent awareness that the burying of waste in traditional landfills can contaminate both air and water has required more carefully constructed and expensive landfills to reduce contamination of water and air around these landfills. The extra cost of modern landfills now makes land applications of many wastes an economically viable alternative. Many wastes also benefit crop production, by supplying essential plant nutrients and by adding organic matter to improve soil physical and biological conditions to enhance crop growth. Because of these beneficial aspects, these materials are often referred to as by-products rather than wastes. A negative aspect is that some by-products also contain elements that can be toxic if applied to soil in excess amounts.

The science of characterizing and managing urban and industrial by-products when applied to soil was the subject of a symposium on this topic at the American Society of Agronomy Annual Meetings held in Cintinnati in 1993. This special publication constitutes the proceedings of that symposium. A total of fourteen chapters deal with topics ranging from the societal, political, and regulatory issues associated with land application to more technical issues such as the sometimes negative effects on soil moisture, salt levels, pH, and other beneficial effects the by-products may have on soil conditions that affect growth of crops.

This special publication provides a compilation of the benefits of applying these by products to agricultural land and the areas of concern and caution in their use. Through continued study on these topics, a more sustainable modern society will be developed.

CALVIN O. QUALSET, *president*
American Society of Agronomy

VERNON B. CARDWELL, *president*
Crop Science Society of America

LARRY P. WILDING, *president*
Sol Science Society of America

PREFACE

America's cities, farms, and industries are generating in excess of 1 billion tons of by-products each year. Municipal biosolids (sewage sludge and solid wastes) and some industrial wates are generally placed in landfills, but landfill capacity is decreasing and disposal costs are rapidly increasing. Animal manure and industrial by-products such as fly and bottom ash from power plants are often stockpiled at the site of generation.

The accumulation of large amounts of municipal, animal, or industrial waste at production sites can result in degradation of soil, water, and air resources. The by-products themselves, their components, or their degradation products can cause odor problems, release gases into the atmosphere which can contribute to the "greenhouse" effect, or contaminate surface water and groundwater resources with nutrients, trace elements, and microorganisms.

Alternative uses have been found for a small fraction of these materials, but our urban areas have an urgent need for developing long-term environmentally safe methods for recycling and utilizing biosolids and industrial wastes. Fortunately, many of these by-products can have substantial value if they are viewed as a resource and properly used within the total agricultural industry. Methods to optimally integrate urban and industrial by-products into agricultural management practices need to be developed in a manner that could enhance the environmental, economic, and social sustainability of agriculture and provide solutions to what is currently viewed as simply a by-product "disposal" problem.

The goal for this publication is to enhance public awareness of how agriculture (which encompasses the art and science of plant and animal production, provision of machinery and materials for that production; and processing, manufacture, and marketing of food, fiber, and other products useful for human activity), can help solve problems associated with the by-products of our increasingly urban and industrial society. Information contained in these chapters was presented in part at a joint symposium entitled "Wastes as Resources" during the 1993 American Society of Agronomy, Crop Science Society of America, and Soil Science Society of America Annual Meetings in Cincinnati, Ohio. The theme for those meetings was "Building Bridges," and one goal was to share visions of how we (as model agronomists, crop scientists, and soil scientists) might use our knowledge and experience to address new challenges in a changing world.

Developing and encouraging the adoption of soil and crop management strategies that utilize urban and industrial by-products as resources rather than considering them wastes provides many opportunities for building bridges between agricultural and urban communities. However, as pointed out by a "farmer-rancher-veterinarian-County Commissioner" in Chapter 1, building these bridges requires credible scientific research on benefits, hazards, and management of these materials, coupled with the development of an informed base of support within the agricultural community. Regulatory and programmatic responsibilities of the USEPA and contributions of non-profit organizations such as the Composting Council toward development of best management strategies for using urban and industrial by-products are discussed in Chapters 2 and 3. The "Farm Co-Composting Project," initiated by the Rodale Research Center provides a specific example in Chapter 4 of the types of bridges that can and must be built between rural and urban partners.

Chapters 5 through 10 focus on properties and potential uses for industrial by-products associated with coal combustion. Basic mineralogy of these materials, their effect on soil physical and chemical properties, the ability to provide nutrients to horticultural and agronomic plants, and their potential environmental impact are reviewed and discussed by several authors.

The use of noncomposted organic materials including paper products, animal manures, and sawmill by-products is discussed in Chapters 11 to 13. Long-term effects of applying municipal sewage sludge to forest and other degraded soils is discussed in the final chapter.

As members of the Tri-Societies and general public, we are indebted to the authors whose work has previously established and continues to contribute to the scientific foundation on which these problem-solving bridges can be built. It is our hope that the information summarized in these chapters will stimulate additional efforts to develop partnerships between urban and rural communities. Economically and environmentally viable recycling across watershed or ecoregion boundaries, our industrial, agricultural, and urban by-products can truly become "resources" rather than "wastes." We wish to thank the reviewers who contributed suggestions for improvement of the manuscripts, and the many people who stimulated discussion at the symposium. We also appreciate the excellent work of the ASA-CSSA-SSSA Headquarters staff provided to ensure a smooth and efficient publication process.

D.L. KARLEN, *coeditor*
USDA-ARS, National Soil Tilth Laboratory
Ames, Iowa

R.J. WRIGHT, *coeditor*
USDA-ARS, Environmental Chemistry Laboratory
Beltsville, Maryland

W.D. KEMPER, *coeditor*
USDA-ARS, National Program Staff
Beltsville, Maryland

CONTRIBUTORS

V. C. Baligar	Lead Scientist, USDA-ARS, Appalachian Soil and Water Conservation Research Laboratory, Beckley, WV 25802-0867
C. R. Berry	Emeritus Plant Pathologist (retired), USDA-Forest Service, Institute of Tree Root Biology, Athens, GA 30606
Jeffrey C. Burnham	President, BioCheck Laboratories, Inc., Toledo, OH 43614
E. C. Burt	USDA-ARS, National Soil Dynamics Laboratory, Auburn, AL 36831-0792
R. B. Clark	Research Plant Physiologist, USDA-ARS, Appalachian Soil and Water Conservation Research Laboratory, Beckley, WV 25802-0867
D. M. G. de Sousa	Research Chemist, EMBRAPA-CPAC, Planaltina, D.F., 73301-970 Brazil
J. H. Edwards	Soil Scientist, USDA-ARS, National Soil Dynamics Laboratory, Auburn, AL 36831-3439
C. M. Feldhake	Soil Scientist, USDA-ARS, Appalachian Soil and Water Conservation Research Laboratory, Beckley, WV 25802-0867
D. T. Hill	Agricultural Engineering Department, Auburn University, Auburn, AL 36831
Gary W. Hyatt	Procter and Gamble Pharmaceuticals, Cincinnati, OH 45241
D. L. Karlen	USDA-ARS, National Soil Tilth Laboratory, Ames, IA 50011
W. D. Kemper	USDA-ARS, BARC-West, Beltsville, MD 20705
R. F. Korcak	Silviculturist, Research Leader, USDA-ARS, Fruit Laboratory, Beltsville, MD 20705-2350
P. P. Kormanik	USDA-Forest Service, Institute of Tree Root Biology, Athens, GA 30602
Terry J. Logan	Professor of Soil Chemistry, School of Natural Resources, The Ohio State University, Columbus, OH 43210
D. H. Marx	Emeritus Scientist, USDA-Forest Service, Institute of Tree Root Biology, Athens, GA 30602
W. P. Miller	Department of Crop & Soil Sciences, University of Georgia, Athens, GA 30602
L. Darrell Norton	Research Soil Scientist, USDA-ARS National Soil Erosion Research Laboratory, Purdue University, West Lafayette, IN 47906
Cary Oshins	Composting Specialist, Rodale Institute, Kutztown, PA 19530

R. L. Raper — USDA-ARS, National Soil Dynamics Laboratory, Auburn, AL 36831

D. W. Reeves — Research Agronomist, USDA-ARS, National Soil Dynamics Laboratory, Auburn, AL 36831-3439

K. Dale Ritchey — Research Soil Scientist, USDA-ARS, Appalachian Soil and Water Conservation Research Laboratory, Beckley, WV 25802-0867

Gerald E. Schuman — Research Leader/Soil Scientist, USDA-ARS, High Plains Grasslands Research Station, Cheyenne, WY 82009

John R. Stulp — Colorado State Board of Agriculture, Prowers County, Lamar, CO 81052

A. A. Trotman — Department of Agriculture and Home Economics, Tuskegee University, Tuskegee Institute, AL 36088

John M. Walker — Physical Scientist, USEPA Municipal Technology Branch, Washington, DC 20460

R. H. Walker — Professor of Weed Science, Agronomy and Soils Department, Auburn University, Auburn, AL 36849-5412

T. R. Way — USDA-ARS, National Soil Dynamics Laboratory, Auburn, AL 36831-0792

R. R. Wendell — Research Affiliate, USDA-ARS, Appalachian Soil and Water Conservation Research Laboratory, Beckley, WV 25802-0867

R. J. Wright — USDA-ARS, Environmental Chemistry Laboratory, BARC-West, Beltsville, MD 20705

S. K. Zeto — Biologist, USDA-ARS, Appalachian Soil and Water Conservation Research Laboratory, Beckley, WV 25802-0867

Conversion Factors for SI and non-SI Units

Conversion Factors for SI and non-SI Units

To convert Column 1 into Column 2, multiply by	Column 1 SI Unit	Column 2 non-SI Unit	To convert Column 2 into Column 1, multiply by
Length			
0.621	kilometer, km (10^3 m)	mile, mi	1.609
1.094	meter, m	yard, yd	0.914
3.28	meter, m	foot, ft	0.304
1.0	micrometer, μm (10^{-6} m)	micron, μ	1.0
3.94×10^{-2}	millimeter, mm (10^{-3} m)	inch, in	25.4
10	nanometer, nm (10^{-9} m)	Angstrom, Å	0.1
Area			
2.47	hectare, ha	acre	0.405
247	square kilometer, km² (10^3 m)²	acre	4.05×10^{-3}
0.386	square kilometer, km² (10^3 m)²	square mile, mi²	2.590
2.47×10^{-4}	square meter, m²	acre	4.05×10^3
10.76	square meter, m²	square foot, ft²	9.29×10^{-2}
1.55×10^{-3}	square millimeter, mm² (10^{-3} m)²	square inch, in²	645
Volume			
9.73×10^{-3}	cubic meter, m³	acre-inch	102.8
35.3	cubic meter, m³	cubic foot, ft³	2.83×10^{-2}
6.10×10^4	cubic meter, m³	cubic inch, in³	1.64×10^{-5}
2.84×10^{-2}	liter, L (10^{-3} m³)	bushel, bu	35.24
1.057	liter, L (10^{-3} m³)	quart (liquid), qt	0.946
3.53×10^{-2}	liter, L (10^{-3} m³)	cubic foot, ft³	28.3
0.265	liter, L (10^{-3} m³)	gallon	3.78
33.78	liter, L (10^{-3} m³)	ounce (fluid), oz	2.96×10^{-2}
2.11	liter, L (10^{-3} m³)	pint (fluid), pt	0.473

Mass

Column 1 SI Unit	Column 2 non-SI Unit	To convert Column 1 into Column 2, multiply by	To convert Column 2 into Column 1, multiply by
gram, g (10^{-3} kg)	pound, lb	2.20×10^{-3}	454
gram, g (10^{-3} kg)	ounce (avdp), oz	3.52×10^{-2}	28.4
kilogram, kg	pound, lb	2.205	0.454
kilogram, kg	quintal (metric), q	0.01	100
kilogram, kg	ton (2000 lb), ton	1.10×10^{-3}	907
megagram, Mg (tonne)	ton (U.S.), ton	1.102	0.907
tonne, t	ton (U.S.), ton	1.102	0.907

Yield and Rate

Column 1 SI Unit	Column 2 non-SI Unit	To convert Column 1 into Column 2, multiply by	To convert Column 2 into Column 1, multiply by
kilogram per hectare, kg ha^{-1}	pound per acre, lb acre^{-1}	0.893	1.12
kilogram per cubic meter, kg m^{-3}	pound per bushel, lb bu^{-1}	7.77×10^{-2}	12.87
kilogram per hectare, kg ha^{-1}	bushel per acre, 60 lb	1.49×10^{-2}	67.19
kilogram per hectare, kg ha^{-1}	bushel per acre, 56 lb	1.59×10^{-2}	62.71
kilogram per hectare, kg ha^{-1}	bushel per acre, 48 lb	1.86×10^{-2}	53.75
liter per hectare, L ha^{-1}	gallon per acre	0.107	9.35
tonnes per hectare, t ha^{-1}	pound per acre, lb acre^{-1}	893	1.12×10^{-3}
megagram per hectare, Mg ha^{-1}	pound per acre, lb acre^{-1}	893	1.12×10^{-3}
megagram per hectare, Mg ha^{-1}	ton (2000 lb) per acre, ton acre^{-1}	0.446	2.24
meter per second, m s^{-1}	mile per hour	2.24	0.447

Specific Surface

Column 1 SI Unit	Column 2 non-SI Unit	To convert Column 1 into Column 2, multiply by	To convert Column 2 into Column 1, multiply by
square meter per kilogram, m^2 kg^{-1}	square centimeter per gram, cm^2 g^{-1}	10	0.1
square meter per kilogram, m^2 kg^{-1}	square millimeter per gram, mm^2 g^{-1}	1000	0.001

Pressure

Column 1 SI Unit	Column 2 non-SI Unit	To convert Column 1 into Column 2, multiply by	To convert Column 2 into Column 1, multiply by
megapascal, MPa (10^6 Pa)	atmosphere	9.90	0.101
megapascal, MPa (10^6 Pa)	bar	10	0.1
megagram per cubic meter, Mg m^{-3}	gram per cubic centimeter, g cm^{-3}	1.00	1.00
pascal, Pa	pound per square foot, lb ft^{-2}	2.09×10^{-2}	47.9
pascal, Pa	pound per square inch, lb in^{-2}	1.45×10^{-4}	6.90×10^3

(continued on next page)

Conversion Factors for SI and non-SI Units

To convert Column 1 into Column 2, multiply by	Column 1 SI Unit	Column 2 non-SI Unit	To convert Column 2 into Column 1, multiply by
Temperature			
$1.00\ (K - 273)$	Kelvin, K	Celsius, °C	$1.00\ (°C + 273)$
$(9/5\ °C) + 32$	Celsius, °C	Fahrenheit, °F	$5/9\ (°F - 32)$
Energy, Work, Quantity of Heat			
9.52×10^{-4}	joule, J	British thermal unit, Btu	1.05×10^{3}
0.239	joule, J	calorie, cal	4.19
10^{7}	joule, J	erg	10^{-7}
0.735	joule, J	foot-pound	1.36
2.387×10^{-5}	joule per square meter, $J\ m^{-2}$	calorie per square centimeter (langley)	4.19×10^{4}
10^{5}	newton, N	dyne	10^{-5}
1.43×10^{-3}	watt per square meter, $W\ m^{-2}$	calorie per square centimeter minute (irradiance), $cal\ cm^{-2}\ min^{-1}$	698
Transpiration and Photosynthesis			
3.60×10^{-2}	milligram per square meter second, $mg\ m^{-2}\ s^{-1}$	gram per square decimeter hour, $g\ dm^{-2}\ h^{-1}$	27.8
5.56×10^{-3}	milligram (H_2O) per square meter second, $mg\ m^{-2}\ s^{-1}$	micromole (H_2O) per square centimeter second, $\mu mol\ cm^{-2}\ s^{-1}$	180
10^{-4}	milligram per square meter second, $mg\ m^{-2}\ s^{-1}$	milligram per square centimeter second, $mg\ cm^{-2}\ s^{-1}$	10^{4}
35.97	milligram per square meter second, $mg\ m^{-2}\ s^{-1}$	milligram per square decimeter hour, $mg\ dm^{-2}\ h^{-1}$	2.78×10^{-2}
Plane Angle			
57.3	radian, rad	degrees (angle), °	1.75×10^{-2}

To convert Column 2 into Column 1, multiply by	Column 1 SI Unit	Column 2 non-SI Unit	To convert Column 1 into Column 2, multiply by
Electrical Conductivity, Electricity, and Magnetism			
10	siemen per meter, S m^{-1}	millimho per centimeter, mmho cm^{-1}	0.1
10^4	tesla, T	gauss, G	10^{-4}
Water Measurement			
9.73 × 10^{-3}	cubic meter, m^3	acre-inches, acre-in	102.8
9.81 × 10^{-3}	cubic meter per hour, m^3 h^{-1}	cubic feet per second, ft^3 s^{-1}	101.9
4.40	cubic meter per hour, m^3 h^{-1}	U.S. gallons per minute, gal min^{-1}	0.227
8.11	hectare-meters, ha-m	acre-feet, acre-ft	0.123
97.28	hectare-meters, ha-m	acre-inches, acre-in	1.03 × 10^{-2}
8.1 × 10^{-2}	hectare-centimeters, ha-cm	acre-feet, acre-ft	12.33
Concentrations			
1	centimole per kilogram, cmol kg^{-1}	milliequivalents per 100 grams, meq 100 g^{-1}	1
0.1	gram per kilogram, g kg^{-1}	percent, %	10
1	milligram per kilogram, mg kg^{-1}	parts per million, ppm	1
Radioactivity			
2.7 × 10^{-11}	becquerel, Bq	curie, Ci	3.7 × 10^{10}
2.7 × 10^{-2}	becquerel per kilogram, Bq kg^{-1}	picocurie per gram, pCi g^{-1}	37
100	gray, Gy (absorbed dose)	rad, rd	0.01
100	sievert, Sv (equivalent dose)	rem (roentgen equivalent man)	0.01
Plant Nutrient Conversion			
	Elemental	*Oxide*	
2.29	P	P$_2$O$_5$	0.437
1.20	K	K$_2$O	0.830
1.39	Ca	CaO	0.715
1.66	Mg	MgO	0.602

1 Social, Political, and Educational Factors Involved in Facilitating Municipal Waste Utilization

John R. Stulp

Colorado State Board of Agriculture, Prowers County Lamar, Colorado

PREFACE[1]

A few years ago, newspapers across the country carried headlines about a barge carrying sewage sludge from New York City that was at sea and was being denied permission to unload its cargo or even dock at many ports on our East Coast. Their problems were compounded by recent laws which prohibited dumping of sewage sludge at sea. This was followed in subsequent months by a series of reports of negotiations between representatives of New York City and other large cities on the East Coast with landowners in states with low population densities who were willing to have this by-product of society applied to their lands under controlled and monitored conditions. However, in case after case state permits necessary to allow use of this by-product from outside the state on the land were denied. The perception of those denying the permits was often that they were protecting citizens of their state from the potential dangers emanating from this by-product of society. The facts were that these big city sludges were being analyzed carefully and those scheduled for disposal would pose no greater hazard than sludges from local communities and cities that were being used on adjacent agricultural lands.

Efforts were made to educate state and county decision makers. However, even when these decision makers understood the facts and wished to allow the farmers to gain the benefits of applying these materials, which contain many essential plant nutrients, public opinion against use of these big city wastes became so strong that a decision to permit use of these big city, out of state, wastes would have been political suicide for the decision makers.

Negotiations to gain permission to use sludge from New York City or "biosolids" to improve production on agricultural land failed in New York, Texas, Oklahoma, South Dakota, and several other states. Even though research has shown that these biosolids were beneficial to the land and posed no health hazards when applied properly, there were always those individuals who perceived hazards in their use, and some of these individuals led strong emotional campaigns that turned the voting public against their use. These rejections, based on perceptions rather than facts prevented

---

[1] The preface was written by W.D. Kemper, USDA-ARS, Building 005, BARC-West, Beltsville, MD 20705.

farm lands from receiving the nutrients and organic matter that they need, and were escalating the costs of disposal of these wastes to the citizens of New York City.

Consequently, society in general has benefited from the decisions by the state of Colorado and Kiowa and Prowers counties to allow their farmers to use "biosolids" from New York City. As these farmers, their neighbors, and state scientists carefully monitored effects of these biosolids on their crops, soils, and environment, they concluded that they brought a substantial net benefit to their farms and community. Additional farmers requested that these "biosolids" be applied to their lands. The demand exceeded the supply and a letter signed by 59 farmers and business people was sent to the mayor of New York City requesting that more of the biosolids be sent to them in Colorado. This request was such a contrast to the many highly publicized refusals that it gained headlines in newpapers of all the big cities across the nation. A material that had been perceived by the public as a noxious waste had been recognized as a beneficial resource that could be used safely.

Many of our nations major wastes have potential for beneficial use in agriculture soils because they can supply needed nutrients and organic matter or improve soil structure and efficiency of water use. Research is showing how those potentials can be achieved. However, those potentials will not be recognized until the decision makers and the public are educated concerning benefits and are convinced that there are controls which can eliminate hazards accompanying their use.

The success in Colorado was a primary factor initiating this symposium. We looked for the factors that had enabled this project to succeed after so many failures in other attempts. While many contributed to this achievement, all those contacted acknowledged John Stulp as the primary leader. He saw the potential, recognized the concerns of his neighbors and county and state officials, encouraged organization of the forums in which those concerns could be aired and misconceptions could be replaced by facts, and extended the benefits broadly.

Sociopolitical factors play such a major role in the ability of society to use its waste products that the organizers of this symposium decided to feature the Colorado success story and asked John Stulp to tell us what enabled them to succeed where so many others had failed.

I have been invited to discuss an ongoing project using bio-solids from New York City on agriculture land in southeastern Colorado. I hope our experiences and my opinions will help other similar projects be successful. This presentation does not represent the final analysis but rather the beginning of an evolving process.

A current television commercial depicts a group of cowboys throwing the cook out of the camp when they discover the salsa dip he is substituting for their regular brand was *made in New York City!* That same bias was common among many who asked, "Why is sludge from New York City being applied in Southeast Colorado?"

In June 1991, six farmers in Kiowa and Prowers counties consented to accept biosolids from New York City. Initially 6900 ha (17 000 acres) of farm land were submitted for permits. A municipal waste management company, Enviro-Gro Technologies (Indianapolis, IN), had contracted with New York

City to handle some of their biosolids from waste water treatment plants due to a pending ban on ocean dumping of municipal wastes.

Kiowa and Prowers counties have a low population with rural residents comprising a density of approximately one person per square mile. When considering total population, the density in these counties is only about 4.4 people per square mile. This small population allowed us to provide information on the benefits and safety precautions of this project to those who had concerns.

Other aspects of southeastern Colorado that were attractive for a biosolids application included our semiarid climate and predominant farming practice which is summer fallowing and winter wheat (*Triticum aestivum* L.). Application of biosolids to the fallow land can be done nearly all year without damaging growing crops or soils. Most soil types in this area are high in clay content and have fairly high pH values. In these dry land wheat areas pH is commonly 7.5 to 8.5. High clay and high pH minimize heavy metal movement in the soil. Railroads with rural sidings for off-loading the regular shipments of biosolids also were a consideration. A rural siding was used for traffic safety reasons. After proper permits were issued to Enviro-Gro Technologies by the Colorado Department of Health, the first shipments arrived in the spring of 1992. Let me first describe the actual physical operation, and then consider the various sociopolitical factors addressed long before the first shipment arrived from New York.

PHYSICAL OPERATION

Sealed cargo type containers with cake sludge (25–30% solids) were off-loaded from rail flat cars onto truck trailers and transported to the permitted fields. At the fields the seals placed on the container in New York City were broken and the contents were dumped on the ground. At this point samples were taken for the monitoring program. The biosolids were then loaded onto spreading units and applied at the predetermined rates that would supply the N needed by the next crop. Modern well cared for equipment was used. Where possible, this equipment was purchased locally.

The average application rate for most fields was 6.7 Mg ha^{-1} (3 tons acre^{-1}) of dry material or 26.9 Mg ha^{-1} (12 tons acre^{-1}) for wet material. A field that had been treated was not easily noticed. The fields were fallowed during the summer of 1992 in a normal manner and planted to winter wheat in late summer or early fall that year. Adequate soil water content throughout the area allowed for good stands of new wheat, and above-average winter precipitation in the form of snow created an excellent potential for above-average yields in most of southeastern Colorado.

Wheat yields in this area are dependent on the blessings of timely rainfall in the spring. However, the same clouds that bring rain to help fill the grain bins sometimes bring hail. Every site in the two county area that had biosolids applications in 1992, received devastating hail shortly before harvest in July 1993. Area fields missed by the hail averaged nearly 2.5 Mg ha^{-1} (40 bushels acre^{-1}). One test plot site was salvageable after an estimated 50% hail loss. The control plots yielded 785 kg ha^{-1} (12.5 bushel acre^{-1}), the bio-solids treated plots yielded 1320 kg ha^{-1} (21 bushel acre^{-1}). A plot receiving two times the recommended agronomic rate of biosolids yielded 1190 kg ha^{-1} (19 bushels acre^{-1}).

The application of biosolids has continued with a growing list of producers willing to try them. Enviro-Gro Technologies has been purchased by Bio-Gro Systems (Indianapolis, IN), another a municipal waste company. They have assumed the same position as Enviro-Gro Technologies had with the project. A dry pelleted form of bio-solids is now being used in place of the 75% moisture cake product.

EDUCATIONAL, SOCIAL, AND POLITICAL FACTORS

Let me now turn to factors in this project, and hopefully future projects, that are important to the overall success and acceptance by farmers, the general public, and municipalities.

The reputations of the land owners planning to accept and utilize the biosolids was important in creating confidence in this project. All of the farmers to first use biosolids are well known in their communities. They are successful farmers and businessmen, represent several generations of farm families, and are recognized by their fellow farmers and city friends as people who take care and pride in their farming operations. They generally have records of public service. For instance, one farmer is a county judge, and another a county commissioner.

Farmers need to be knowledgeable about any products they are utilizing, whether they are biosolids, commercial fertilizers, crop protection chemicals, or any new agriculture products. In a time of increasing environmental awareness and concerns, those of us in production agriculture must be able to articulate and defend our usage of the various products and techniques vital to the safe and economic production of food and fiber.

Some of the resources that were extremely useful to our education about utilization of biosolids and have contributed to the success of the project include the following:

1. Colorado State University Department of Agronomy personnel, and in particular Dr. Ken Barbarick, have been conducting field research with biosolids on dry land wheat farms in Colorado for more than 12 yr. The long-term data he has produced using biosolids from the Denver area waste treatment facilities and his willingness to make presentations in southeastern Colorado helped establish the benefits that could be obtained from using bio-solids. This helped put the minimal potentials for negative effects in their proper perspective.

2. The Colorado Department of Health is the regulatory agency in Colorado that issues permits for land application of bio-solids. Phil Hegeman is the director of municipal sludge permitting. New York City requires regular chemical and pathogen analyses of the biosolids produced at each

of their treatment plants. After careful study and review of records from New York City treatment plants, Hegeman and coworkers excluded 4 of 14 facilities considered as possible sources for shipping biosolids to Colorado. One of the excluded facilities had high Cd levels, and the other three did not have adequate pathogen reduction records. The New York City biosolids permitted into Colorado were very similar to biosolids being produced in most of the sewage treatment plants in Colorado.

3. The USEPA was helpful in providing a federal overview of the project. Bob Bropst provided reams of research papers and documents pertaining to biosolids. He also was available for answering many questions by concerned citizens. A 12-min video produced by the USEPA discussed safety and biosolid utilization in several areas (including some Colorado locations).

4. Enviro-Gro Technologies was represented by Mike Scharp, a field representative, and Kipp Parker, their Colorado agent, who also was a private agriculture consultant. Any waste management company must have excellent credentials to make a project like this acceptable. Their track record in waste management was good. The references they gave were supportive of their experience claims. The application they made to the Colorado Department of Health was professional and thorough. Representatives of the company met on a regular basis with farmers participating in the project, the various regulatory agencies, and county commissioner groups. Their willingness to listen to concerns by the water department of the city of Lamar resulted in some eligible tracts being pulled from the original permits, but contributed to acceptance of the project by the Lamar residents. Local purchases of some basic vehicles and equipment enhanced their stature as a good neighbor. A special demonstration test plot project using various levels of biosolids was funded through the local soil conservation district by Enviro-Gro. When Bio-Gro Systems took over the operation as the new owner, much of the previously established good will had to be reestablished by the new management.

Political entities that were considered included more than just two sets of county commissioners. Several neighboring counties also were provided with information on the project in anticipation that expansion might include land in these parts of southeastern Colorado. The city of Lamar and their water department, which pumps their municipal water from wells, were kept informed of the application sites. Water is a very precious commodity in the western USA. Potential containimation fears concerning mobilities of heavy metals and microorganisms were addressed using data from other studies.

At the state government level, in addition to the health department, the department of agriculture, and the governor's office were advised of what was going on in the Kiowa and Prowers counties. Again the credibility of the people involved played an important role in gaining interest in the governor's office and the department of agriculture. A field visit by Governor Romer and his cabinet confirmed local acceptance of the project. A local aide to one of our U.S. senators had kept a file of newspaper reports and articles about the New York City Sludge project. This was useful to the senator's office when a concerned constituent claimed the project was being done in secrecy and should be stopped. Continuing communications between the farmers involved in this project and the area state legislators and staffs of congressional and senatorial offices, helped keep them informed on what we were doing and minimized misunderstanding. Since 1992 was an election year, all local and regional candidates were informed of the project so as to reduce the possibility of any of them taking a negative position on this issue. Public meetings are important but should be well structured so all parties are able to participate on a non-emotional, fact-compoaring level. A panel format for presenting information about the various aspects of biosolid application is helpful. Representatives of the previous mentioned entities—the Colorado State University Agronomy Department, the Colorado Department of Health, the USEPA, the municipal waste management company, the local county commissioners, and the conservation districts should be included in such a panel.

Testimonies from entities using biosolids are extremely helpful, but not always available when a project is new. Comparing the heavy metal analysis of biosolids to the analysis of common multivitamin and mineral products such as Centrum (A–Zn) helped define concentrations at which elements such as Cu, Fe, Mg, P, Zn, Cr, Mn, Mo, Ni, Se, and Va are found as being within the healthy nutrition ranges rather than in the toxic ranges.

Questions and answers and debate are best left until after the informational presentations in meetings, since many of the questions are then answered in logical sequences before they are asked. It is important that public meetings be conducted by a fair minded, yet firm, moderator or the situation can deteriorate to a nonfactual emotionally charged situation where logic and reason are not given a fair chance to prevail.

A recent article in a major farm publication was very negative toward use of sludge on soils on the basis of reports from a single out of state incident. Consideration of this article was placed on the agenda of our next public meeting. A more accurate set of facts was obtained from the state regulatory agencies handling this incident prior to our meeting. Presentation of these facts and the following discussion placed this incident and article in their proper perspective.

General public education beyond public meetings is very important. Informal discussions with main street merchants and community leaders resulted in a couple of unsolicited letters to the editor of the local newspaper endorsing the biosolids project.

Copies of the Colorado Department of Health regulations and the USEPA regulations related to use of biosoils were sent to local public libraries. The regional USEPA supplied many research articles and technical papers to local libraries and individuals requesting them. Copies of a short (12-min) video tape discussing bio-solid utilization and safety, which had been produced by the USEPA, also were placed in local libraries and schools. Copies of the tape were given to anyone requesting it. The local farmer elected Agriculture Stabilization and Conservation Service (ASCS) and Soil Conservation District (SCD) supervisors were kept informed as the project progressed.

One-on-one contact with area citizens is very important. Concerned persons often ask questions individually that they would not ask in a group. Neighbors to the treated fields need to be educated even if they are not participating in a biosolid program. One person expressed a concern about contracting human immunodeficiency virus (HIV) from wheat flour products that were grown in fields utilizing biosolids from New York City. This concern may seem unrealistic to scientists educated in these subject areas. However, it was a sincere concern and time spent in giving this individual a personal short course in virology and agronomy was well invested.

Contacts from other areas that are utilizing biosolids included those with Gary Wegner, a grain farmer from near Reardon, Washington. Gary has worked with biosolids for over 5 yr and had good data and a short publication summarizing his experience which we used frequently and distributed to our group of concerned citizens.

Milwaukee, Wisconsin, provided some of the most interesting information on long-term biosolid use. They have been using bio-solids as a beneficial product from their municipal treatment facilities since 1926. Milorganite is a trademark biosolids product marketed across the USA by Milwaukee. They helped by providing a historical, as well as a highly successful program perspective for new biosolid users.

As previously mentioned, the ongoing study by Colorado State University Department of Agronomy was very applicable to southeastern Colorado. When we were able to document the similarities between the New York City and Colorado sludges, the results of the 12-yr study became directly pertinent and added to the credibility of using New York City biosolids as a fertilizer and soil amendment.

The local media gave this project good coverage. A local talk radio program invited people opposed to the project as well as those supporting the project to several programs. Local and regional publications were useful in explaining the project to the citizens of the area. The Colorado Department of Health and USEPA responded to some letters to the editors on some of the published concerns. It takes patience and diplomacy to explain in an inoffensive manner that a spill of biosolids will not eat a hole in the asphalt and have to be treated like a nuclear disaster site!

Medical professionals in the area are an important resource. However, it should not be assumed that an M.D. or D.V.M. will automatically have a thorough knowledge of biosolids. Residents of rural areas do look to the medical profession and veterinary profession for assurance on new projects concerning their personal and animals' health. Consequently literature left with doctors and veterinarians and discussions with them would be good investments.

The cost and trail of liability are part of any new project. New York City is the end financier of the entire project. As the generator of the biosolids, New York City has contracted with various municipal waste management companies to properly handle their biosolids. The applicant (Bio-Gro) pays a fee based on dry tonnage to the Colorado Department of Health as well as a monthly fee to the local county governments. These fees are used by the Colorado Department of Health and counties to pay for their own monitoring programs. The applicant has paid for repair of some road surface damage which resulted from increased truck traffic hauling the biosolids. One of our objectives was to manage this project so there was no extra cost to local government.

The monitoring program put in place and agreed to by New York, Bio-Gro (and Enviro-Gro), the Colorado Department of Health, and Kiowa and Prowers counties was extremely helpful in gaining necessary approval for this project. New York City, as the generator of the biosolids, monitors the quality of their product. Bio-Gro, as a contractor with New York City, also monitors the bio-solids in order to know where and how they manage the biosolids. Bio-Gro has been using a Cu screening test on each container. Copper has been the metal ion that was nearest toxicity levels and consequently it was often used as an index of the biosolid quality. Before leaving New York City by rail, each container is sealed in order to detect any tampering with the shipment. On arrival in Colorado, the containers are inspected for the seals being intact. If tampering with the seals is suspected, that load is rejected and sent back to New York City.

A representative of the Colorado Department of Health has been on site 5 d wk $^{-1}$ to inspect shipments, observe application, take samples, and represent the health department's interest in the project. The County Commissioners in Kiowa and Prowers counties have contracted with a local agriculture consulting company to provide additional monitoring and product sampling on a random basis. The same agriculture consulting business owned by a local soil scientist has contracted with the local soil conservation district to conduct the necessary soil tests on a test plot utilizing bio-solids at various rates. The cost of the test plot study has been funded by Enviro-Gro and Bio-Gro.

This arrangement of monitoring on four levels utilizing different laboratories has resulted in a comfort level by all involved. The laboratory reports have been generally consistent with each other. Rare differences have been traced to laboratory or typographical errors that were corrected.

Any new project has to overcome natural suspicions. Honesty, openness, and willingness to invest in communication and continuing education are principles that help relieve our concerns regarding the unknown.

The success of this project to this point has occurred because the parties involved with it have adhered to these principles. Everyone from the farmers, to the contractors, local leaders and citizens, and the regulatory agencies involved, is important to this success. We cannot make general assumptions that all of these are being addressed. Ongoing, specific attention to detail at all levels is critical. Wherever the weakest link is will possibly determine the overall success or failure.

Primary factors contributing to the acceptance of this project have been the availability of credible scientific research on benefits, hazards, and management of biosolids coupled with articulate individuals, well educated and supportive of this type of environmental recycling.

Postscript

The Kiowa County Board of Commissioners are no longer supporting the biosolid project. Two of the three commissioners are new to the three-member board. A very vocal and somewhat emotional group of Kiowa County citizens are opposed to the project. A recent Kiowa County incident where dry biosolids were blown by high winds from a field loading site onto another unpermitted property caused a crisis. Bio-gro admitted their personnel were at fault for applying biosolids under those weather conditions and have offered compensation to these parties not wanting the biosolids on their rangeland.

Prowers County Commissioners and farmers are still quite supportive on this ongoing project. The waiting list for bio-solids application is still large.

2 United States Environmental Protection Agency's Roles in Waste Utilization: The Need for Partnership

John M. Walker

USEPA
Washington, District of Columbia

The U.S. Environmental Protection Agency (USEPA) is charged by the U.S. Congress under a number of different, and sometimes conflicting, statutes to regulate the management of wastes to protect human health and the environment. This chapter will discuss several aspects of this difficult task.

This chapter begins by listing examples of the comprehensiveness and pervasiveness of USEPA's roles. It then briefly considers what might constitute a good standard of protectiveness for environmental sustainability associated with additions of wasteborne pollutants to soil, e.g., standards either based on scientifically acceptable change or policy-determined no change above background levels in soils. The paper then addresses some of the challenges facing USEPA in developing sensible scientifically based rules and the need for assistance of the agricultural community in accomplishing these tasks. Finally, it examines some of the remarkable agricultural and environmental benefits that can result from knowledge gained from research and tailor making and using waste products, either singularly or in combination, to achieve these benefits.

HOW UNITED STATES ENVIRONMENTAL PROTECTION AGENCY'S ACTIVITIES IMPACT OUR EVERYDAY LIVES

The USEPA rules impact the air inside and outside our buildings, the electricity we use, the food we eat, the water we drink, and the wastewater we flush down the drain. The air regulations govern not only the air emissions and ash from power generation, heating, cooling, manufacturing, automobiles and other forms of transportation, but also the nature of the volatile compounds that are permissible in homes such as from glues used to manufacture and fasten down our carpets, solvents in our paints, and aerosolizing agents in containerized sprays.

The USEPA regulations govern certain aspects of the manufacture of gasoline and automobile combustion and exhaust systems to control the emissions. They regulate the collection and disposal of top and bottom ash from thermal and air pollution control systems. They regulate the handling of fluorocarbons used in air conditioning systems. There are regulations pertaining to the safe level of Rn in our homes.

The USEPA categorizes and separately regulates hazardous and nonhazardous wastes including cradle to grave control of hazardous wastes. They regulate the clean-up of spills and contaminated soils, as well as the assessment of costs for cleanup as absolute liability (liability without determination of fault for liability).

Other USEPA rules have far-reaching implications on the production of foods and the use and disposal of food wastes. For example, USEPA rules govern various aspects of the manufacture and disposal of fertilizer manufacturing wastes. They also regulate the manufacture, use, and disposal of herbicides and pesticides. The USEPA soon will be involved in regulating the comprehensive management of nutrients from fertilizers, animal wastes, food processing wastes, urban biosolids and municipal wastes and other industrial wastes.

The USEPA regulates the treatment of the water we drink and the treatment and disposal of wastewater from municipal and industrial systems as well as the use and disposal of the residues from treatment. The USEPA regulations also impact the encapsulation and/or removal of asbestos and lead paint in buildings in which we live. Still further, USEPA regulations control the Pb permitted in solder, paint, gasoline, and plumbing. The USEPA regulates the management of wastes from the manufacture of drywall products used in our homes and the treatment of wood with various preserving materials. Its regulations impact the manufacture of paper and paper mill products and the disposal and utilization of those manufacturing wastes. The USEPA regulations also impact the use and disposal of yard wastes, other household wastes, and municipal wastes. The USEPA regulations also have a direct impact on the possible uses of various forms of land, especially wetlands.

It can be seen from this partial listing that USEPA rules and regulations have a far more comprehensive, extensive, and pervasive impact on our lives than you might have imagined, even more than the Internal Revenue Service. It is incumbent on the agricultural community to be deeply involved so that its expertise is fully utilized and its points of view carefully considered to help assure that USEPA's regulatory efforts are well thought out and based on sound science and policy.

WHAT STANDARD OF PROTECTIVENESS IS ACCEPTABLE?

Should our standard of protectiveness from wasteborne pollutants be based on scientific risk assessment, or a no-change policy? If the medium to which wastes will be added is the soil, a risk-assessment approach would permit an increase in wasteborne pollutants provided the scientifically determined acceptable level for pollutants in soil was not exceeded. A no-change policy might limit any increase of wasteborne pollutants in soil to some fraction of background pollutant levels.

Some European countries, provinces in Canada, and scientists in this country have adopted or urge, as a matter of policy, a protectiveness standard that permits minimal to zero change in soil pollutants as a result of waste addition to soils. The USEPA's new biosolids rule (40 CFR Part 503) determined a scientific risk-based standard that permits an acceptable change in soil pollutant contents.

Dr. Margaret Maxey, Professor of Environmental Ethics at the University of Texas in Austin, spoke at a recent opening session of the Water Environment Federation. She reminded the thousands of attendants that nature is not fragile and that a standard of zero risk followed to its ultimate end means very limited production of wealth. The wealth, which comes from our ability to produce quality foods and goods, enables us to conduct research and institute environmental controls that enhances our environment and takes the drudgery out of our lives.

Dr. Maxey said that the "good old days" were in fact the "bad old days." She said just imagine the manure on our city streets today if there were no automobiles and other forms of modern transportation. Think of having to wash clothes with a washing board, storing foods only with salt or ice, and cooking and heating with wood stoves. She made no secret of her resolve to firmly resist giving up modern conveniences based on the incorrect perception of environmental fragility that necessitates no change and zero risk. Her message was strong—do not sit back and allow the advocation of zero risk to drive society back to the bad old days.

COOPERATIVE ROLES FOR ENGINEERS, SCIENTISTS, AND OTHER CONCERNED WASTE MANAGEMENT PROFESSIONALS OUTSIDE THE USEPA

While the USEPA's final rule governing biosolids utilization was risk based and used sound science, it was not that way when proposed. The proposed rule used data from salt and greenhouse pot studies rather than field experiments where soils were amended with biosolids. Therefore, the proposed rule did not account for reduced pollutant bioavailability due to the biosolids chemical matrix. In addition, the proposed rule used certain assumptions and models that were later found, during review by experts, to cause a significant overestimation of risk. Hence, the proposed rule contained regulatory requirements that were very limiting, e.g., the proposed limit for Cu was equivalent to background soil levels.

Experts with agronomic and biosolids expertise assisted the USEPA in correcting the science. In so doing they developed a new level of understanding and ability to select and manage data for conducting risk assessment (Chaney,

1993; Ryan & Chaney, 1992; Chaney & Ryan, 1992, 1994). The efforts of this expert team helped simplify the final biosolids rule, including the adoption of an "exceptional quality" (EQ) biosolids that can be used without further regulatory control once certified as having met EQ requirements. The adoption of these changes will potentially save millions of dollars and encourage the production of high-quality biosolids. The production and use of such high-quality products are consistent with the goal of environmental sustainability.

States are now being asked to take on the responsibility of implementing the Federal Part 503 biosolids rule. The final Part 503 rule was less stringent than when proposed. And in many cases less stringent than rules governing the use of biosolids already promulgated by states. This was surprising to many people.

While states may have special situations that require more stringency than federal rules, it is important that any decision for increased stringency be soundly based. If scientists, professionals, and rule makers in each state would learn more about what was really involved in the federal USEPA risk assessment process, sound science and not ignorance would be the basis for new and revised state rules for managing biosolids and other nutrient-containing waste materials.

The speakers in plenary session of the 1993 American Society of Agronomy Meetings (ASA) urged that ASA members get involved as a participating partner with the USEPA in managing wastes. Rob Wolcott from the USEPA's policy office pointed out the upcoming role of USEPA in holistic management of nutrients applied to soils. The cooperative involvement of agricultural scientists with the USEPA would significantly contribute to successful completion of this task.

OPPORTUNITIES FOR DESIGNER/TAILOR MADE WASTES

There are a number of opportunities for creative uses of wastes. Knowledge gained from basic research as well as practice is paving the way for creative uses of waste materials that arise either directly from various municipal and industrial waste treatment processes or from combinations or special treatment of waste materials to give them desirable characteristics. This knowledge also leads directly to improved methods of overcoming other environmental problems. The proper use of these waste products helps achieve the goal of sustainable farming. The production of these designer waste products creates opportunities for new jobs in both rural and urban areas. Several examples follow:

1. Metal bioavailability can be managed. The bioavailability of metal pollutants in a waste material such as wastewater sludge biosolids is reduced due to matrices components like oxides of Fe, A, and Mn, humid acids from the biosolids organic matter, and phosphates (Chaney & Ryan, 1992). Dewatering biosolids with an iron salt such as ferric chloride enhances its ability to reduce bioavailability of metal pollutants.

2. On-going research is suggesting that biosolids, which are added to soils contaminated by Pb-containing paints or past automotive emissions, can reduce the soil lead bioavailability and hence help protect children who ingest soils (James Henningen, personal communication).

3. Research by Hoitink et al. (1992) has demonstrated, and the potting industry has adopted the use of specially prepared compost to suppress plant root disease.

4. Research with municipal solid waste compost as a soil amendment has resulted in increased yield and quality of field and fruit crops (Peverly & Gates, 1993).

5. Research by Falahi-Ardakani et al. (1987) has shown how composted biosolids can be used in potting media as a substitute for peat and at the same time reduce the need for chemical macro- and micronutrient fertilization.

6. Recent cooperative studies at the Rodale Institute involve the comparative root and plant growth responses to inorganic fertilization and organic composts. This research is being conducted to examine how plant root structure and effectiveness is impacted in soils due to the stimulation of fungi and other microorganisms, better physical structure, and other related factors (T. Schettini, I. Lynch, personal communication).

7. Many studies and actual operations have shown that municipal solid waste, biosolids, plant and animal wastes added to soils can be an important slow-release source of N and other nutrients (Wright, 1994).

8. There are a number of cases where the use of waste materials like biosolids has been highly successful for renovating drastically disturbed sites that are generally devoid of organic matter, very high in acid content, and low in nutrients (Sopper, 1993).

9. Blending biosolids with kiln dust provides a more balanced fertilizer material that also has usefulness as a lime substitute (Logan & Burnham, 1994).

ROLES

Just as the current administration is redefining government, I have taken the liberty to become philosophical and suggest that we continue to define and redefine our roles for managing wastes in partnership with others including the USEPA. Outlined below are a series of partnership roles that, if assumed by scientists, would help achieve improved regulation, management, and utilization of waste resources.

1. Continue to conduct basic and applied research that increases our understanding and enables better management of wastes.

2. Provide counsel and training and a sound scientific bridge among the customers for your environmental knowledge. These customers include waste generators, the public, other scientists, farmers, bankers, and state and federal regulators. As your provide counsel and advice on environmental is-

sues, clearly separate science from policy and indicate distinctly which is which.

3. Advise policy makers about the waste management risks arising from one policy decision compared with another.

4. Advise waste managers and regulators on whether rules developed for biosolids can apply to other waste products such as composted municipal solid wastes.

5. Determine and advise regulators on the real world impacts of rules. For example: (i) if a regulatory limit for Cd in a sludge biosolids is 39 mg/kg, the actual mean content of Cd in the biosolids would have to be less so that the 39 mg/kg Cd limit would not be exceeded. For example, the actual mean content of Cd might be 10 mg/kg less than the 39 mg/kg regulatory limit (at 29 mg/kg) to avoid exceedance, but exactly how much less would depend on the variability of Cd in the biosolids; and if a rule is in error, there is ample time to make corrections because at agronomic rates of application it takes several hundred years to reach most of the cumulative pollutant loading limits for the biosolids.

6. Be prepared to speak out against nonscience-based rules and policies that force uneconomic practices or disposal rather than beneficial use. For example, I believe that it was important to challenge the October 1993 biosolids-related lending policy of the Farm Credit Bank of Springfield, Massachusetts. That policy would have singularly required farmers to seek indemnification from generators and land appliers of biosolids backed by insurance for the low-risk use of biosolids on their farms. The farmers would have been unable to obtain loans from the Farm Credit Bank if they used biosolids on their farms without having obtained indemnification backed by insurance. The Farm Credit Bank had no similar indemnification requirements for use of other wastes, chemical fertilizers, herbicides and pesticides.

A useful outcome of the challenge to the Farm Credit Bank lending policy was a mutually agreeable indemnification statement that could be given by biosolids generators and land appliers stating that the biosolids being land applied were of the quality stated and that they were being land applied in accordance with applicable rules for land application at the time of the indemnification. While the problem still exists that no similar statement is required by generators of other wastes that are being applied on farmland, the guarantee being given is what any good biosolids generator and land applier would be willing to do anyway.

7. Carefully examine instances where alleged problems have occurred from the land utilization of waste materials. Provide counsel on what really has occurred and why.

8. Provide solutions to factors that limit the beneficial utilization of wastes.

9. Help answer questions like: (i) Do we really need a two- to five-mile buffer between composting facilites and the nearest neighbor?, and (ii) Is their a time-bomb effect, i.e., the potential for subsequent increased bioavailability of metals to plants and increased metal leaching to groundwater such that there is significant environmental risk?

10. Evaluate the economics of sludge vs. fertilizer use considering such issues as the comparative economic value of using nutrients and organic matter from waste materials and other sources for enhancing soils and producing quality crops.

11. Provide counsel to other researchers on experimental design of waste studies, helping them avoid past mistakes and take advantage of already accumulated knowledge.

12. Provide assistance in the determination of what is acceptable change due to waste addition to land.

13. Become accountable for your research. Ask important questions. Encourage oversight by your customers. Show them where your research can lead.

14. Train new people.

SUMMARY AND CONCLUSIONS

The USEPA has both regulatory and programmatic responsibilities for utilization of wastes. The regulatory role provides rules, regulations, and enforcement activities that govern the production and utilization of wastes. The programmatic role has involved funding as well as technical assistance and guidance for the design, construction, and operation of facilities that use sewage sludge biosolids and other forms of waste materials. The regulatory role is a USEPA priority with far-reaching impacts on all aspects of waste utilization practices. Individuals responsible for developing the regulations often have had little opportunity to either manage wastes or learn extensively about the environmental impacts of their management or mismanagement. It is important, therefore, that the agricultural community, with its extensive knowledge of environmental science and expertise in waste utilization, work closely with USEPA. This cooperative effort would help assure that, regulations which are developed, are scientifically sound and implementable, rather than excessively stringent. Such cooperation should result in agronomic practices that benefit from the creatively designed, economically viable and sustainable practices that use urban and rural waste products in agriculture.

REFERENCES

Chaney, R.L. 1993. Risks associated with the use of sewage sludge in agriculture. p. 7–11. *In* Proc. 15th Federal Conv. Australian Water and Wastewater Assoc., Gold Coast, Queensland. 11–13 April. Australian Water Wastewater Assoc., West End, Queensland, Australia.

Chaney, R.L., and J.A. Ryan. 1992. Heavy metals and toxic organic pollutants in MSW-composts: Research results on phytoavailability, bioavailability, fate, etc. p. 451–506. *In* H.A.J. Hoitink et al. (ed.) Science and engineering of composting. Design, environmental, microbiological and utilization aspects. Renaissance Publ., Worthington, OH.

Chaney, R.L., and J.A. Ryan. 1994. Risk based standards for As, Pb and Cd in urban soils. DECHEMA-Fachgespräche Umweltschutz, Frankfurt/Main.

Falahi-Ardakani, A., F.R. Gouin, J.C. Bouwkamp, and R.L. Chaney. 1987. Growth responses and mineral uptake of vegetable transplants grown in a composted sewage sludge amended medium. Part 2. As influenced by time of application of nitrogen and potassium. J. Environ. Hortic. 5:112–116.

Hoitink, H.A.J., M.J. Boehm, and Y. Hadar. 1992. Mechanisms of suppression of soilborne plant pathogens in compost-amended substrates. p. 601–621. *In* H.A.J. Hoitink et al. (ed.) Science and engineering of composting. Design, environmental, microbiological and utilization aspects. Renaissance Publ., Worthington, OH.

Logan, T.J., and J.C. Burnham. 1994. Alkaline stabilization with accelerated drying process (N-Viro): An advanced technology to convert sewage sludge into a soil product. *In* D.L. Karlen et al. (ed.) Agriculture utilization of urban and industrial by-products. ASA Spec. Publ. 58. ASA and SSSA, Madison, WI.

Peverly, J.H., and P.B. Gates. 1993. Utilization of municipal solid waste and sludge composts in crop production systems. p. 189–199. *In* C.E. Clapp et al. (ed.) Sewage sludge: Land utilization and the environment. Soil Sci. Soc. Am. Misc. Publ. ASA, CSSA, and SSSA, Madison, WI.

Ryan, J.A., and R.L. Chaney. 1992. Regulation of municipal sewage sludge under the Clean Water Act Section 503: A model for exposure and risk assessment for MSW-compost. p. 422–450. *In* H.A.J. Hoitink et al. (ed.) Science and engineering of composting. Design, environmental, microbiological and utilization aspects. Renaissance Publ., Worthington, OH.

Sopper, W.E. 1993. Municipal sludge use in land reclamation. Lewis Publ., Ann Arbor, MI.

Wright, R. (ed.). 1994. Agricultural utilization of industrial byproducts, agricultural, and municipal wastes, USDA-ARS, Beltsville, MD.

3 Economic, Scientific, and Infrastructure Basis for Using Municipal Composts in Agriculture

Gary W. Hyatt

Procter & Gamble Pharmaceuticals
Cincinnati, Ohio

Agricultural, social, economic and political science bear on the development of solid waste compost usage for crop production in the USA. Following the introduction, this chapter begins immediately by answering a fundamental question: What is the economic benefit to the farmer of using compost as a production input? The financial models project that the value of residual N from composts applied over a series of years in a structured management program can be cost competitive with conventional N fertilizer. Suggestions for optimizing compost N value for even more favorable economics are demonstrated.

Following the financial analysis are summaries of basic and applied research projects that are currently sponsored by the Composting Council, a round–table association broadly representing the commercial and municipal stakeholders and participants in the composting industry. These studies are still underway, but as the majority approach completion, important findings on the safe use of composts in agriculture are forthcoming. When brought together in one place, it is easy to see there is a large body of supporting information challenging us to make broader use in agriculture of the organic materials we routinely discard (e.g., see TVA, 1975; Composting Council, 1992; and below).

THERE IS NO LACK OF INFORMATION ON COMPOSTING

This is at once encouraging and frustrating. Encouraging because we have volumes of data on which to base positive decisions for compost use in a wide variety of agricultural, horticultural and land reclamation applications; frustrating because decision makers are not receiving the information, or when they receive it, action is not taken. It seems society quickly forgets this historic work, and a good deal of replication is required—about once per generation, on average—to refine and re-establish our information and knowledge base.

How much information is available? The Composting Council (1992), through an agreement with the University of Washington, published a comprehensive review of the composting literature. This review, in book form, includes annotated references on many aspects of compost science, including preprocessing, processing, curing, refinement and usage. It refers not only to municipal solid waste (MSW) compost, but also to sludge biosolids, food, drug and paper manufacturing residuals and other composts. Several hundred academic, popular and government agency technical publications are summarized.

Computerized data bases also exist. The Composting Council, in cooperation with Cornell University, maintains and distributes a diskette-based bibliography of over 7000 annotated references, each retrievable by keyword search. Various institutions (e.g., the University of Florida, Auburn University) have composting data bases which they maintain and respond to inquiries from files.

The USDA compiled a list of ARS-authored publications in June 1993. The 373 titles (published 1971–1993) all focus on agricultural use of urban and rural organic and inorganic wastes. The USDA is further working to establish a national consortium, with an advisory board coordinated out of Beltsville, to research and develop sustained agricultural uses for urban and rural wastes. The expressed purpose of the consortium is to identify and prioritize research needs on an annual basis, solicit and review proposals, award grants, and monitor/publicize research results. As a first step, a team of authors has developed a document reviewing the agricultural use of municipal, animal and industrial wastes (Wright, 1993).

If publication is an indicator, scientific and popular interest in composting has grown in the past two decades. From 1970 to 1993 the number of citations on the subject of compost has grown to 11 353 (Fig. 3–1), including both U.S. and international references. The trend in publication is clearly upward, with a slight dip, for whatever reason, during the 1980s. The banner year was 1992 with 1316 citations—about 12% of total publications over the 23-yr period. Also notable, but not graphed, are patent citations from the Dialog Claims database for the early 1990s. In 1991 there were 35 U.S. patents issued for compost-related technologies, processes, and products. In 1992 there were 60. For 1993, as this was written in late September, there are already 51.

Clearly, good science and information are available. And excellent studies, described later, are presently underway. The action step for us, as scientists and farmers, is to make this information known, and to hold decision makers accountable to use it and build upon it. We need to stop reinventing this wheel of knowledge every 20 yr, and get on with returning our organic resources to the land from where they were derived, instead of locking them away in landfills or burning them.

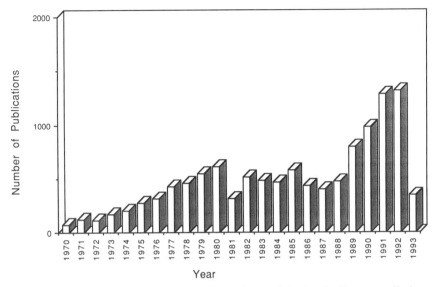

Year

Fig. 3-1. Distribution by year of 11 353 title and/or keyword citations for "compostx," where "x" stands for "-er," "-ing," "-able," etc. The author searched the following data bases: *Agribusiness, Agricola, Agris International, Chemical Abstracts, Conference of Paper Industries, Enviroline, Environmental Bibliography, Paper Chemistry, and Prompt.* Dialog's *Claims* database also was searched for compost patent information cited separately in the text.

FOR A REASONABLE COST AND EFFORT, COMPOST USAGE COMPLETES A BASIC ECOLOGICAL CYCLE

Today, with few exceptions, the movement of nutrients from soil-to-food/fiber-to-urban dweller is an open cycle. It is a one-way trip, which is ultimately unsustainable.

Once agricultural products are used in our cities, how can the nutrients, now converted to a variety of "solid wastes," be returned to the farm or forest? Part of the answer is the production and land application of composts made from the secondary organic resources we call "solid wastes." Returning composts to agriculture and silviculture closes the loop and sustains the organic C cycle. At the same time, part of our nation's solid waste challenge is solved. We increase crop yields, conserve resources, improve soils, and reduce nonpoint source pollution. It sounds compelling and simple. It is, in fact, complex and challenging, but nevertheless achievable.

A closed nutrient cycle including commercial-scale composting has many discrete operations. These begin at the household where organic residues are generated and discarded and continue through the collection and separation of compostable materials from other wastes, design and management of commercial-scale compost facilities, compost product curing, refining and distribution, and the ultimate application to the land. Each operation contributes associated cost. Each operation can have unique impacts on compost and crop quality, the environment, and the social/political processes of citzens and local governments. Both costs and impacts have been ana-

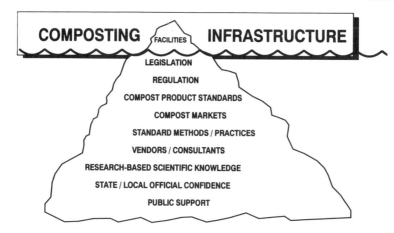

Fig. 3-2. The composting infrastructure conceptualized as an iceberg. Much activity must go on below the waterline in order to support the production and use of compost at the tip. Coordinating all of it is a singularly tough social, technical, scientific and political challenge.

lyzed in the past, and are under analysis today (see, e.g., Hyatt & Richard, 1992).

Compost production is only the tip of the iceberg representing the whole composting infrastructure (Fig. 3-2). All the activities below the "waterline," from legislation to research to public support, are needed to enable compost manufacturing facilities to exist, and to enable compost to be used. Getting them all started and coordinated is a singularly tough social, political, technical and scientific challenge.

A goal of the Composting Council, a national round–table association, is to build the infrastructure required to assure the success of composting in the USA. The Council is playing a vital and active role in all the discrete operations and activities cited above. The Council's primary objectives are to encourage the production of safe, high-quality compost derived from organic secondary resources, and to encourage its beneficial use in order to complete the C cycle of returning organic matter to the land. Council leadership has established six standing committees addressing key issues in composting infrastructure development. Agendas of the committees include fundamental research, development of technical guidelines, compost market development, enlistment of government support, and providing public education. A thorough description of the Composting Council, its organization and activities is given by Leege (1992).

SCOPE OF OPPORTUNITY

Does compost use have anything to do with sustainable agriculture? Compost production and use can be agronomically effective, economically attractive, environmentally sound, and socially acceptable: all goals of sus-

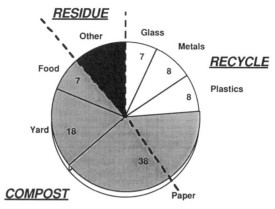

Fig. 3-3. The composition of the U.S. municipal solid waste stream. The heavy lines delineate the fractions which may be composted and recycled, and the fraction which must be treated in some other way, such as landfill or incineration.

tainable agriculture. Achieving these goals requires focus on the things commonly dealt with as throwaways in daily life: yard debris, food residuals, sludge biosolids, residential wastes and other organic materials commonly discarded as useless.

The use of urban and industrial wastes in agriculture is not a new idea. But societal changes have long since made collection of village "honey buckets" untenable. Today new challenges exist. Not only have the absolute amounts of discards increased, but their compositions also have changed.

Consider municipal waste. This material, by far, represents the greatest opportunity to recycle organic C back to the farm. In the 1990s, about 181 Mg (200 million tons)/yr^{-1} of municipal waste are discarded. Household and light commercial waste fractions in the USA may differ a percentage point or two locally, but typically 50+% of the MSW in the USA is compostable (Fig. 3-3). All of the food residue, yard debris, and a fraction of discarded paper is compostable (about 40% of the paper fraction, or 16% of MSW, consists of paper which is soiled beyond practical use in repulping operations or otherwise unrecyclable). When separated from the rest of the stream, these organics are the raw materials for MSW composting facilities.

Besides residential waste, commercial food residues (from restaurants, supermarkets or food/drink manufacturing), agricultural residues, barks, manures, and sludge biosolids/septage are all raw materials for composting. In a comprehensive, national scale analysis of compost production and usage, the Battelle Institute (Slivka et al., 1992) calculated that the USA, if it processed all of these compostable materials, could produce upwards of 45.4 Mg (50 million tons) of composts per year. In the same report, they calculated potential usage in the dozen or so major sectors at about 453.5 Mg (500 million tons). All 45.4 Mg of production, and much more, could easily be used in the agriculture sector alone.

In addition to the Battelle study, which examines compost use on a national scale, there are numerous local and regional usage studies. Many of these are cited and reviewed by the Community Environmental Council of Santa Barbara (1993) and by Shiralipour et al. (1992b). Most of these studies find that composts produced within the locality or region could easily find nearby outlets.

FOCUS ON AGRICULTURAL MARKETS: ECONOMICS, QUALITY AND BEHAVIORS

What's In It for the Farmer?

A "market" implies there is a material or commodity with an agreed upon value, for which users are willing to pay money. Supply and demand control composts, as with any other goods or services in a free market society. Composted materials are yet to enjoy broad penetration of the agricultural market. What are the issues facing compost producers as they attempt to establish agricultural markets?

Close behind the demonstration of safety and efficacy, a primary challenge is cost justifying the use of compost as a production input—as a substitute for standard commercial fertilizers. In a perceptive analysis by Parr and Hornick (1993), compost is viewed as having comparative, agronomic and economic value. Comparative value is calculated as the value of composts as substitutes for commercial fertilizers. This is based on the market value of the N, P, and K they contain, and ignores any soil conditioning, water holding, trace element, disease control, or other valuable properties composts may confer. Parr and Hornick (1993, Table 5) tabulate dollar value for the N, P and K in manures, biosolids and MSW. Comparative values range from $3.66 Mg^{-1} for MSW to $23.47 Mg^{-1} for cattle manure.

Agronomic value is related to the increase in crop yield or quality, while economic value is reflected in the value, in cash, of that increase to the farmer. There are many studies documenting agronomic benefits of composted materials (Shiralipour et al., 1992b; Appendix 3-1). Crop yields and qualities are known to increase. Shiralipour et al. (1992b) cite studies where tomatoes (*Lycopersicon esculentum* L.) are bigger and more numerous, melons (*Cucumis* sp.) are sweeter, flowers are taller and on and on. Increasing compost application rate shows diminishing returns to plant growth, however, with maximum rates of 50 to 100 Mg ha^{-1} being optimum.

Composts are low in total N (0.5–2.0%, generally), and traditionally have not compared well economically with chemical N fertilizers. But, unlike chemical N, compost N is "slow release" (in organic form), and keeps on yielding available N for several years after application. Thus compost economic value can accrue from the ability to produce yield responses in future crops. This is practiced by applying a relatively high dose (25–50 Mg ha^{-1}) in the 1st yr to improve soil physical properties and establish a nutrient level, followed by smaller annual applications (5–10 Mg ha^{-1}) for maintenance. Parr and Hornick (1993) cite Barbarika, et al. (1980) and point out that this residual effect needs to be accounted for to determine the true economic value of composted materials.

Cost Justification Based on Residual N for Compost Beneficial Use

Research has shown compost has valuable residual N (organic N that is mineralized slowly and available to plants from one growing season to the next). But what is that residual value, and what variables affect it? What would a best management strategy of applying composts require to take advantage of their great residual effects on soil? Such computations for compost have not been developed, although an attempt has been made to analyze value based on historic records of corn (*Zea mays* L.) yields under various compost treatments (Collins, 1991).

What is needed is a net present value (NPV) calculation to help farmers derive the financial advantages of using composts long term in preference to alternatives. In other words, given the costs of chemical N and its application, what is the value of avoiding these costs in the out years by using compost in a managed application program beginning today, counting on compost's residual N to provide that value? This is the economic equivalent to a future payment at a particular "discount rate" (The discount rate is the interest rate used in the discounting process. Discounting is the inverse of compounding. It is the process used to determine the present value of a cash flow.) I offer the following as a "jump ball" to start us thinking, and encourage readers to grapple further with the issue of compost cost justification.

The model reviewed below was developed predominantly through energetic conversations with David A. Miller, Director, Feed Grains, Soybean and Wheat Department, American Farm Bureau Federation, Chicago. He is an agricultural economist and added valuable insight to the problem. Our model pits the use of compost for the value of its initial and cumulative residual N against the annual use of synthetic fertilizer N. Nitrogen is the common anchor, mainly because it has an easily quantifiable price. The model does *not* include the value of P, K, or other macro/micronutrients in compost, or more difficult to assess parameters such as the value of time, the value of agronomic benefits (e.g., increased yield), soil conditioning or disease control benefits, or the advantages of reduced runoff and irrigation because of increased water retention by compost-amended soils. We have been fairly criticized for not including these latter items, since they do have value, and ultimately should be estimated and included in any cost-benefit analysis. However, our position is the most conservative approach, because adding these other variables will almost certainly increase the value of compost. The argument is stronger if economic benefit can be projected using only one nutrient, as we have done.

Some of the basic assumptions and parameters of the model (Table 3–1) were programmed into a common spreadsheet, using common financial equations. "What if...?" scenarios were then entered and solved. Mineralization rates, available N analysis, weight of required available N, and several

Table 3-1. Assumptions, parameters and initial values used in the compost economics
   model. Each parameter is variable.

Initial value	Parameter
10%	1st yr N mineralization rate
5%	2nd yr N mineralization rate
2%	3rd yr N mineralization rate
2%	4th yr N and later mineralization rate
$0.51 ($0.23)	Applied commercial N equivalent value in dollars kg^{-1} (dollars lb^{-1})
1.2%	Available N analysis of compost (%)†
$3.09 ($1.25)	Cost of compost application in dollars kg^{-1} ha^{-1} ($ ton^{-1} acre^{-1})
$0.00	Cost of compost
8%	Discount rate

† The weight of available N applied per acre also was a variable. Values of 112, 168 and
   224 kg ha^{-1} (100, 150 and 200 lb acre^{-1}) were used.

cost parameters were varied. The management program time basis was arbitrarily set at 25 yr, which, as you will see, is an adequate period to begin
understanding cost/future value behavior.

The numbers serving as our initial entries for the model (Table 3-1),
reflect "typical" values. The $0.51 kg^{-1} ($0.23 lb^{-1}) cost of chemical N includes application. The $3.09 ha^{-1} ($1.25 acre^{-1}) compost application cost
is fully burdened in order to make it comparable to applied conventional
N. With this set of assumptions, over the 25-yr period, the model tells us
compost has a total net return of $14.60 ha^{-1} ($5.91 acre^{-1}) greater than
chemical N.

Would U.S. farmers care about $14.60 hectare^{-1} over 25 yr) Maybe-
maybe not. Clearly, compost would *not* be an option if it turned in a negative value over the management period. But under this set of fairly conservative conditions it breaks even, plus a few dollars. To our knowledge this
is the first projection suggesting that compost can yield financial advantage
over conventional fertilizer, judged only on the value of residual N.

But, what changes in assumptions might produce a better or worse financial picture? We questioned the model further, producing the following
results. For the graphs that follow, points above the zero-axis represent economic benefits to farmers, while points below the axis represent economic
loss.

What Effect Does Mineralization Rate and Initial Nitrogen Content Have on Net Present Value of Compost?

Holding mineralization rate at 10% the 1st yr, 5% the 2nd yr and 2%
in each of the remaining 23 outyears, Fig. 3-4 shows the monetary value
of residual N when the compost N content is varied from 0.5 to 1.5%. The
graph shows, for three different required available N needs (112, 168, 224
kg ha^{-1} or 100, 150, 200 lb acre^{-1}), that farmers would break even against
using chemical N if the N content of the compost was about 1.2% (assuming all other assumptions are held constant as in Table 3-1, i.e., chemical
N value of $0.51 ha^{-1} or $0.23 lb^{-1}, etc.). As expected, this graph shows
that a low N analysis compost asked to deliver high available N is not very

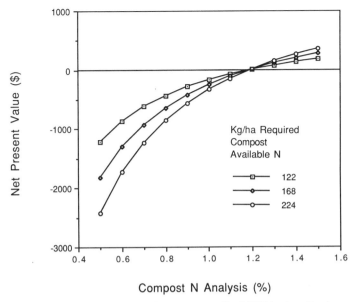

Fig. 3-4. Effect of a range of compost N contents at a fixed (10%) mineralization rate on net present value (NPV) for three different available N values.

cost effective. However, the break-even point against chemical N (i.e., 1.2%) is a reasonable value for the N present in typical MSW/biosolids composts.

Figure 3-5 shows a similar graph where the mineralization rates are 20% the 1st yr, 10% the 2nd yr and 2% in each of the remaining 23 outyears. As the mineralization rate goes up, there is a strong interaction bewteen it and available N that generates economic value. The break-even point in this scenario is about a 0.87% compost N content. A 100% increase in mineralization rate gives about a 30% reduction in required compost available N content. These are realistic mineralization and N values for MSW or biosolids composts, and econonmic parity as compared to chemical fertilizer could exist with careful, long-term management.

A family of curves was developed for varied mineralization rates, given a constant 112 lb ha^{-1} (100 lb acre^{-1}) of required available N yield (Fig. 3-6). A similar family of curves could be generated for any required available N demand. Even a slow mineralizing compost (5% in the 1st yr, 2% in the remaining outyears) could achieve cost parity over a management cycle if its available N content is high enough (approaching 1.4% in this case).

Carrying this logic one step further, compost available N percentage vs. mineralization rate is depicted (Fig. 3-7). The curve represents a family of break-even points (isovalue points), indicating equivalent tradeoffs between mineralization rates and available N contents. This graph says that a farmer would theoretically be indifferent to using any compost which falls on the curve, since each point on the curve represents a mineralization rate/available N parity to chemical N.

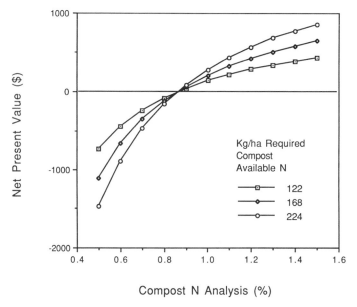

Fig. 3-5. Net present value at 20% mineralization over a range of required available N values
and compost analyses. Compared to Fig. 3-4, a higher mineralization rate shifts the compost
N analysis requirement down to about 0.87% to deliver parity against chemical N. Said another
way, doubling the mineralization rate reduces the N analysis requirement by about a third.

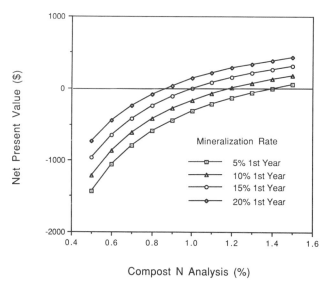

Fig. 3-6. Net present value for a range of mineralization rates for 122 kg ha $^{-1}$ (100 lb acre $^{-1}$)
available N. Compared to Fig. 3-4 and 3-5, this graph shows how mineralization rates and
N analyses affect compost NPV. The graph shows that even a slow (5%/yr $^{-1}$) mineralizing
compost could achieve cost parity with chemical N if its available N content is high enough
(i.e., 1.4%).

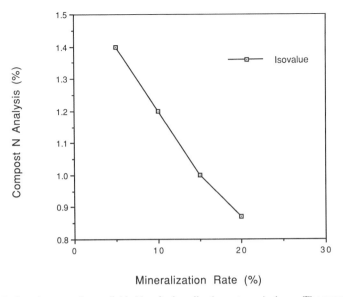

Fig. 3-7. Isovalue curve for available N and mineralization rate equivalency. The curve represents a family of breakeven points indicating equivalent trade offs between mineralization rates and available N contents.

How does Net Present Value of Compost and Available Nitrogen Vary with Time Across a 25-Year Management Horizon?

Figure 3-8 shows NPV of the added N, the net value of the compost, and the NPV of the available N in the compost through 25 yr. The assumptions are 10% mineralization rate the 1st yr, 168 kg ha^{-1} (150 lb acre^{-1}) required available N and 1.2% N analysis. Value of the added N reduces over time, since not as much compost needs to be added in succeeding years. The value of residual available N achieves a positive payback vs. chemical N in the 7th yr. The net value of added compost (lower curve) is never positive the year it is added, because of the cost of application. However maintaining the compost application regime keeps the NPV of available N positive for the duration of the management period. Figure 3-8 suggests a farmer would need to own, or at least have a 7-yr lease on the land for this scenario to be viable.

Figure 3-9 shows the relationship between compost residual N NPV, and mineralization rate across 25 yr. Composts with higher mineralization rates break even and maximize on average 5 to 6 yr ahead of those with slower mineralization rates. Composts of any mineralization rate, once reaching parity with chemical N, maintain a positive, slowly decreasing value throughout the 25-yr management period. According to this analysis, compost needs to approach a 20% mineralization rate to be practical/feasible for most leased landholding operations. Composts with lower mineralization rates may be more attractive to a land owner, rather than a short leaseholder.

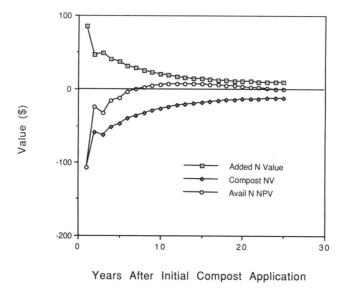

Years After Initial Compost Application

Fig. 3-8. Relation with time of compost net value, value of added N and NPV of available N assuming 10% mineralization, 168 kg ha^{-1} (150 lb acre^{-1}) available N and 1.2% N in the compost.

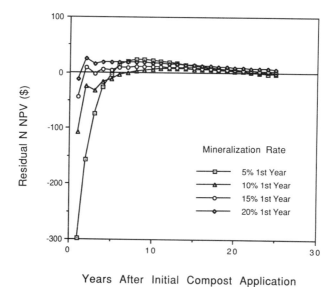

Years After Initial Compost Application

Fig. 3-9. The effect of mineralization rate on compost NPV over time [assuming 168 kg ha^{-1} (150 lb acre^{-1}) required available N]. Composts with higher mineralization rates break even and maximize on average 5 to 6 yr ahead of those with slower mineralization rates. The 5% rate reaches parity in 5 yr and maximizes in Year 9, the 10% rate achieves parity in Year 7 and maximizes in Year 11, the 15% mineralization rate sees parity in Year 1, and maximum in Year 8, the 20% rate Year 1 and Year 2, respectively.

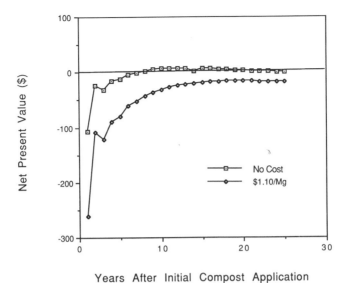

Years After Initial Compost Application

Fig. 3-10. The impact of cost to the farmer is shown in this graph. Even a modest cost of $1.10 Mg^{-1} ($1.00 ton^{-1}) prevents compost N from breaking even over the 25-yr management cycle. Assumptions are 1.2% compost N content, 168 kg ha^{-1} (150 lb acre^{-1}) available N applied and 10% mineralization rate, 8% discount. This graph raises the public policy question: Should municipalities subsidize agriculture to use composts?

The 5% mineralization rate demonstrates interesting behavior between Years 5 and 13. Under this scenerio, large amounts of compost need to be applied initially, due to the low mineralization rate. The value of residual N surges in the early to midyears due to the 2% outyear mineralization of the larger amounts applied up front. The overall NPV is negative, however, because of the high front end expense (Years 1–5).

What Impact Does Price of Compost Have on Net Present Value?

Figure 3-10 graphs the relationship of NPV to the price a farmer might have to pay per ton of delivered compost. At a cost of a dollar per ton ($1.10 Mg^{-1}) to the farmer, all other things being equal, the present value never reaches a positive payback. One thing this suggests is that municipalities, if they intend to approach agriculture as an outlet for their MSW and biosolids composts, need to build distribution costs into their tipping fees so there is no expense to farmers (assuming that value to the farmer is based solely on the compost's N content).

Here is an opportunity for the composting industry, agribusiness, economics and public policy to come together. Clearly, if the paybacks calculated above are attractive to farmers and they would pay for the agronomic benefits, then the waste generator should pay for the processing and handling costs. Specifically, delivery of compost to the user is no different from collection and "disposal" for the generators. After all, if the organics are

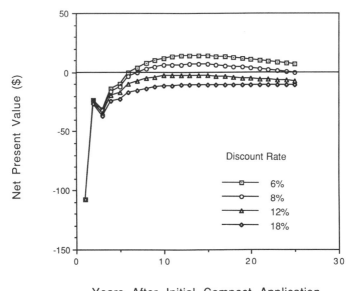

Years After Initial Compost Application

Fig. 3–11. Net present value vs. discount rate for four discount rates typical of those seen over the past decade or so. As the time value of money goes down, compost use becomes more attractive. Assumptions are as in Fig. 3–10. Sums of values over 25 yr for each discount rate are as follows: 6% = +$289.82 ha^{-1}, 8% = +$14.60 ha^{-1}, 12% = −$262.59 ha^{-1}, 18% = $440.88 ha^{-1}.

not composted, the generators pay the cost of delivery to the landfill or incinerator, as well as capital, operating and environmental costs associated with these facilities. These are benefits to society that the farmer should not have to pay for.

What is the Effect of Discount Rate?

Figure 3–11 graphs NPV across 25 yr, driven by four discount rates. The discount rates reflect actual values seen over the past 15 yr or so. There is little difference in Years 1 to 4, but higher discount rates take away all the incentive for using compost in the middle years of the management program. This graph suggests that compost use is much more attractive during times of low inflation and low discount rates. As the time value of money goes down, compost use becomes more viable. Compost use is more favorable in stable macroeconomic times. Similar to bonds, as the discount rate declines, long-term investment in agricultural productivity becomes much more attractive. When discount rates are high, the advantage lies in the short term. Under the assumptions of this figure, a farmer would best invest in compost if the discount rate was just below 9%.

What Happens if the Cost of Chemical Nitrogen Goes Up or Down?

The effect of synthetic fertilizer N price volatility on compost residual N value was analyzed (Fig. 3–12). The available N released per year is as-

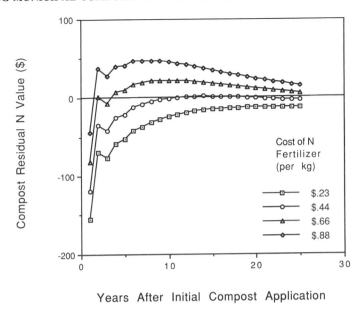

Fig. 3–12. The effect of chemical N cost on compost residual N NPV. Net present value is very sensitive to the cost of chemical N. A positive payout (not graphed) can be achieved in the 1st yr if the cost of chemical N is $1.17 kg ha^{-1} ($0.53 pound^{-1}) applied. Assumptions as in Fig. 3–10.

sumed to be 168 kg ha^{-1} (150 lb acre^{-1}) at all prices. The price of chemical N fertilizer does not have to rise much to make the NPV of compost residual N skyrocket. In the example, a doubling of chemical N price from $0.44 to $0.88 kg^{-1} ($0.20 up to $0.40 lb^{-1}) results in NPV for compost residual N rising over sevenfold from *negative* $158.44 ha^{-1} to *positive* $995.27 ha^{-1} (−$64.12– +$402.78 acre^{-1}). Under the same set of assumptions (not graphed), a positive payout can be achieved in the 1st yr if the cost of chemical N is $1.16 kg^{-1} ($0.53 lb^{-1}).

What Happens if We can Drive Down the Cost to Apply Compost?

Figure 3–13 shows the effect of compost application cost on return. The other assumptions are as in Fig. 3–10. Clearly, if application cost can be reduced, there is a major impact on the value of compost residual N vs. commercial fertilizer N. Like chemical N prices, compost application cost has a big effect. The graph shows that if application expense can be reduced from $3.09 ha^{-1} ($1.25 acre^{-1}) to zero, the NPV of residual N per acre rises 90 times. Indeed, by reducing application costs to $1.23 ha^{-1} ($0.50 acre^{-1}), compost N will pay out in the 1st yr against chemical N. This seems like another public policy opportunity.

What Happens if a Farmer's Wildest Dreams Come True

Figure 3–14 shows the net present value picture if everything is going right. The assumptions leading to this curve are: 168 kg ha^{-1} (150 lb acre^{-1})

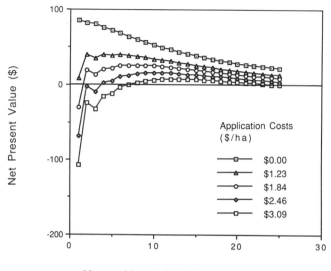

Years After Initial Compost Application

Fig. 3–13. The effect of application cost on NPV. At no cost to apply, the NPV is about 100 times greater than at $3.09 ha^{-1} ($1.25 acre^{-1}). Assumptions as in Fig. 3–10. Like chemical N cost, application cost has a big effect. At no cost to apply, the aggregate NPV over 25 yr is $1326.78 ha^{-1}, at $1.23 the NPV is $801.91 ha^{-1}, at $2.46 it is $277.04, at $3.09 it is $14.60.

available N applied, 6% discount rate, 1.5% available N in the compost, 20% mineralization rate, commercial N at $0.88 ha^{-1} ($0.40 lb^{-1}) applied, and no cost to the farmer for the compost or its application. With this set of assumptions, over the 25-yr period, the use of compost has a total net return of $2998.16 ha^{-1} ($1213.34 acre^{-1}) greater than chemical N, all the value accruing only from the investment worth of residual N in the compost. For comparison, Table 3–2 shows the aggregate values for 112 or 224 kg ha^{-1} (100 and 200 lb acre^{-1}) available N application.

The model was challenged to show the effects of alternative year applications under similar basic assumptions. The resulting curves were saw-toothed, trending generally in the same directions as annual application curves. Although we didn't pursue this analysis, if the case could be made for a lower application cost because of volume advantages (such as in out-years of the management plan, where smaller amounts of compost are applied annually), then further exploring alternative year application might be worthwhile.

A Plan for Understanding Compost Users

Currently, there is a three-sector model emerging to guide the infrastructure needed for sustainable markets and use of composts. The sectors are Potential Markets Definition, Standards for Production, and Specific User Segment Needs Definition. Activities in each sector are at various stages of

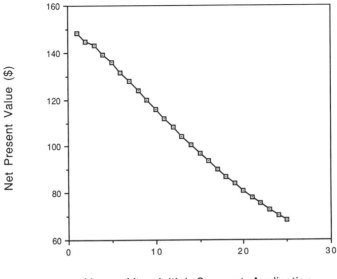

Years After Initial Compost Application

Fig. 3-14. This figure projects NPV under "ideal" conditions: 168 kg/ha^{-1} (150 lb acre^{-1}) available N applied, 6% discount rate, 1.5% available N in the compost; 20% mineralization rate; commercial N at \$.88 kg^{-1} (\$0.40 lb^{-1}) applied; no cost to the farmer for the compost or its application. The aggregate NPV for the 25-yr management period is \$2998.16 ha^{-1} (\$1213.34 acre^{-1}). See also Table 3-2.

completion and refinement by the Marketing and Standards Committee of the Composting Council.

In the first sector, Potential Markets Definition, a significant analysis of *potential markets* exists, built on demographic information and inferences from markets for similar materials, such as topsoil and peat. The study was commissioned by the Composting Council and performed by the Battelle Institute (Slivka et al., 1992). Battelle investigated the potential for use of composted products in the USA. Battelle was asked to address a range of composted products, including yard debris, sludge biosolids and other or-

Table 3-2. The value of residual N for three different available N needs. See also Figure 14. The "ideal" conditions yielding these values are: 6% discount rate, 1.5% available N in the compost; 20% mineralization rate; commercial N at \$.88 ha^{-1} (\$.40 lb^{-1}) applied; no cost to the farmer for the compost or its application. These are "ideal" conditions, but Figures 4-14 demonstrate that varying any one of them can yield positive payback within the bounds of a reasonable management program.

Mg (tons) compost applied		N available		NPV†	
ha^{-1}	(acre^{-1})	kg ha^{-1}	(lb acre^{-1})	\$ ha^{-1}	(\$ acre^{-1})
37.4	(16.7)	112	(100)	\$1998.79	(\$808.90)
56.0	(25.0)	168	(150)	\$2998.16	(\$1213.34)
74.6	(33.3)	224	(200)	\$3997.56	(\$1617.79)

† No net cost application scenario (see text and Fig. 3-14).

Potential Production Versus Applications

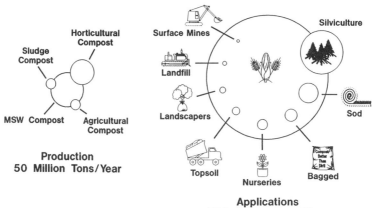

Fig. 3–15. Graphic representation of potential compost supply and potential demand in the USA. The circles are to scale and represent the relative magnitudes of each source and usage segment. The largest circle on the right represents the agriculture segment (from Slivka et al., 1992).

ganic secondary resource composts. Through trade associations, census data, and numerous other sources, they compiled what is generally accepted to be a reasonable estimate.

Battelle confined their estimates for compost outlets to nine user segments. Outlets range from agriculture to surface mine reclamation. These segments, along with the U.S. potential for compost production from all types of secondary resources, are shown graphically in Fig. 3–15 (with some segments combined to eliminate clutter). The size of the circles is in proportion to the size of the source or the outlet. The estimate for cubic years of capacity for application is shown on the right for the nation as a whole. Agriculture and silviculture are far and away the leading usage opportunities in the USA, with sod production running a distant third. Potential production from all sources is shown on the left. If you think in cubic yards, the conversion factor is about 2 yd of compost per ton (about 1.5 m^3 Mg^{-1}).

One assumption used in making these estimates is important. The potential usages had to be within 80 km (50 mi) of a population center of greater than 100 000 people. The distance constraint was added because of the perceived limit on the economic viability of shipping farther. Thus, Fig. 3–15 is a conservative estimator of outlets, it does not include outlets conceivably farther than 80 km from population centers of less than 100 000.

The significant finding from this study is as follows. The USA may someday, at full capacity, composting all its secondary resources, including 50 to 60% of its MSW, produce about 45.4 million megagrams (50 million tons), of compost per year. *However, the potential use for composted products in the USA is approximately 10 times the potential production of composts in any given year.* Current production is less than 20% of the total possible production. Today we make around 8.2 million megagrams (9 million tons) of composts per year from all sources.

Net, even allowing for competition among all products, such as biosolids compost, natural top soil and peat, there should be no doubt that composts from secondary resources, manufactured to realistic safety standards, have beneficial use in the nation in an array of outlets, which are yet to be fulfilled. What we have is a market development challenge, *not* a lack of outlets.

The full text of the Battelle report is available from the Composting Council at the address provided in the Acknowledgments. The full report contains very useful appendices which break down the potential outlets for compost on a state-by-state basis, providing useful perspectives for macro-planning.

In the second sector, Standards for Production, guidelines for the production of composts are under development by the Standards Committee of the Composting Council. The total ability to satisfy customers is embodied in using the best manufacturing practices. Best production methods involve product quality verification protocols, science-based process management, recognition of the concerns with facility siting, and it involves management of feedstocks to fit the ultimate consumer's usage criteria. The standards encompass quality criteria that are process and product safety related, and which therefore should be under regulatory control, as well as user criteria, which should be market driven and unregulated. How these criteria relate to the composting process and usage is described by Leege (1992) and diagrammed (Fig. 3–16). The figure shows *General Use Compost* complying with regulatory standards to protect public health, safety and the environment. *General Use Compost* is suitable for general distribution and use as a soil amendment. *Designated Use Compost* does not comply with regulated safety standards, and its distribution and use are subject to regulatory control. Notice that this composting model is bisected horizontally, with safety issues above the line, and marketing issues below. As diagrammed, *General Use Compost* is further amended, refined and/or cured to meet nonregulated individual user needs.

The third sector, Specific User Segment Needs Definition, is the greatest opportunity area for any of the Council's Committees. The current texts on quality in manufacturing define quality as, "fitness for use," or "conformance to requirements" (see, e.g., Deming, 1986; Ishikawa, 1985). That there are about 8.2 million Mg (9 million tons) of composts from all sources being used in the USA in various applications today (Slivka et al., 1992), suggests that, on a local level at least, compost marketers are understanding their consumers. There is a good deal of compost that conforms to user requirements, and users are acknowledging this in the marketplace with their pocketbooks.

But of the estimated 45.4 million megagrams (50 million tons) that could be produced, there are still 36.9 million megagrams (41 million tons) left that will need to be made, sold and used. As the compost supply increases, this is sure to lead to more vigorous competition among suppliers. Even though the potential uses far outrun the potential supply, producers need to be sen-

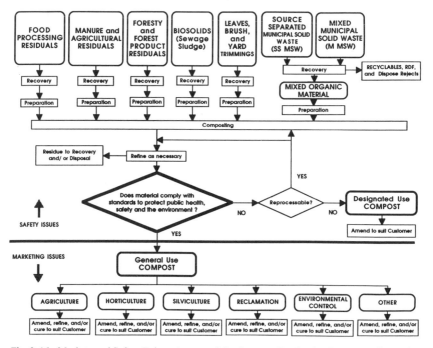

Fig. 3-16. Market- and Safety-Driven Aspects of the Compost Production Process as diagrammed
by the Standards Committee of the Composting Council [From Leege (personal communication, 1993)].

sitive to the fact that customers lost to poor quality may *never* return, and,
more important, may take a dozen or more other customers with them. Local
producers, unsophisticated in the ways of the consumer, will come and go
at the whim of the user.

A national market infrastructure needs to be built of stronger stuff. What
may be in place locally may not be good enough to assure beneficial use of
a future national supply. The Composting Council, through its Marketing
Committee, is building guidelines for successful use of compost, which
represent the voice of the consumer. Table 3-3 outlines the basic model being used with compost consumers to capture their needs.

THE COMPOSTING COUNCIL'S NATIONAL APPLIED
RESEARCH PROGRAM

The use of municipal composts in crop production is widespread throughout the USA. Appendix 3-1 broadly summarizes results of 61 applied research
projects ongoing in 17 states as of October 1993. Past uses, benefits and chemical and physical characteristics of MSW compost, and a variety of other
composts, have been reviewed elsewhere (Shiralipour et al., 1992a, b).

Table 3-3. Compost Council Marketing Committee model for capturing specific user segment needs definitions.

Section 1: Description of type use
Section 2: Description of material/compost required
Section 3: Method of use or application
Section 4: Guidelines for use of related materials
Section 5: Long-term maintenance
Section 6: Environmental considerations

Shiralipour's (1992b) paper contains a useful annotated table (Table 1, p. 268-269) of plant responses to MSW compost application.

Using the metaphor of the composting infrastructure iceberg cited above, if the applied research program is placed up at the tip there is a significant group of submerged activities needed to support it. Basic research, regulatory consistency and standards are among them. Without identifying individual projects, what follows is a synthesis of scientific findings. Some of the science and infrastructure basis for the work follows. Appendix 3-1 provides some details beyond this synthesis. Readers are encouraged to contact individual principal investigators with questions and for published results of their projects.

Through winter and spring of 1992, 23 research/demonstration projects, sponsored in part by Composting Council members and state and local government agencies, began around the USA, using composts as soil amendments on commercially important crops. Essential goals of the work are to: (i) demonstrate and develop data to document agronomic benefits, (ii) identify social, economic and political issues inherent in the safe and effective use of MSW compost, and (iii) communicate the results first to users, regulators, and public decision makers and then to the general public. The proximate emphasis is on performing and recording good science. The ultimate emphasis is on community-oriented actions, which can be expanded for state and regional adoption. The Composting Council commissioned an aggressive and impressive group of principal investigators who designed an array of experiments and demonstrations to meet the goals.

The projects focused on crops and soils common to the Northeast (7 projects in Connecticut, Maryland, Massachusetts, New Hampshire, New Jersey, and New York), Southeast (11 projects in Florida), and Midwest (3 projects in Minnesota). Two projects are in final planning and start-up stages in California and Washington. As this chapter is written, the crops in the majority of these projects have finished their 2nd yr.

The composts were obtained from commercial-scale composting facilities, and include composts made from separated MSW, MSW/biosolids, and yard trimmings ± biosolids. All composts meet individual state regulatory requirements for safety. The crops represent a range of species and agronomic categories of economic importance to each region; from flowers to trees, food to decorative foliage and turfgrass, fruits and vegetables for human consumption to forage for animals.

Besides the Composting Council, numerous groups and organizations are serving as cooperators on the projects. For example, universities, private industry, growers/farmers, state agencies and state and local governments are all active in planning and execution of the projects. All projects include public outreach and education components to help communities and users manage the changes in practice and perception inherent in the introduction, interstate transport, and use of composts. The USA can expect these projects to provide not only good science on the use of compost in various agricultural settings, but also to open channels of communication to make it easier to introduce, transport, and use secondary resource composts in various applications. During the 1992 growing season, all Council projects successfully grew crops, even though a few faced bad weather (including Hurricane Andrew), delayed harvests, and logistical delays during project start ups. Final results are expected in 1994–1995.

Besides the 23 projects described above, Appendix 3-1 includes a compendium of 38 additional projects in other states not sponsored by the Composting Council. *All results are preliminary*, and are subject to change through project completion and final analysis and evaluation. Some general treatment results are as follows.

Where cited in Appendix 3-1, yield data generally show about a 30% increase for compost-treated crops vs. controls. There is a general increase in available N, P, K. Compost used as mulch shows mixed results in weed control in two of the projects.

Regulated trace element concentrations in soils are well below USEPA CFR (Code of Federal Regulations) 40 Part 503 levels (for biosolids applied to the land), and they are restricted to upper soil levels. To date, 2 yr after application, regulated elements show no migration through the soil. There is no uptake of regulated trace elements by corn or grapes (*Vitis* sp.), although Zn levels were elevated vs. controls in corn root tissue in two projects.

Groundwater quality in Connecticut and Long Island, New York, showed no change vs. controls on compost-treated plots. There was an elevation of nitrate in soil pore water on corn plots in Seneca Falls, New York (on high rate application sites), and a concern for inorganic P levels after 2 yr of compost application in one of the Alabama projects. To date, there are no other documented physical or chemical concerns regarding plant nutrients or regulated trace elements.

Product maturity and physical contaminants are important agronomic and aesthetic variables. Of the 23 Council-sponsored projects, 10 confirmed that immature compost adversely affects germination and plant vigor. For example, grass seed germination and muskmelon maturity were delayed a couple of weeks, and early growth of tomato and pepper (*Capsicum frutescens grossum*) was reduced because of "nitrogen rob." Four projects mention physical contaminants, like glass, and soluble salt content (another potential cause of reduced germination). Three projects mention trace elements as a concern. These findings were shared with compost producers and suppliers for their action and continued improvement.

Final results from these projects are becoming available (e.g., Maynard, 1993; Stilwell, 1993a, b) and will be published individually by the principal investigators. What is clear is that there is a balance of substantial agronomic benefit, with comparatively minor negatives of inerts, delayed germination, early growth suppression from immature products, and macronutrient surpluses. Actual carry-over effects have not been quantified, but it is clear that yields of second crops are usually improved, reserves of organic N in the soil will increase with continued application, and will be available for slow release. We should not lose sight of the fact that these benefits would not occur if organic residues are landfilled or incinerated, instead of being recycled into compost. Nor should we lose sight of the fact that composts and composting processes can be improved to eventually eliminate the negatives.

THE COMPOSTING COUNCIL'S NATIONAL BASIC RESEARCH PROGRAM

Wholesome, affordable and safe foods for the nation's population is the goal of farmers and regulatory agencies. Currently, there is no scientific evidence indicating that use of composts meeting the rules and guidelines for human and environmental safety regarding regulated elements and other parameters is incompatible with that goal. Yet, several factors still remain to be verified to more fully support broader and ecologically appropriate uses of composts in agriculture. Some of the factors are scientific, others are social/perceptual.

Developing the Science Base for Use of Composts in Agriculture

One of the first things the Composting Council addressed was the critical gap in basic scientific knowledge about composting. In addition to all the good work of the past, what priority items were missing and needed to be known about composting? What was preventing composting from being accepted as a waste management option?

Through interviews with top compost scientists and manufacturers around the nation, a list of project topics was compiled and prioritized. Requests for proposals were written and distributed to qualified scientists for response. In June 1991 17 projects were funded. Seven were literature review and synthesis projects, covering volatile organic compounds, trace elements, lead sources in MSW, paper compostability, beneficial use and compost markets. The seven literature reviews were commissioned because the relevant information existed, it just needed to be brought together into one place. Five papers resulting from the reviews were published, along with several other papers on related topics, in a special issue of *Biomass and Bioenergy* (Hyatt & Richard, 1992). These reviews are available in bound form from the Composting Council. The other 10 projects are the fundamental research studies summarized below. As the majority of the projects, and others sponsored by the Council, approach completion, some important findings and publications are forthcoming from the work. What follows is a top line summary of the work. Please refer to cited publications for data and details.

Volatile Organic Compounds

The presence, characterization and quantification of manmade, or xenobiotic, volatile organic compounds (VOC's) in MSW composts and composting plants were not well characterized. Understanding VOC's was a primary interest of the Council. Briefly, the work on VOC's performed at the Connecticut Agricultural Research Station is highly significant in that it characterizes and quantifies for the first time the types of VOC's present in and around active, commercial-scale MSW composting plants (Eitzer, 1992). Eight plants were sampled. Air was sampled ($N = 161$ individual samples) from various locations within and outside the facilities. A suite of 67 target VOC's was chosen for analysis by thermal desorption gas chromatography/mass spectrometry.

Over half ($N = 35$) of the target compounds were never detected. The rest were identified and quantified. Results showed that, regardless of the sampling location, concentration levels remained well below (2–4 orders of magnitude below, on average) permissible exposure limits for workplace air established by the American Conference of Governmental Industrial Hygienists. Concentrations are also well below Office of Safety and Health Administration (OSHA) safety recommendations. The only location where concentrations were close to guideline levels was at the tipping floor itself, where fresh MSW is delivered to the facilities. But even here, adequate ventilation controls the elevated levels. Measurements of type and quantity of VOC's are in rough agreement with those cited by Kissel et al. (1992) in a review of VOC's near landfills. Another finding was related to the variability between samples. Results indicate great similarities in VOC species and quantities across facilities from different locations, running under different operating conditions and feedstocks. These findings suggest that VOC's present in MSW are fairly consistent regardless of the method used to do the composting. Interestingly, greater variability appears among samples taken in different locations within the same facility. Highest to lowest VOC levels were detected in the tipping piles, near the shredders, in the active composting areas and in curing piles, respectively.

Environmental Fate and Plant Uptake of Regulated Trace Elements and Xenobiotic Organic Compounds

There are four terrestrial research groups sponsored by the Council. Two groups are performing work on trace metals: one project on fate and transport in soils (Ohio State University), and another on uptake by plants (University of Washington). Two parallel projects are examining fate and transport (University of Iowa) and uptake by plants (North Carolina State University) of organic compounds. These four groups have built and demonstrated an important cooperative synergy, and represent an investment of expertise for the composting industry.

The University of Washington researchers deserve considerable credit for taking on the difficult job of compost supplier to the other three terrestrial research groups. This important task was performed as follows. Separated, size-reduced organics from MSW from the Bellingham, Washington, compost facility were delivered to the University of Washington site at Eatonville. Each batch of MSW was custom enriched with regulated metals (to 503 regulated levels, using acetate forms of Se, Hg, Zn, Cu, Cd, Ni, Mo, and chloride form of Cr) or organics. Then compost was carefully made in small batches, screened, tested and shipped to the other campuses as the standard material for experimentation. Thus, all the fate and uptake studies are standardized on one compost source.

To date, these studies of mobility and uptake of metals and organics support the safety of MSW composts used on the land, even when the compost is artificially enriched to full regulatory limits for metals, or "spiked" with organics. The data are beginning to illustrate the low probability for long-term hyperconcentration in soils of regulated metals or organics, as organic C metabolizes. Organic C is stable and regulated metals and organics are not in bioavailable forms.

Plant studies of regulated metals at the University of Washington show no effects of application rates or type of compost on plant uptake. Seldom were plant concentrations above control levels. The single exception is Cd in "spiked" compost.

For the organics fate study, a polychlorinated biphenyl (PCB) (Arochlor 1248), benzo(a)pyrene, bis-ethylhexyl phthalate and chlordane (1,2,4,5, 6,7,8,8-Octachloro-2,3,3a,6,7,7a-hexahydro-4,7-methano-1H-indene) were chosen as "model" organics. In Iowa, field sites were instrumented, planted and sampled periodically to measure the movement of organics through the tilled soil and into the deeper soil profile. The plants being investigated are fescue (*Festuca* sp.), corn and poplar (*Populus* sp.).

Findings show these compounds are strongly sorbed to compost organic matter and do not migrate in the short term (Fannin, 1993; Hsu et al., 1993). The higher the percentage of compost in the soil, the more tightly bound the organics are held. Biotransformation rates for PCB and chlordane increased slightly in the presence of compost. Benzo(a)pyrene was far more tightly bound to the soil than expected.

Fannin (1993) states: "The combination of biotransformation and compost-enhanced soil holding properties shows promise as a remediation method. . . . Partitioning into soil and/or compost does not destroy harmful compounds, but it can extend the time available for other removal mechanisms to operate. Moreover, results indicate that trace organic chemicals in land-applied compost will not infiltrate to groundwater or migrate in runoff water unless the soil/compost particles become entrained via soil erosion." Field data closely fit computer model (USEPA Pesticide Root Zone Model II) predictions for the fates of these organics. Longer-term modeling suggests (through 5 yr) the organics stay in the rooting zone and slowly biode-grade (Hsu, 1992; Hsu et al., 1993).

In North Carolina, the organics uptake study is still underway. Analytic methods for soil and plant PCB have been developed and validated, with recoveries of PCB averaging 90%. Greenhouse results to date show no adverse germination or growth effects from test organic-amended compost. Plant species are potato (*Solanum tuberosum*), lettuce (*Lactuca sativa*), wheat (*Triticum aestivum*), peanut (*Arachis hypogaea*), petunia (*Petunia* sp.) and fescue.

An important benefit of the terrestrial studies is the potential for the data to be used to support a risk-based picture for the use of MSW composts. The USEPA invested over a decade in developing protocols and methods for beneficial use. These were tested on biosolids sludge and resulted in the 503 rules. The USEPA is now in a position to use this body of knowledge and methods to test other beneficial use materials, such as MSW composts. The relevant, field-scale methods to test other beneficial use materials, such as MSW composts. The relevant, field-scale data, forthcoming from these terrestrial studies have the potential to demonstrate how trace elements behave in MSW-treated as compared to biosolids-treated soils. These long-term tests are needed to see if assumptions for a risk analysis (e.g., changes in uptake coefficients) for MSW compost are justified. The data will point to usage standards which are "protective" in the long term, and will be useful to support any future changes in regulations.

Compost Stability and Maturity

The maturity projects (University of Cincinnati, University of Illinois and Ohio State University) provided workable measures of compost stability. The objectives of this work were to: (i) develop and validate a set of practical maturity assays which could be used by compost suppliers and customers routinely to assess compost product maturity on the spot at facilities, rather than at specialized laboratories, and (ii) identify the critical levels for each maturity assay or, ideally, a single "maturity index" based on the results of several assays, which will provide compost users with a quantitative measure of maturity.

Three potentially useful assays have been developed: colorimetric (M. Cole, personal communication, 1993), dissolved O_2 respirometry (Frost et al., 1992; Iannoti et al., 1993) and enzymatic (Herrmann & Shann, 1993). Reactive C (via colorimetry) is a good measure of maturity, but not as good a measure as plant growth (i.e., soluble salts or root pathogens may be present in otherwise mature compost, which may hamper growth). Dissolved O_2 respirometry shows reliable descending curves for percentage average saturation, which correlate with increasing compost stability. It also predicts maturity in terms of the potential for growth of rye grass (*Lolium* sp.). CO_2 absorption respirometry is insensitive for any but the most unstable compost samples. The enzymes alkaline phosphatase, acid phosphatase, esterase (FDA), lipase (caprate C-10), glucosidase, and cellulase have been tested in vitro. Lipase is presently considered the best candidate for a maturity indicator, since it appears beyond 100-d curing time (Herrmann & Shann, 1993).

Mobility and Bioavailability of Regulated Elements on Applied Research Sites

The unrestricted distribution and use of MSW compost requires data to show that trace levels of regulated metals are safe for the environment or organisms. Compost is most often analyzed for total nutrients and metals, with little emphasis on bioavailability. Speciation of metals in soils or soil amendments controls the bioavailability of the metals to plants, and determines the potential to contaminate ground or surface water. Among the questions which come to mind are: How much of a given element is leached with water? What forms are the elements in? How much is bound to the soil organic matter, and how tightly? How deeply will surface-applied elements penetrate the soil?

To come to grips with some of these questions, soil profile samples were collected from the high application, low application and control plots of each Council-sponsored applied experiment in the Northeast (Connecticut, Maryland, Massachusetts, New Hampshire, and New York plots). The agronomic projects opened a unique window to collect samples, and to analyze the potential bioavailability and other physicochemical characteristics of trace elements and micronutrients under actual field conditions. Especially important was that for all experimental sites, MSW/biosolid compost from a single source in Delaware was applied at similar rates, at approximately the same time, at approximately the same latitude, to 10 distinctly different Northeast soil types (identified in Appendix 3–1). *The significant variable is the soil types.* The intent is to develop a database for the fate and speciation of metals in different northeastern soil types amended with the same MSW compost.

All totaled, 668 samples were collected from 77 1-m-deep profile pits, at nine Northeast sites. Results to date show, for all soils, regulated metal totals are confined to the top 10 to 15 cm of soil, as illustrated for the concentrations of Zn across an 80-cm profile (Fig. 3–17). The differences between the amendment and control plots is generally less than 10 mg kg^{-1} at depths greater than 15 cm. Zinc levels range from 35 to 707 mg kg^{-1} in the 0- to 10-cm layer, depending on loading rate. The other regulated metals show similar curves (data not plotted). Metals were not observed to be mobile at these sites, even after 6 mo of weathering in the soil.

Controls compared to amended totals for regulated metals in the 0- to 10-cm horizon only, contrasted with the analysis of the Delaware compost applied to all plots, shows maximum metals levels are found at the South Deerfield site, where compost was surface applied as a mulch (Table 3–4). Where compost was tilled into the soil, metal concentrations reflect loading rates.

There was especially close agreement between the control and amended levels between the two sites with 112 Mg ha^{-1} (100 tons acre^{-1}) of compost applied (Winding Brook and Kingman Farm) and the two sites with 56

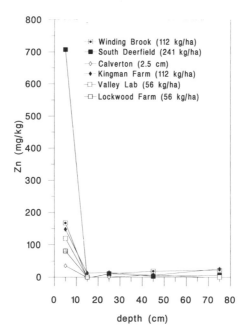

Fig. 3–17. Concentration of total Zn in soil profiles collected at six differnet northeast locations. Presence of Zn essentially drops to zero at 15-cm depth. This study is still in progress (McAvoy, Kerr and Hyatt, unpublished data), but indications are that other regulated metals behave about the same. See Table 3-4 for supplementary information. Information given with icons includes name of study site and compost application rate.

Mg ha^{-1} (50 ton acre^{-1}) (Lockwood Farm and Valley Lab). All these soils are sandy loams or loamy sands. Further, compare these with the heavier clay soil found at Seneca Falls. We do not have enough data to postulate a reason, but Seneca Falls seems to yield higher totals than the sandy soils even when the compost application rate on sandy soil (Kingman Farm) is higher than Seneca Falls. Complete data are in preparation and will be published in late 1994 (McAvoy, Kerr & Hyatt, unpublished data). This work will include information on metals fractionation and speciation, and an assessment of metals bioavailability.

Final results from all the fundamental research projects are being published as they become available from the principal investigators (e.g., He et al., 1992; Mielke & Heneghan, 1991; and others cited above). Results to-date demonstrate there is no science-based reason not to proceed with composting as a method of recycling organic resources, and there is no reason not to use composts as soil amendments in agricultural applications.

OTHER KEY SCIENTIFIC, SOCIAL AND REGULATORY SUPPORTS

A range of important, yet underdeveloped issues are important to the agricultural use of composts. These issues span many disciplines from en-

Table 3-4. Analysis for total regulated metals from eight Northeast soils.†

Sites	Cd	Cr	Cu	Hg	Ni	Pb	Zn
				mg kg^{-1}			
DE‡	3.73	36	368	--	172	376	990
WB	0.072	36.08	14.63	--	21.45	13.81	60.45
112	0.862	46.37	93.73		50.32	136.15	227.90
SD	0.228	30.20	57.47	--	25.32	15.46	71.10
241	3.273	89.47	319.12		114.33	250.52	778.42
MD	0.042	0.00	4.95	--	2.97	2.97	11.87
44	0.804	21.41	59.36		22.38	56.44	140.13
SF	0.174	8.33	11.58	--	22.19	18.33	60.78
83	1.300	49.78	177.66		73.21	141.54	358.25
CV	0.095	19.48	41.87	--	7.79	24.34	25.32
1″	0.202	24.73	70.24		13.85	38.58	60.35
KF	0.181	24.55	9.82	--	17.67	29.45	47.13
112	0.655	49.01	76.46		43.13	84.30	195.07
VL	0.115	15.82	11.87	--	19.78	8.90	33.63
55	0.464	33.45	59.03		34.44	55.10	153.48
LF	0.167	20.86	16.88	--	0.00	14.90	35.75
55	0.448	34.66	57.43		25.75	43.57	116.85
APL	23§	1200	1500	17	420	300	2800

† The top number is the control measure. The bottom number is the experimental measure. Soil types, crops and other details are cited in Appendix 3-1 for the respective studies.
‡ DE = Analysis of MSW/biosolids compost from the Delaware Reclamation Authority, Newcastle, Delaware. This compost was used on all the following experimental sites. Numbers under the site identifiers are Mg compost applied: WB = Winding Brook Turf Farm, Lyman, Maine (part of University of New Hampshire project); SD = South Deerfield, Massachusetts (Univ. Massachusetts); MD = Salisbury, Maryland (Univ. Maryland Agric. Exp. Stn.); SF = Seneca Falls, New York (Cornell Univ.); CV = Calverton, New York (State Univ. New York, Stony Brook); KF = Kingman Farm, New Hampshire (Univ. New Hampshire); VL = Valley Laboratory (Connecticut Agric. Exp. Stn.); and LF = Lockwood Farm (Connecticut Agric. Exp. Stn.).
§ Part 503 levels for regulated metals (*Federal Register*, 1993).

gineering to economics to politics. Some are discussed below, indicating that solutions are within reach.

Biological Process Control

In order to be meaningful as a national management option, composting must be performed on a commercial scale. Making compost, whether from complex feedstocks like MSW or from more homogeneous feedstocks like leaves, is not just a matter of stacking material and letting nature takes its course. If the composting process is not in control, the quality of the end product is not in control and is unpredictable at best.

Commercial-scale composting must be seen and approached as a sophisticated manufacturing process, with all the exactness of biological process control and engineering methods applied to make a useful product. In today's mix of composting technologies, some facility designers and builders manage these fairly well, others haven't caught on yet. These latter cause problems for the whole industry, which manifest themselves as odor nuisance, useless compost and plant closings. They give commercial composting, a fragile industry in the USA, to begin with, a bad name.

A better understanding of biological process control is needed to assure more consistency in compost quality. Among the process variables that are critical to successful aerobic composting are pH, temperature, moisture and aeration. The last is key to controlling the three others. Simplistically, traditional wisdom says to reduce the size of the organic feedstock, establish a critical mass of moist material in a pile or rotating drum, and blow lots of air through it to keep the aerobic microbes operating at peak efficiency.

However, there is some evidence from the literature (Finstein et al., 1980; Nakasaki et al., 1992; Stutzenberger et al., 1971) and from our laboratory data (Pettigrew, Kain, & Hyatt, unpublished data) to suggest there is another way to proceed. In fact, it might be better to limit aeration early in the process. By doing so, pH is kept low, and ammonia N is converted to the less volatile ammonium form. Nitrogen is retained in the mix, resulting in two major advantages: (i) less ammonia odor, and (ii) lower C/N ratio, which offers the opportunity to produce a higher-quality finished compost. The exact control parameters, and the best technology design(s) to achieve them, are yet to be demonstrated. More work needs to be done to determine the best control methods and specifications. This is a great opportunity area for the composting industry.

Regulatory Consistency

Today, only a handful of states regulate compost production and use. These states differ in their classifications of composts, testing methods and frequency, and in their definitions and specifications for safety and use.

To help move things along, the Composting Council developed and made available to state governments, guidance on legislation and regulations. These models are being used in formulating and/or modifying individual laws and rules in California, Florida, Maryland, New York, North Carolina, Ohio, Pennsylvania, Texas and Washington.

Another landmark is the publication of the national sludge rule, CFR 40 Part 503 (*Federal Register*, 1993), which provides guidance on allowable trace element levels to state regulators. The USDA spent over a decade constructing a risk-based approach for the use of sludge biosolids on the land. Extensive agronomic research and analysis, across a range of crops and soil types, coupled with laboratory studies of bioavailability/human health, and ecological and toxicological analysis of 14 food chain pathways, have yielded the CFR 40 Part 503 rule based on the risk analysis data base for sludge-containing products (including composts) applied to the land. Details of the studies, and the risk analysis methodology, are well described by Ryan and Chaney (1992) and Chaney and Ryan (1992). The rule covers land filling, incineration, distribution and marketing of biosolids materials, and includes their application to agricultural and nonagricultural lands. A summary (Table 3-5) of the key elements of the rule and their meanings regarding levels of regulated metals acceptable for land application was drafted with input from Dr. Chaney and Dr. Ryan (personal communication, 1993).

The regulation also covers trace elements, organic chemicals and pathogens. It considers individuals within an exposed population, as well as the general population as a whole (aggregate risk analysis). The contaminant loading of the most restrictive exposure pathway [i.e., the pathway affecting the Most Exposed Individual (MEI)] becomes the upper discharge limit for each contaminant. The logic is, if the MEI is protected, then the rest of the population is protected.

Table 3-5. Definitions of "Ceiling," "Cumulative Load," "APL," and "APLR" as used in the National Sludge Rule (CWA-503) signed on 25 Nov. 1992.†

Ceiling (pollutant ceiling concentration) is the highest permissible level of a given trace element measured in milligrams of element per kilogram of sludge, above which sludge biosolids may not be *applied to land*. Below the Ceiling level, material may be *applied to land by site-specific permit*.

For example, sludge biosolids containing at maximum 4300 mg kg^{-1} Cu (which is above the APL and therefore is not suitable for unlimited distribution and marketing—see APL) may be land applied as long as a site specific permit is acquired. More than 4300 mg Cu per kilogram sludge it may not be applied to land under any circumstance.

Cumulative Load is the highest permissible total final level *applied to land* of a given trace element, measured in kilograms of element per hectare of land, *with a site specific permit*. *Bulk* material may be applied until Cumulative Load is reached, as long as the material is below Ceiling.

For example, sludge biosolids containing Zn between 2800 (APL) and 7500 (Ceiling) mg kg^{-1} may be applied to the land until the Cumulative Load reaches 2800 kg Zn ha^{-1}. Records and files must be kept of the application activity and reports prepared for USEPA on an annual basis.

APL, or Alternative Pollutant Limit, is the highest level of a given trace element, measured in milligrams of element per kilogram of sludge, which may be marketed *without restrictions*, based on trace element content. The sludge biosolids may be used on vegetable gardens or accidentally ingested.

For example, sludge biosolids containing up to 1200 mg Cr kg^{-1} or 300 mg Pb kg^{-1} may be used in food crop culture, in any amounts, without ill effects due to lead or Cr toxicity, and without concern for environmental degradation.

APLR or Annual Pollutant Loading Rate is the highest annual rate of *application to land*, in kilograms of element per hectare per year, for sludge biosolids containing trace elements at or above the APL. The APLR is generally for use with bagged materials. If a product contains levels of elements greater than APL, but less than Ceiling, then it may be sold in bags if, by labeling, it is constrained to loading applications less than APLR.

For example, sludge biosolids containing 4000 mg kg^{-1} Zn (= 4 kg Mg^{-1}; higher than APL, but less than Ceiling) must be labeled for application of not more than the APLR equivalent of 140 kg ha^{-1} yr^{-1} (2800 kg ha^{-1} divided by 20 yr). In the case of the 4000 mg kg^{-1} material, the annual loading rate is 35 Mg ha^{-1} yr^{-1} (140 kg ha^{-1} yr^{-1} divided by 4 kg Mg^{-1}).

† All weights are "dry weight." For "bulk" and "bag" definitions, there is a Boolean rule for permissible distribution criteria which simplifies the distinction: BULK = ("ceiling" *AND* cumulative LOAD) *OR* APL; BAG = ("ceiling" *AND* APLR) *OR* APL.

Continued work is needed to assure composting and compost use is viewed as a broad, national solution to waste management. The CFR 40 Part 503 provides a good, science-based, national structure with which to frame state rules. For free and open commerce in composts to occur across state boundaries, the states will need to develop a consistant regulatory approach.

Universal Standards

A word describing the state of compost standards today is "diverse." For example, the USEPA does not have a mandate to develop standards except for sludge biosolids. Internationally, there is an apparently arbitrary, or at best inconsistent, approach to standards. At the state level in the USA, some regulators survey the diversity and arbitrarily choose numbers and adopt them as their own rules. Or, they attempt to base rules on water standards or state soil remediation standards, which have little relevance to compost. This leads to a further extension of an arbitrary vs. a science-based approach to standards. This results in public policy resting on arbitrariness and a lack of common quality verification protocols. Furthermore, some watchdog organizations do not accept or acknowledge the science and inject fear, indecision, and a debilitating conservatism as they critique the industry.

To move the industry ahead, the Composting Council is trying to lend credibility to the reapplication of sludge limits to all composts, including MSW compost (see the basic research work on metals reviewed above). To this end, the Council supports the USEPA recommendations for metals and pathogens, except where the USDA has provided the scientific basis for the modification of a few of the metals limits. The USDA has written a letter of exception regarding Cd and Mo to the USEPA. The Composting Council's Standards Committee has reviewed and supports the USDA justifications for the exceptions, and has suggested that the USDA recommendations be accepted in the case of the lower limit for Cd.

To the 503 baseline, the Standards Committee of the Composting Council is adding additional (unregulated) biological, chemical and physical limits, including biological stability, pH, inerts/sharps, and soluble salts. Because O_2 uptake correlates with biomass reduction status, the stability measure is included to specify when samples should be analyzed to determine compliance with weight-based contaminant concentration limits. Compost pH was included because it can relate to metal and nutrient mobility and availability, apparent compost stability, and phytotoxicity. Manmade inerts/sharps include glass and metal fragments, needles, pins and films that could pose a human or animal hazard through unprotected exposure or through ingestion. Soluble salts, while not a safety issue, is a phytotoxicity attribute which affects marketability. The Council chooses to include these nonregulated limits in order to enable broadbased distribution without additional regulatory control in the use of composts.

A characteristic of the composting industry is that products leave the plants both as a function of technical capabilities and of customer applications. This attribute must be reflected in sampling methods so that testing remains valid across a range of customer/technology variations. To establish protocols for sampling and testing, the Standards Committee has undertaken a cooperative effort with the Minnesota Office of Waste Management and its contractors to monitor compost products on a monthly basis from eight facilites in Minnesota during a 2-yr period. Sampling methods are being developed that will account for the diversity in customer demands and processing technologies. Both sampling and analytic test methods are being refined, proven, and documented. A report on recommended sampling and test methods is planned for publication in early 1996.

DISCUSSION AND CONCLUSIONS

This chapter in no way reviews all the available literature, and it hasn't begun to do justice to the Council-sponsored work. But I hope the macrosources cited above, namely the Composting Council's bibliography and diskette database, Appendix 3–1 and the commercially available databases listed in the caption of Fig. 3–1, give interested readers plenty of entry into the field. The perception of product safety, economics and markets comprise the major challenges to the successful use of composts in agriculture and elsewhere.

Product Safety

A broadly acceptable, federal rule for safely applying biosolids to the land is on the books (*Federal Register*, 1993). A similar rule for MSW compost (if one is found to be necessary) does not exist yet. While all the data on MSW composts are not in, there is emerging a pattern of low probability of bioaccumulation, and low probability of migration of regulated trace elements or organics in soils from the application of MSW composts.

Studies of total metals, including the survey across northeast soils, show that regulated metals don't migrate beyond the plow layer. Metals fractionation studies show strong binding of cations to the organic matrix, leaving regulated elements unavailable for plant uptake or leaching to groundwater.

Most composting facilities operating in the USA today produce MSW-containing composts which meet federal limits (Table 3–4, bottom row) for regulated elements in biosolids. The metal closest to surpassing the federal limit is lead. Waste producers, through source reduction, and compost production facilities, through process control, can do more to reduce the levels of regulated elements, especially lead, in finished compost (Epstein et al., 1992).

The Composting Council's Standards Committee continues to develop guidelines for aesthetic and safety variables which attempt to go beyond any state's regulatory requirements. The point is to develop a self-regulated industry, since it is fairly clear that if the industry does not regulate itself, it will surely stimulate regulation by others.

Economics

A topic that has great opportunity for exploration is the cost of applying safe compost in agricultural settings in comparison to the use of conventional fertilizers. The economic analysis in this chapter lays out for the industry several opportunities and hurdles that might be encountered in moving compost into commercial crop production. Parr and Hornick (1993) imply a break-even point at which the agronomic and economic values of compost vs. commercial fertilizer balance. Modeling presented in this chapter confirms this relationship and suggests that there is value in working further to establish the impact and value of compost residual N, and adding to that the potential value of other macro/micronutrients, improvement in soil tilth and other parameters that are more difficult to quantify.

"Traditional wisdom" says that compost is "not a fertilizer." Yet, the above demonstration of a positive NPV vs. chemical N, under a variety of fairly conservative conditions, define it as an economically viable alternative to chemical fertilizer. If safety and economic conditions are met, farmers may well use composts to meet their crops' N needs. Surely, the crops don't care where their N comes from.

Preliminary guidance coming out of the modeling suggests we should focus on using composts with a measured mineralization of at least 10 to 15%, with 1% or more available N (the good news is that higher mineralization rates and available N are routine characteristics of composts made with sludge biosolids). A management cycle of at least 12 yr should be committed to. With the set of assumptions made for the modeling above (Table 3-1), the economics of using MSW and sludge composts are fragile, and compost that has to be paid for by the farmer solely for its N benefit may find difficulty being marketed successfully. The analysis might be better if the full suite of macro/micronutrients is included. It seems logical that including P and K in the analysis would make compost more competitive because these elements add value without adding cost—P and K are present in the compost as applied, whether or not they are figured into the value.

At this early stage of compost cost exploration, computation and justification there are still many alternatives. First, as illustrated, is to base the justification on the residual effects of N in compost applied year after year (the "net present value" approach). Alternatively, the cost of spreading and incorporating could be borne by citizens in the form of higher tip fees at the composting plant, or a few cents more on their water bills in the case of biosolids, especially if it is judged that it is in the public interest to recycle organic residues rather than landfill or incinerate them. A third alternative might be to provide low interest loans or develop a system of shared equipment ownership for delivering and applying composted materials. These kinds of public policy decisions can have a lot to do with reducing the cost to the farmer. A fourth is more broadly to investigate ways to apply larger amounts of material less frequently, thus saving application costs (the lime model). A fifth alternative is to include the cost avoidances which accrue to the disease control, water conservation, and soil conditioning benefits of using composts. Building on these last criteria, the question might well be asked whether it is possible to get better *overall results* with 36 kg (80 lb) of N delivered via compost vs. 45 kg (100 lb) of N delivered via chemical fertilizer. Working these factors into the model may significantly change the value picture, most likely for the better. Therefore they deserve rigorous and immediate research attention.

Markets

There has been some informal conversation with the Composting Council that the estimates of absolute amounts of agricultural waste materials (manures, especially) fall short of the mark defined by Slivka et al. (1992). Their documented (Anonymous, 1972) estimate is 1451 million wet megagrams (1600 million wet tons) per year of cow, poultry, sheep and pig feces and urine, which would yield about 163 dry megagrams (180 million dry tons) of compostable material. In the absence of data to the contrary, it may not be worth the effort to refine the estimate, since, even if the total supply of composts from all sources is underestimated by one-half, and the total demand is overestimated by the same, there is still a healthy potential market. The majority of the 1451 million megagrams is directly land applied, as well.

Two quality issues have emerged from the applied research. Both could improve through better communication between users and producers. Users and applied research scientists have pointed out and voiced dissatisfaction with the level of maturity and with the amounts of non-composted inerts present in some delivered composts. Said bluntly, compost producers need to understand the simple idea that FARMERS DO NOT NEED TO ACCEPT COMPOSTS WHICH DO NOT MEET THEIR STANDARDS FOR MATURITY OR FOR INERTS CONTENT. The flip side is that farmers need to define what their standards are, and to hold producers to them. Users who *specify* and producers who *deliver* will be the ones to enjoy the advantages of compost over their competitors.

In conclusion, it is commonly estimated that over half of the rich loam topsoils in some of the prairie states have been lost since the sod was busted a scant century ago. It takes decades for nature to make an inch of topsoil. With the use of composts, the inch can be achieved more quickly. Charles Cannon, Executive Vice President of the Composting Council has said it is a twisted logic which chooses to capture glass, plastic, Fe, and Al for recycling, yet allows our precious organics to be entombed in landfills or burned.

This sentiment is captured more poetically in the following words of Victor Hugo "These heaps of garbage at the corners of the stone blocks, these tumbrils of mire jolting through the streets at night, these hybrid scavengers' carts, the fetid streams of subterranean slime which the pavement hides from you, do you know what all this is? It is the flowering meadow, it is the green grass, it is marjoram and thyme and sage, it is game, it is cattle, it is the satisfied low of huge oxen at evening, it is perfumed hay, it is golden corn, it is bread on your table, it is warm blood in your veins, *it is health, it is joy, it is life. . . .*"

ACKNOWLEDGMENTS

A very large group (literally hundreds) of people have helped, not only personally with the preparation of information shared in this chapter, but organizationally with our present effort to develop the scientific and social basis for compost use in the USA. I would like to recognize and thank each of them here, but the list is too long. My colleagues at Procter & Gamble: Marty Cannon, Margy Conditt, Kay Dorsey, Bob Greene, Bill Greggs, Bruce Jones, Larry King, Phil Leege, Drew McAvoy and many others brought an immnense amount of intellectual and interpersonal energy to Procter & Gamble's national and international commitment to composting as a means of solid waste management.

I especially would like to thank all the Council-sponsored basic and applied researchers: Allen Barker, Vince Breslin, Mike Cole, Brian Eitzer, George Estes, Rob Harrison, Chuck Henry, Ron Herrmann, Harry Hoitinck, Terry Logan, Abbie Maynard, Mike Overcash, John Peverly, Jerry Schnoor, Jodi Shann, and Sam Traina. They, and others, made my job so enjoyable for a couple of too-brief years. We spent many delightful days in laboratories, in meeting rooms, and in the fields. Each project, besides its considerable scientific merit, has a personal side that I value as well. I am looking forward to the publications coming from their work, and to ongoing personal and professional relationships.

I am grateful to Rufus Chaney of the USDA, and Jim Ryan and John Walker of the USEPA, for stimulating conversations and for the profound knowledge they bring to their areas of expertise. All these scientists represent a critical mass of compost expertise for the USA. Their synergy will become more valuable as public and regulatory attention to composting increases.

I would like to recognize the Northeast Agricultural Experiment Station Directors for their help in mobilizing the applied research sponsored by the Composting Council. Not only did they identify the principal investigators who ultimately became involved in compost research, but also the lobsterfest following their 1992 annual meeting in Newport (RI) was an event to remember. And I thank Professor (Emeritus) William Mitchell, University of Delaware, for his unwavering, enthusiastic aid and advice, and for the significant body of scientists he introduced to me, and Dr. Charles Frink, Vice Director (Emeritus) of the Connecticut Agricultural Experiment Station, who likes to say he taught me everything I know about agronomy (and which I insist to him is a dubious credit!).

I am grateful to Anzillia Murray of Procter & Gamble's Winton Hill Technical Center library for providing technical library and data base search services. I am grateful to Dr. Wayne H. Smith of the University of Florida for thoughtful conversations and for compiling and updating the information in Appendix 3-1. I thank Jim Parr, USDA, for sharing with me an early draft of his 1993 paper on uses of municipal wastes, and for conversations which helped me develop my thoughts on compost economics. I especially thank Dave Miller, American Farm Bureau Federation, for insight, ideas and conversations which helped develop the economic modeling presented in this paper.

Finally, I appreciate very much the staff of the Composting Council, Charlie Cannon, Connie Kunzler, Jan King, Martin Simpson and Randy Monk, for superb hospitality at their association offices, for great coffee and bagels, and for a quiet place to meet and get things done. Continuing updates on the status of the described research, as well as other compost-related reports and public information, are available from the: Composting Council, 114 S. Pitt St., Alexandria, VA 22314; 703-739-2401 and FAX-739-2407.

APPENDIX 3-1

Compost utilization research and demonstration projects
in several states, October 1993.[1]

Location:	Auburn, Alabama
Investigators:	Drs. C.C. Mitchell and C.E. Brown
Soil series:	Marvyn loamy sand
Crops:	Greenhouse bioassay with sorghum [*Sorghum bicolor* (L.) Moench] and sudangrass (*S. bicolor* var. sudanense)
Compost types:	Broiler litter and composted poultry mortalities
Source:	Poultry broiler producers
Results:	Fresh litter and composted poultry mortalities resulted in 33 to 47% of total N recovered in herbage (available N) over an 8-wk period. Well-composted litter resulted in only 15% recovery. Phosphorus recovery ranged from 8% for the aged compost to 23% for mortality compost. Potassium recovery ranged from 50–87%.
Concerns:	Composts (aged) are not a good source of nutrients for crops

Location:	Crossville, Alabama
Investigators:	Drs. C.W. Wood and R.P. Flynn
Soil series:	Wynnville sandy loam
Crops:	Corn (*Zea mays* L.)
Compost types:	Broiler litter composted with plant material and paper mill sludge
Source:	Biogrow, Inc., Luveme, Alabama
Results:	Experiments used compost rates to supply O, 36.3, 72.6 or 108.9 kg (0, 80, 160 and 240 lb) of total N. Compared compost to fresh broiler litter and ammonium nitrate fertilizer. Treatments were applied for 2 yr and a third growing season was added to evaluate residual effects of the fertilizer sources. Grain and stover yield, stalk nitrate, soil nitrate, P, and C were measured as well as micronutrients. No significant difference in corn yield between sources was observed until the residual year. Corn yield was greatest from plots receiving compost supplying at 72.6 kg (160 lb) of N per year. Corn

[1] These reports were compiled by Wayne H. Smith, University of Florida, to inform readers about the projects. The data are preliminary and not completely analyzed. Reference may be made to the project, but no data or interpretation should be cited without the written permission of the principal investigator(s).

stalks from compost treated soil did not exhibit nitrate accumulation while the ammonium nitrate and fresh litter treatments did. Ammonium nitrate treatments also had greater nitrate concentration in the soil profile after 2 yr of application followed by fresh litter and then compost.

Concerns: Inorganic P significantly greater in soil after 2 yr of application

---

Location: Santa Barbara, California
Investigators: Dr. Aziz Shiralipour
Soil series: Goleta loam, Todos-lodo clay loam, Amold loamy sand
Crops: Strawberry (*Fragaria* spp.), broccoli (*Brassica olerucea* var. buligtis L.), lettuce (*Lastuco* spp.), greenhous/nursery ornamentals
Compost types: Municipal solid waste (MSW), municipal sewage sludge (MSS), yard waste (YW)
Source: Truman, Minnesota, and Kellogg, California
Results: Work plan drafted. Greenhouse studies in progress. Truman material immature causing delay of germination and stunting (volatile organic acids). YW/MSS showing excellent results vs. controls. Growth data forthcoming.
Concerns: Compost immaturity, phytotoxicity

Location: Town of Hamden and Windsor, Connecticut
Investigators: Dr. Abigail Maynard
Soil series: Merrimac sandy loam, Cheshire fine sandy loam
Crops: Woody nursery stock—*Acer* spp., *Pinus* spp., *Quercus* spp., mulch and incorporation used
Compost types: MSW/MSS, Source separated MSW
Source: Delaware MSW Fairfield, Connecticut (MSW/MSS)
Results: Second-year canopy and height measurements being made. Growth data available in early 1994. Six to eight times fewer weeds on mulched plots. No differences in water nitrate between treatments in 18 mo of sampling.
Concerns: None to date

Location: Sanford, Florida
Investigators: Dr. Richard Beeson
Soil series: Greenhouse media
Crops: Virburnum (*Viburnum* spp.), Ligustrum (*Ligustrum* spp.), and Azalea [*Rhodendron calewlarcum* (Michaux) Torrey]
Compost types: MSW & YW
Source: Amerecyle, Reuter Recycling, & Enviro-Comp
Results: At final harvest, only the ligustrum plants grown in 20% Reuter were taller than controls. The 20 and 30% Amerecycle and 20% Reuter compost media produced larger plants than the control media. Since there were no dry weight differences, this suggests the larger plants had lower moisture stress (more succulent, larger plants). At 40%, Amerecycle both shoot dry weight and canopy sizes were less than controls. With the fashion azaleas, full flowering appears to be more frequent at the 10 and 20% compost treatment. Harvest will be in September. The formosa azaleas in the lysimeter pots are growing well and expected to be at marketable size by November. Some plants at 40% Amerecycle compost show symptoms of Fe deficiency. Chelated Fe will be applied as needed.

Physical properties of the media are being monitored during production.

Concerns: None to date

Location: Homestead, Florida
Investigators: Drs. Herb Bryan, Bruce Schaffer, and Jonathan Crane
Soil series: Krome gravely loam
Crops: Tomato (*Lycopersicon* spp.), squash (*Cucurbita* spp.)
Compost types: MSW/MSS, MSW
Source: Bedminster: Texas, Tennessee; Reuter Recycling
Results: Project establishment was delayed by Hurricane Andrew. In March, site preparation, treatment with Reuter compost [4 repetitions at 30 and 60 tons acre^{-1}], Bedminster [2 repetitions each, Texas 91/Tennessee 93 at 15 and 30 tons acre^{-1}] and planting to 'Sunny' tomato was accomplished. Mature Reuter and Bedminster composts resulted in higher yields of large tomato and total market yield. Immature compost seriously reduced yield. Plant height was similarly effected by mature and immature composts. The site was replanted to 'Dixie' squash on 22 Sept. 1993.
Concerns: None to date

Location: Ft. Lauderdale and West Palm Beach, Florida
Investigators: Drs. John Cisar and George Snyder
Soil series: Hallandale fine sand
Crops: Turfgrass/Landscape, St. Augustine grass [*Stenotaphrum secundatum* (Walter) Kuntu]
Compost types: MSW
Source: Reuter Recycling
Results: Leaf nutrient concentrations were determined for St. Augustine grass grown in media containing 0, 10, 20, 30, 40 and 50% compost or at 0, 33, 67 and 100% by volume. Nitrogen, P and K levels were increased by compost treatments—N and K as ratios increased and P up to 20% compost. These increases in nutrient levels generally relate to the higher turf quality ratings and more chipping growth with increasing levels of Reuter compost. Leaf samples taken following two dry-downs to wilting showed that compost treatments sustained increased N and K. Pot weight loss measures showed that increased compost rates resulted in greater cumulative water loss before the onset of turf wilting, although the days-to-wilt was not affected as much as it was during the intense evaporative period reported earlier.
Concerns: Soluble surfactantlike organics leached from compost treated soil. Identification role in transport unknown

---

Location: Bradenton, Florida
Investigators: Drs. Gary Clark, Craig Stanley, and Donald Maynard
Soil series: Eau Gallie fine sand
Crops: Bell peppers (*Capsicum annuum* L. var. annuum), tomato
Compost types: MSW
Source: Reuter Recycling
Results: Reuter compost applied in 1992 at 0, 30 and 60 t acre^{-1} decreased pepper yield (7 and 14%) due to immaturity and N-rob so compost rates increased. Tomato was then planted on the same plots that

received irrigation by drip and subirrigation procedures with 160 and 230 lb acre^{-1} of N fertilizer. Addition of compost to the drip irrigated plots increased yield of extra large fruit and total marketable fruit. For example, at the low N treatment total fruit yield was increased from approximately 2100 boxes 11.3 kg (25 lb) to over 2600 boxes of fruit per acre. Only yield of extra large fruit was affected by fertilizer in the subirrigated plots. Even though there was considerable variability in soil water measures where compost was applied, soil water content was higher in proportion to compost rates. In 1993 the soils data indicate a stabilization of the compost. The plots were rebedded, fumigated and fertilized in August 1993 in preparation for planting to pepper in September 1993.

Concerns: Glass in compost, immature compost, high C/N ratio

Location: Gainesville, Florida
Investigators: Dr. A.E. Dudeck
Soil series: Arredondo fine sand
Crops: Turfgrass/sod, St. Augustine grass
Compost types: MSW, MSW plus sewage sludge, YW
Source: Amerecycle, Vital Earth, Resources Recycled, Wood Products
Results: St. Augustine grass was grown under a rain-out shelter in three composted solid wastes at 10, 20 or 30% by volume in the top 15 cm of sand soil and received fertigation every 3, 6 or 9 d. Composts incorporated at 30% enhanced early grass establishment with less water stress. Grass growth in 30% MSW was better than grass growth in MSW + SS and in yard trash composts at the same rate of incorporation.
Concerns: None to date

Location: Ft. Lauderdale, Florida
Investigators: Dr. George Fitzpatrick
Soil series: Margate fine sand
Crops: Philodendron (*Philodendron* spp.), Lantana (*Lantana cumara* L.), Schefflera (Araliaceae spp.), Oleander (*Neriam oleander* L.), Jessamine (*Castrum nocturnum* L.), West Indian Mahogany [Swietenia mahogani (L.) Jacq.], live Oak (*Quercus* spp.)
Compost types: MSW, sludge, YW
Source: Reuter Recycling, Palm Beach Solid Waste Authority (SWA) Broward County
Results: Biomass data were complied for four species grown in containers with stand-alone composts, or where compost replaced peat in a control commercial mix (4 peat, 5 pine bark, 1 sand). For lantana in stand-alone, Reuter and sludge/yard trimmings composts, plant biomass was greater than in the control medium. Other compost treatments produced biomass levels similar to the controls. In dwarf schefflera and philodendron, the Reuter compost and the sludge/yard trimmings compost, (both as a stand-alone medium or blended) produced plants with greater biomass than those in control medium. With orange-jessamine, only the sludge/yard trimmings compost produced plants with greater biomass than the control medium. West Indian mahogany trees were replanted in the landscape site in preparation for follow-up work on the effects of com-

post topdressing on transplant stress. Due to high rainfall, withholding irrigation did not cause transplant stress.

Concerns: Compost preprocessing, postprocessing, and maturity

---

Location: Ft. Lauderdale, Florida
Investigators: Dr. George Fitzpatrick
Soil series: Greenhouse media
Crops: Ficus (*Ficus* spp.), Areca Palm, Jessamine, Dwarf Oleander, Hibiscus (*Hibiscus* spp.), Philodendron, Schefflera
Compost types: Mixed waste paper, sewage sludge, Refuse Derived Fuel (RDF), RDF Residuals, yard trash
Source: Palm Beach SWA
Results: Plant material has been acquried, but media not deliverd to date
Concerns: None to date

Location: Gainesville, Florida
Investigators: Dr. Ray N. Gallaher
Soil series: Bonneau fine sand
Crops: corn, vegetables
Compost types: YW
Source: Wood Resource Recovery
Results: Yard trimmings (grass, shrubs, tree parts) from urban homesites comprised the feedstock. The objective of the study is to determine the effect of applying the yard waste compost on soil properties, corn nutrition, phytoparasitic nematodes and yield. Application of 120 t acre^{-1} of yard waste compost increased soil organic matter and should provide long-term nutrient and water benefits to soil properties. Planting in Year one soon after incorporation of compost induced N deficiency which had to be corrected by modified fertilization. Compost suppressed three of four phytoparasitic nematodes, especially root knot. This yard waste compost (120 ton t; 108.8 Mg) contained fertilzier elements at an estimated purchase value of $500. An expanded experiment is in progress.
Concerns: High C/N ratio, plowdown timing, mulch vs. incorporate

---

Location: Gainesville, Florida
Investigators: Dr. Don Graetz
Soil series: NA
Crops: NA
Compost types: MSW
Source: Reuter Recycling and Bedminster Amerecycle
Results: Compost stability evaluated using CO_2 evaluation showed that the Reuter compost was more reactive than Amerecycle and Bedminster composts. Reuter compost generally had higher C/N ratios than the other composts. Little C mineralization occurred with the Amerecycle and Bedminster products, indicating stability. With Reuter compost, CO_2 evaluation was proportioned to rate. At 106 d, the samples were stirred and moisture adjusted, causing CO_2 evolution to be stimulated briefly and then stabilize. This suggests that immature compost becomes stable in soil in about 100 d. Water content analyses showed that stable compost retained less water than the immature composts. An error was found in earlier data for Ni

so none of the data should be used. The source of error is being sought.

Concerns: Instability of flow-through system

Location: Gainesville, Florida
Investigators: Dr. Don Graetz
Soil series: NA
Crops: NA
Compost types: Mixed waste paper, sewage sludge, RDG and RDF residuals, yard trash
Source: Palm Beach SWA
Results: Samples of the compost applied in Stoffella experiments were received and entered into the analysis sequence for water soluble nutrients and total metals.
Concerns: None to date

Location: Lake Alfred, Florida
Investigators: Dr. James H. Graham
Soil series: Candler fine sand
Crops: Citrus (*Rutaceae* spp.)
Compost types: MSW
Source: Reuter Recycling
Results: Orlando tangelo trees (R. X *Tangelo* J. Ingram & H.E. Moore) on Cleopatra mandarin rootstock were inoculated with *Phytophthora parasitica* or remained noninoculated and 48 d later were planted at 6.1- by 14-m (20- by 6-ft) spacing into plots amended with 100 t acre^{-1} compost or into nonamended plots. At transplant, an additional 0.04 m^3 (0.05 yd^3) was added to the backfill of the planting hole in amended plots. Carbon and N content of the soil were both increased fourfold, thus C/N ratio remained unchanged. At 7 mo after transplant, compost had no effect on root mass density. *P. parasitica* had spread to noninoculated plants in each plot. Populations of *P. parasitica* were not different in noninoculated plots and inoculated plots with and without compost. Evaluations of root growth, *Phytophthora* population and tree productivity will continue.
Concerns: None to date

Location: Hastings, Florida
Investigators: Dr. Dale R. Hensel
Soil series: Ellzey fine sand
Crops: Potato
Compost types: YW
Source: Southland Environmental Systems, Jacksonville, Florida
Results: Potato yields were increased from 330 to 360 cwt acre^{-1} when compost was increased from 0 to 60 t acre^{-1}, respectively (where cwt acre^{-1} 100 lb acre^{-1} = 112 by ha^{-1}). When 29.3 kg (65 lb acre^{-1} of extra N was applied to the basic rate of 45.9 kg (210 lb acre^{-1} more potatoes were produced when the extra N was added. Nematode samples taken in August showed no difference between treatments. They were relatively low numbers.
Concerns: High C/N ratio, cost of transportation from source, Long-term effects

---

Location: Gainesville and Tampa, Florida
Investigators: Dr. Dennis McConnell
Soil series: Greenhouse media
Crops: Dracaena (*Dracaena* spp.), Azalea, Peperomia, Shumard Oak (*Q. shenardii* Backl), Red Maple (*Acer rubrum* L.), Pickerelweed (Pontederin cordata L. var. cordata), Schefflera, Anthurium (*Anthurium* spp.), Begonia (*Begoniceae* spp.), Boston Fern (Nephrolepsis exaltata cv. Bostoniensis), Burford Holly (*Ilexcornata* cv. Burfordii), Chlorophytum (*Liliaceae* spp.), Crape-Myrtle (*Lagerstronia indica*), Daylilly (*Hemerocalles* spp.), Dieffenbachias (*Aroceae* spp.), Ficus Benjamina (*F. nitida* Thumb.), Golden Pothos (*Epipremnum aureum*), Juniper (*Juniperus* spp.), Orchids (*Orchidaceae* spp.), Yaupon Holly (*Ilexviritoria* Aut)
Compost types: MSW, YW
Source: Agrisoil Bedminster, Wood Resource Recovery
Results: Overall, composted wastes can be used as a component of potting mixes with acceptable plant growth rates. In most cases, plants grow as well or better in potting mixes containing composted wastes, even in mixes containing 100% compost. Certain plants seem to exhibit a greater number of problems than other plants. Azalea and oak had reduced growth rates in potting mixes containing composted yard wastes or MSW when compared to commercial mixes. Composted yard waste can be used as a partial or complete substitute for pine bark and a partial substitute for peat. Plant growth was reduced when oxidative settling decreased potting mix pore size.
Concerns: Toxic compounds, nutrient binding, decomposition and settling, N-rob by immature compost

Location: Immokalee, Florida
Investigators: Dr. Thomas Obreza
Soil series: NA
Crops: Tomato, watermelons (*Citrullus lanatus*)
Compost types: MSW
Source: Reuter Recycling, Bedminster
Results: In 1992, Reuter compost was applied at 29.9 and 45.4 Mg (33 and 50 dry tons) acre^{-1} and Bedminster compost at 5.8 and 10.9 Mg (6 and 12 dry tons) acre^{-1}. A standard fertilizer control and a chicken manure compost treatment were included. The plots were planted with tomato and then followed with a watermelon crop and irrigated at 1.25, 1.00 and 0.75 evapotranspiration (ET). Because of compost immaturity, the tomato yield was not increased. However, total marketable watermelon yield was 54% greater where the high Reuter compost was applied. The lower Reuter rate increased yield 33%. The Bedminster compost had no residual effect. Compost treatments generally increased soil organic content, water-holding capacity, mineral concentrations and pH in proportion to compost rate through 300 d after treatment. The high rate Reuter compost soil contained about 33% more water than the control. The "dry bed side" was wetter 60 d after planting tomato. The site was retreated with compost in September 1993 for planting in October 1993.
Concerns: Glass in compost, high C/N ratio

Location: Lake Alfred, Florida
Investigators: Drs. Larry Parsons and Adaire Wheaton
Soil series: Astatula fine sand
Crops: Citrus
Compost types: MSW
Source: Reuter Recycling
Results: Hamlin orange trees (*Citrus sinensii* cv. Hamlin) were planted in soil treated with 0, 50 and 100 mt compost ha^{-1}. At another site, 49 mt compost ha^{-1} was soil incorporated, planted with Ambersweet (*Citrus* spp.) trees and irrigated with treated wastewater. After 9 mo, the Hamlin trees showed no growth increase while the Ambersweet trunk diameters were increased by the compost in the first 9 mo but not afterwards. Root study revealed that the 50 and 100 t ha^{-1} compost treatment increased total root length and the concentration of roots at the 158.4- to 304.8-mm (6- to 12-in) soil depth. Extractable minerals and pH were increased in the top 152.4-mm (6-in) soil layer and the pH increased in the 152.4- to 304.8-mm (6- to 12-in) depth.
Concerns: Economics, consumer acceptance of crops, possible heavy metals

---

Location: Gainesville, A.C.F., Florida
Investigators: Dr. Hans Riekerk
Soil series: (Spodosol)
Crops: Slash pine trees (seedlings and 7-yr old trees) (*Pinus elliothi*)
Compost types: MS
Source: Reuter Recycling
Results: Soil moisture and water chemistry measures in the seedling plots and the 7-yr old stand treated with compost were difficult during this period because of a dry summer. Analysis of previous samples showed ammonium proportional to compost rate, especially in soil water in the forested plot, but not groundwater. Nitrate was high in soil water below the plow layer in the seedling plot. At the highest compost rate nitrate was 18 mg kg^{-1} (ppm) in the groundwater. The highest phosphate-P levels were in soil water of the control plots. Elemental levels (Cl, K, Na) increased in samples generally increased with compost rate. Heavy metals remained below detection levels. Seedling survival was 93% in the control plots and decreased to 40 to 58% when compost was applied. The reduction was associated with observed massive weed growth. A growth response in the 7-yr old trees has not been detected.
Concerns: Deer browse on seedlings, weeds on seedling site, umbrella effect in older site

---

Location: Gainesville, Florida
Investigators: Drs. James Stephens and Stephen Kostewicz
Soil series: Arredondo fine sand
Crops: vegetables
Compost types: YW, animal manures + litter
Source: Wood Resource Recover, Red Rooster Inc., and other local sources
Results: A database of information and visual materials including written fact sheets are to be developed as part of a thrust to educate the public as to the potential savings of energy and benefit to the environment. A slide set of appropriate techniques involved in use

and management of organic amendments and composted materials in the home vegetable garden. The materials are to be used as resource material in extension programs at the local level.

Concerns: None to date

Location: Gainesville, Florida
Investigators: Drs. James Stephens and Stephen Kostewicz
Soil series: Arredondo fine sand
Crops: vegetables (organic culture)
Compost types: YW, animal manures + litter
Source: Wood Resource Recovery, Red Rooster, Inc.
Results: Yard wastes and other high C/N composted materials provide optimum results only when fertilizers (organic or standard type) are added with the composts. Yield response to applications of organic soil amendments/fertilizer was evaluated in cucumber, mustard green (*Brustia junicea*), southern pea (*Pisum saturum* L.)
Concerns: None to date

Location: Boynton Beach and Ft. Pierce, Florida
Investigators: Dr. Peter Stoffella
Soil series: Myaka sand
Crops: Tomato, peppers (*Capsicum* spp.), cucumber (*Cucumis sativas*)
Compost types: Mixed waste paper, sewage sludge, RDF, RDF residuals, yard trash
Source: Palm Beach SWA
Results: The yard waste/sewage sludge compost was applied 19 and 20 Aug. 1993 at 0 and 60 t acre^{-1} compost, rorotilled, fertilized, bedded and mulched with clear plastic over drip irrigation emitters. Tomato and pepper plants and cucumber seeds were planted or sowed at weekly intervals beginning 24 August until 27 September. Cucumber seeds were planted in compost or a commercial mix in flats outdoors. In a greenhouse, seeds were planted in compost, field soil or commercial media. Plant growth measures of transplants were not different regardless of planting data. Field emergence of germinated cucumber seeds was 74% with compost and 6% without, probably related to moisture during the non-irrigated period. There were no phytotoxic or growth suppression symptoms from the compost. Seed germination was equal or better than in the other media.
Concerns: None to date

Location: Moscow, Idaho
Investigators: Drs. Robert Tripepi and Alton Campbell
Soil series: NA
Crops: Poinsettia (*Euphorbia pulcherrima*), Chrysanthemum (*Chrysanthemum* spp.), and nursery stock
Compost types: pulp and paper sludge
Source: Newsprint Mill
Results: Poinsettia—fair growth in paper sludge, but plants were more susceptible to diseases. Chrysanthemum good growth in sludge media. Plants appear unaffected by sludge compared to plants in commercial mixes. Nursery stock—four species of nursery stock grew well in composted sludge and bark mixes.
Concerns: Composted sludge contains phytotoxic compounds

Location: Moscow, Idaho
Investigators: Drs. Robert Tripepi and Alton Campbell
Soil series: NA
Crops: Tomato plants
Compost types: Bluegrass residues (straw, seed heads, etc.)
Source: Bluegrass (*Poa* spp.) seed fields in northern Idaho
Results: Tomato plants grew best in medium that was made from composted bluegrass residue (BG) plus alfalfa (*Medicago sativa* L.) seed screenings. Plants in this medium grew better than those in a mix of 75% peat + 25% perlite. Plants potted in media made from BG + cattle manure or BG unamended grew much less.
Concerns: Phytotoxic compounds in BG or N immobilization problems.

Location: Moscow, Idaho
Investigators: Drs. Robert Tripepi, Joe McCaffrey, and Jack Brown
Soil series: NA
Crops: Amur Maple (*Acer Ginnala*)
Compost types: Rapeseed (*Brustica hapas*) meal mixed with bark
Source: Rapeseed crushing plant
Results: Experiment was just started in June 1993. No results as yet to report.
Concerns: Phytotoxicity of rapeseed meal to plant roots

Location: Salisbury, Wyoming, and College Park, Maryland
Investigators: Dr. Frank Goulin and John Bouwkamp
Soil series: Norfolk loamy sand and Wye silt potting mixes
Crops: Greenhouse plants, poinsettia, melon, tomato, spinach (*Spinacia elerucea*)
Compost types: MSW/MSS
Source: Delaware Ferst
Results: Melons with higher sugar content than controls. Vegetable crops yields above controls. Nitrogen stress on all crops due to immature compost. Compost concentration increased to one-third by volume in potting mixes.
Concerns: Phytotoxicity, delayed root development and germination due to N stress, maturity index needed for field/pot crops

Location: College Park, Maryland
Investigators: Dr. Frank Gouin
Soil series: NA
Crops: Bedding plants
Compost types: MSW
Source: F & E Resource System Technology
Results: Compost concentrations above 25% by volume cause stunting of plants and excessive growth of mushrooms.
Concerns: Compost maturity

Location: Belchertown and Deerfield, Massachusetts
Investigators: Dr. Allen Barker
Soil series: Winooski very fine sandy loam
Crops: Strawberry, blueberry (*Vaccinium*), apple (*Malus* spp.), landscape flowers, wildflowers, and sod
Compost types: MSW/MSS, YW, chicken manure, and cranberry wastes
Source: Delaware and Massachusetts

Results: With surface-applied or shallowly incorporated composts in field plots, drying and NH_4-N in composts and weeds in soil limited grass and wildflower sod productivity. Problems overcome by producing sods in composts layered on plastic. Composts applied to fruit crops had no effect on quality or productivity in the 1st yr.

Concerns: Control of soil-borne weeds, need for irrigation, high cost to transport composts, adequate curing of compost

---

Location: St. Paul, Minnesota
Investigators: Dr. Paul R. Bloom and Bruce Cook
Soil series: NA
Crops: NA
Compost types: MSW/MSS, YW, chicken manure, and cranberry wastes
Source: newspaper, rabbit food, composted cow manure (innoculum)
Results: Initiating studies on the fate of herbicides atrazine (6-chloro-N-ethyl-N-(1-methylethyl)-1,3,5-triazine-2,4-diamine), trifluralin [2,6-dinitro-N,N-dipropyl-4-(trifluoromethyl)benzonarmine] and 2,4-D (2,4-dichlorophenoxylacetic acid) during composting. Mineralization, volitilization and percentage material/metabolite measurements will be used to determine if composting enhances the rate of herbicide degradation, and if it could be used as an inexpensive, low-technology method for the detoxification of herbicides in unused solutions, soil, water, and yard waste.
Concerns: None to date

Location: Staples, Lamberton, and Becker, Minnesota
Investigators: Dr. Thomas Halbach and others.
Soil series: Hubbard loamy sand, Verndale sandy loam, and Temial loam
Crops: Corn
Compost types: MSW
Source: Swift Co., Truman St. Cloud, Mora, Fillmore Co., Thief River Falls
Results: Compost with C/N <20 had positive yield effect on corn. Compost with C/N >30 decreased yields.
Concerns: C/N, soluble salts

Location: Cloquet Forestry Center, Minnesota
Investigators: Dr. T.J. Nichols
Soil series: Omega loamy sand, Cloquet sandy loam
Crops: Forestry (timber and Christmas trees)
Compost types: MSW
Source: Mora, Thief River Falls
Results: Weeds flourished in compost-treated soils at rates to 140 tons acre^{-1} (TPA). High over-winter tree mortality on site with finer textured soil.
Concerns: Metals, salts, aesthetics of glass and plastics on surface

---

Location: St. Paul, Minnesota
Investigators: Dr. Bert T. Swanson and J.B. Calkins
Soil series: Growth media
Crops: Horticultural
Compost types: MSW
Source: Swift Co., Truman St. Cloud, Mora, Fillmore Co., Thief River Falls

Results: Compost outperformed controls.
Concerns: High soluble salts, visible plastics, odors, shards

---

Location: St. Paul, Minnesota
Investigators: Dr. Bert T. Swanson and J.B. Calkins
Soil series: Growth media
Crops: Nursery crops, turf
Compost types: MSW, manure
Source: MSW plants and manure compost facility
Results: Unavailable
Concerns: None to date

Location: St. Paul, Minnesota
Investigators: Dr. Bert T. Swanson and Beth Jarvis
Soil series: Growth media
Crops: Nursery crops
Compost types: YW, MSW, rubber tire chips
Source: MSW plants, municipalities
Results: Not available
Concerns: None to date

Location: Nebraska
Investigators: W. Woldt
Soil series: Not available
Crops: Not available
Compost types: Not available
Source: Not available
Results: Working on the development of a computer based multi-media educational program that includes composting.
Concerns: Not available

Location: New Hampshire (Durham, Madbury, and Lyman, Maine)
Investigators: Dr. George Estes
Soil series: Warwick and Hoosic fine sandy loam, Deerfield loamy fine sand, and Ninigret fine sandy loam
Crops: Silage corn, sod, roadside right-of-way
Compost types: MSW/MSS
Source: Delaware
Results: Turf, corn and road side plantings completed. Corn yield increased 33%; tissue analyses not significant for Cu, Cd, Ni, Cr, Pb.
Concerns: None to date

Location: New Jersey Agricultural Experiment Station Snyder Research Farm, Pittstown, New Jersey
Investigators: Daniel Kluchinski, Dr. Joseph Heckman, and Joseph Mahar
Soil series: Quakertown silt loam
Crops: Corn, soybean [*Glycine Max* (L.) Merr.]
Compost types: Uncomposted leaves
Source: Local municipalities
Results: Uncomposted leaves applied each fall at 0, 76.2, and 152.4 mm (0, 3, 6 in.) depths (0, 10, 20 dry tons acre^{-1}) with manure spreader, incorporated with chisel plow. Field corn and soybean grown in rotations. Studying crop growth and yields; N deficiency; weed, dis-

ease and pest occurrence; soil pH, nutrients, organic matter, moisture retention; economics.

Concerns: None to date

Location: New Jersey Agricultural Experiment Station Farms and Field Stations, Private Nurserys in New Jersey
Investigators: Multiple cooperators
Soil series: Nixon loam, Freehold sandy loam, Quaker Town silt loam, Downer loamy sand
Crops: Container ornamentals, field nursery stock, corn, turfgrass, and landscape plants
Compost types: MSW, MSW/MSS
Source: To be determined
Results: None to date
Concerns: None to date

Location: Calverson, New York
Investigators: Dr. Vincent Breslin
Soil series: Riverhead sandy loam
Crops: Sod
Compost types: MSW/MSS and MSW
Source: Delaware, Pembroke Pines
Results: Metals analysis shows both composts meet New York state criteria, except Pembroke Pb. Soil profile analyses show compost metals restricted to 0 to 5 cm. No significant differences in water quality between pre- and postapplication. Increased compost rates decreased grass cover early on, later growth indistinguishable from controls.
Concerns: Phytotoxicity, delayed germination of grass seed

Location: Ithaca, New York
Investigators: Dr. Eric Nelson
Soil series: Not available
Crops: Turfgrass
Compost types: MSS, YW, animal manures, industrial sludges
Source: Various
Results: Various composts provide high degree of suppression of turfgrass diseases. Active components include bacteria, fungi and actinomycetes. Composts applied as soil amendments, topdressings, or applied as aqueous extracts are all effective.
Concerns: Phytotoxicity

Location: Seneca Falls, Interlaken, New York
Investigators: Dr. John Peverly
Soil series: Schoharie silty clay loam, Honeoye silt loam
Crops: Corn, vineyard
Compost types: MSW/MSS, YW
Source: Delaware, Springfield, Massachusetts
Results: Positive treatment effects for nitrate–N, P, and K in corn. About 30% increase in grain yield. No metals uptake by corn grain. N content of roots, stalks, grain unaffected. No indication of treatment effects in regard to metals in grape tissues. No consistent yield effects in grapes.

Concerns: Elevated Zn (Delaware Reclamation Authority > Springfield) in corn roots bears watching; nitrates in soil pore water, even at low rates.

Location: Ithaca, New York
Investigators: Joseph Regenstein
Soil series: To be determined
Crops: To be determined
Compost types: Zebra mussels (mollusk), poultry manure, and sawdust or peat
Source: Great Lakes, New York
Results: To be examined are compost stability at several ratios of mollusks, pH effects of shells on compost, pH effects of shells and compost on acid soils. Length of potential liming ability study yet to be decided.
Concerns: Sharpness of shell fragments and mollusk absorbed toxin effects

---

Location: Centre County, Pennsylvania
Investigators: Dr. Douglas Beegle
Soil series: Hagerstown silt loam
Crops: Corn
Compost types: Mushroom compost
Source: Not available
Results: Significant yield response to adding compost observed up to at least 4 yr after application of compost. Some but not all response due to N in compost. Response from improvement in physical properties likely.
Concerns: Very heavy rates used to simulate farmer practice

Location: Centre County, Franklin County, Lancaster County, & Montgomery County, Pennsylvania
Investigators: Dr. Douglas Beegle
Soil series: Hagerstown silt loam, Hagerstown silty clay loam, Hagerstown silt loam, and Lehigh Channery silt loam, respectively
Crops: Corn
Compost types: Dairy manure, dairy manure, poultry manure, and packing house waste, respectively
Source: Not available
Results: New experiments, none available
Concerns: Wet, immature compost; wet compost, very heavy rain immediately after application; high residual fertility on the farm where this plot is located; and wet, lumpy compost with a unique feed stock, respectively

---

Location: Liberty, South Carolina
Investigators: R.K. White, R.W. Miller, Jr., and J.D. Ridley
Soil series: Cecil sandy loam
Crops: Apple trees
Compost types: Biosolids
Source: West Richland County
Results: Five years of application at three rates with chemical treatment and control. Applications annually and biannually. Significantly (at 95% level) reduces death loss to trees.
Concerns: None to date

Location: Greenville, South Carolina
Investigators: R.K. White
Soil series: Not available
Crops: Daily landfill cover
Compost types: Biosolid and yard waste
Source: City of Greenville and Western Carolina Regional Sewar Authority
Results: Evaluating material as to its suitability for daily landfill cover.
Concerns: None to date

Location: Pullman, Washington
Investigators: Theresa Beaver and Dan Caldwell
Soil series: Palouse silt loam
Crops: Barley (*Hordeum* spp.), tomato
Compost types: Separated cow manure and coal ash
Source: Washington State University
Results: Separated cow manure and coal ash were composted together with the following rates of coal ash: 0, 5.4, 8.7, 18.8, adn 31.8%. After curing, the compost was applied to field plots planted to barley. A greenhouse experiment also was set up using the compost as a partial substitute to potting soil. Results will not be available till late summer.
Concerns: Heavy metal content of coal ash

Location: Portage, Wisconsin
Investigators: Bill Casey and Tracy Benzel
Soil series: Not available
Crops: Corn
Compost types: MSW/Sludge from Columbia County
Source: Rotating drum and windrows
Results: On-going, full-scale agricultural compost and spreading program. No results to date.
Concerns: Inert materials, N availability, public acceptance

Location: Arlington, Wisconsin
Investigators: Drs. Lloyd A. Peterson and Rickie P. Voland
Soil series: Greenhouse potting mixes
Crops: Vegetable seedlings
Compost types: MSW and YW
Source: Various includeing Columbia County, MSW
Results: Project started 1 July, 1993 to assess suppression of seedling damping-off diseases and compost maturity indices in Wisconsin composts. Assays to include microbial activity, microbial biomass, near-infrared spectroscopy and chemical analyses.
Concerns: None to date

Location: Portage, Wisconsin
Investigators: Dr. Aga Razvi
Soil series: Plainfield loamy sand and Parr or Plano silt loam
Crops: Corn, soybean
Compost types: MSW, Sewage sludge
Source: City of Portage compost facility
Results: (i) Determining the potential for groundwater contamination when composts are applied on agricultural land, using suction lysimeters

and gravity lysimeters, (ii) Looking at the potential for crop growth, yield and accumulation of harmful materials in plant tissue using corn and soybean.

Concerns: Very wet season

Location: Portage, Wisconsin
Investigators: Drs. Aga Razvi and W.W. Stevens
Soil series: Plainfield loamy sand
Crops: Corn
Compost types: MSW/Suldge from Portage, Wisconsin
Source: Rotating drum and outdoor windrows
Results: Compost applicatoin rates to compare growth, soil nutrition, leachate nutrient and metal concentrations. Three years of data not available.
Concerns: Environmental and food chain quality

Location: Columbia County, Portage, Wisconsin
Investigators: Dr. Richard Wolkowski
Soil series: Plainfield loamy sand
Crops: Corn
Compost types: MSW/Sludge from Columbia County
Source: Rotating Drums/Windrows
Results: No data available yet. Comparing growth in compost × fertilizer applications. Applied in spring 1993. Various stages of compost maturity.
Concerns: N availability, metal uptake

Location: Sawyer County, Spooner, Wisconsin
Investigators: Dr. Richard Wolkowski
Soil series: loamy sand
Crops: Corn
Compost types: MSW
Source: Filmore County, Minnesota
Results: Fertilizer × compost trials compared growth and yield. Various stages of compost maturity. 0, 2, 4, 8, 16, 32 dt acre^{-1} × three fertilizer treatments.
Concerns: N availability and metal uptake

Location: Spooner, Wisconsin
Investigators: Dr. Richard Wolkowski
Soil series: Pence sandy loam
Crops: Corn
Compost types: MSW
Source: 1991, Filmore County, Minnesota; 1992, Prairieland Facility, Minnesota; 1993, Columbia County, Wisconsin
Results: Design-Compost applied at 0 to 32 t acre^{-1} dry matter. Separate plot received equivalent nutrients as fertilizer. Measure yield, growth, nutrient, content, soil NO_3. Yield of 1991, positive 30% response to 16 and 32 t acre^{-1}. Whole plant nutrient and metal content slightly increased. No change in grian. More soil NO_3 with fertilizer N. 1992 Compost immature. Yield reduced 50%. Data for 1993 not available.
Concerns: 1992 compost very immature

Location: Pardeeville and Arlington, Wisconsin
Investigators: Dr. Richard Wolkowski
Soil series: Boyer sandy loam, Saybrook silt loam
Crops: Corn
Compost types: MSW
Source: Columbia County, Wisconsin
Results: 1993 Design, 0-40 T acre^{-1} dry matter compost. Separate plots received 0 to 72.6 kg (0 to 160 lb acre^{-1}) N. Compost of three ages (each rate): fresh (just out of vessel), 6 wk in turned pile, 36 wk (1st half in turned pile, remainder in static pile). Will measure yield (grain and silage), growth, nutrient and metal content, soil nutrient, soil NO$_3$ to 0.9 m (3 ft).
Concerns: Compost high in inerts ($>10\%$), especially glass

REFERENCES

Anonymous. 1972. From agricultural waste to feed or fuel. Chem. Eng. News. 29 May, p. 14.

Barbarika, A., D. Colacicco, and W.J. Bellows. 1980. The value and use of organic wastes. Maryland Agri-Economics Coop. Ext. Serv., Univ. Maryland, College Park, MD.

Chaney, R.L., and J.A. Ryan. 1992. Heavy metals and toxic organic pollutants in MSW-composts: Research results on phytoavailability, bioavailability, fate, etc. p. 451-506. In H.A.J. Hoitink and H.M. Keener (ed.) Science and engineering of composting: Design, environmental, microbiological and utilization aspects. Renaissance Publ., Worthington, OH.

Collins, A.R. 1991. How much can farmers pay for MSW compost? BioCycle. October, p. 66-69.

Community Environmental Council. 1993. Compost market development: A literature review. Gildea Resource Center, Santa Barbara, CA.

Composting Council. 1992. A review of composting literature. 2nd ed. Composting Council, Alexandria, VA.

Deming, W.E. 1986. Out of the crisis. Center for Advanced Engineering Study, Massachusetts Inst. Technol., Cambridge, MA.

Eitzer, B.D. 1992. Volatile organic emissions from MSW composting plants. In Proc. Composting Council 3rd Annu. Conf. Washington, DC. 11-13 Nov. 1993.

Epstein, E., R.L. Chaney, C. Henry, and T.J. Logan. 1992. Trace elements in municipal solid waste compost. Biomass Bioenergy 3:3-4, 227-238.

Fannin, S.A. 1993. Fate of xenobiotic organic compounds following land application of municipal solid waste compost. M.S. thesis. Univ. Iowa, Iowa City.

Federal Register. 1993. 40 CFR Parts 257, 403 and 503. Final rules: Standards for the use of disposal of sewage sludge. Fed. Reg. 58:32, 9248-9415.

Finstein, M.S., J. Cirello, D.J. Suler. 1980. Microbial ecosystems responsible for anaerobic digestion and composting. J. Water Pollut. Control Fed. 52:11, 2675-2685.

Frost, D.I., B.L. Toth, and H.A.J. Hoitink. 1992. Compost stability. BioCycle 33:11, 62-66.

He, Xin-Tao, S.J. Traina, and T.J. Logan. 1992. Chemical properties of municipal solid waste composts. J. Environ. Qual. 21:318-329.

Hermann, R.F., and J.R. Shann. 1993. Enzyme activities as indicators of municipal solid waste compost maturity. Compost Sci. Technol. 1(4):54-63.

Hsu, S.M. 1992. Prediction of fate and transport of xenobiotic organic compound in the root and vadose zones following land application of municipal solid waste compost. M.S. thesis. University of Iowa. Iowa City.

Hsu, S.M., J.L. Schnoor, L.A. Licht, M.A. St.Clair, and S.A. Fannin. 1993. Fate and transport of organic compounds in municipal solid waste compost. Compost Sci. Technol. 1(4):36-48.

Hyatt, G.W., and T.L. Richard. (ed.) 1992. Guest editorial: Aerobic composting and compost utilization. Biomass Bioenergy 3:121-125.

Iannotti, D.A., T. Pang, B.L. Toth, D.L. Elwell, H.M. Keener and H.A.J. Hoitink. 1993. A quantitative respirometric method for monitoring compost stability. Compost Sci. Utiliz. 1:3, 52-65.

Ishikawa, K. 1985. What is total quality control? The Japanese way. Prentice-Hall, Englewood Cliffs, NJ.

Kissel, J.C., C.L. Henry, and R.B. Harrison. 1992. Potential emissions of volatile and odorous organic compounds from municipal solid waste composting facilities. Biomass Bioenergy. 3:3-4, 181-194.

Leege, P.B. 1992. Composting infrastructure in the United States. p. 168-184. *In* H.A.J. Hoitink and H.M. Keener (ed.) Science and engineering of composting: Design, environmental, microbiological and utilization aspects. Renaissance Publ., Worthington, OH.

Maynard, A.A. 1993. Evaluating the suitability of MSW compost as a soil amendment in field-grown tomatoes. Part A: Yield of tomatoes. Compost Sci. Utiliz. 1:2, 34-36.

Mielke, H.W., and J.B. Heneghan. 1991. Selected chemical and physical properties of soils and gut physiological processes that influence lead bioavailability. Chem. Spec. Bioavail. 3:3-4, 129-134.

Nakasaki, N., A. Watanabe, and H. Kubota. 1992. Effects of oxygen concentration on composting organics. BioCycle 33:6, 52-54.

Parr, J.F., and S.B. Hornick. 1993. Utilization of municipal wastes. p. 545-559. *In* F.B. Metting (ed.) Soil microbial ecology: Applications in agricultural and environmental management. Marcel Dekker, Inc., New York.

Ryan, J.A., and R.L. Chaney. 1992. Regulation of municipal sewage sludge under the clean water act section 503: A model for exposure and risk assessment. p. 422-450. *In* H.A.J. Hoitink and H.M. Keener (ed.) Science and engineering of composting: Design, environmental, microbiological and utilization aspects. Renaissance Publ., Worthington, OH.

Shiralipour, A., D.B. McConnell, and W.H. Smith. 1992a. Physical and chemical properties of soils as affected by municipal solid waste compost application. Biomass Bioenergy 3:3-4, 261-266.

Shiralipour, A., D.B. McConnell, adn W.H. Smith. 1992b. Uses and benefits of MSW compost: A review and assessment. Biomass Bioenergy 3:3-4, 267-280.

Slivka, D.C., T.A. McClure, A.R. Buhr, and R. Albrecht. 1992. Compost: National supply and demand potential. Biomass Bioenergy 3:3-4, 281-299.

Stilwell, D.E. 1993a. Elemental analysis of composted source separated municipal solid waste. Compost Sci. Utiliz. 1:2, 23-33.

Stilwell, D.E. 1993b. Evaluating the suitability of MSW compost as a soil amendment in field grown tomatoes, Part B: Elemental analysis. Compost Sci. Utiliz. 1:3, 66-72.

Stutzenberger, F.J., A.J. Kaufman, and R.D. Lossin. 1971. Cellulolytic activity in municipal solid waste composting. Can. J. Microbiol. 16:553-560.

Tennessee Valley Authority. 1975. Composting at Johnson City. Final Rep. Joint USEPA-TVA Composting Project with Operational Data, 1967 to 1971. USEPA Rep. SW-31r2. USEPA, Washington, DC.

Wright, R.J. (ed.) 1993. Agricultural utilization of municipal, animal and industrial wastes. U.S. Gov. Print. Office, Washington, DC.

4 Strategies for Encouraging the Use of Organic Wastes in Agriculture

Cary Oshins

Rodale Institute
Kutztown, Pennsylvania

Today's society produces a number of "wastes" that could be benefically used in agriculture. Mutual benefits could accrue to both the farm and non-farm sectors through this utilization. Encouraging the use of these materials requires an understanding of both the barriers to and the opportunities for their use. Composting is one technology that can facilitate the use by farms of nonfarm materials. In 1991, the Rodale Institute Research Center[1] initiated its Farm Co-Composting Project[2]. The goals of the project are: (i) Transforming rural and urban wastes into agricultural resources; and (ii) Converting rural/urban conflicts into enterprise opportunities. The Project has identified a variety of barriers to the cocomposting of agricultural and non-agricultural wastes on farms, and is implementing strategies to reduce or overcome these barriers.

Composting is the controlled biodegradation of putrescible materials to a stable condition. The process and the material it produces, compost, are seeing a surge of interest from both agricultural producers and those responsible for management of solid waste. Agricultural composting of putrescible materials, such as livestock wastes, in a way that is integrated into an agricultural operation is receiving increased interest because of a number of forces. These include a rise in costs and regulation of solid waste management, increased momentum to move to a more sustainable agriculture, and increased recognition of downstream and off-site costs caused by pollution stemming from agricultural activities, most notably overfertilization and erosion.

---

[1] The Rodale Institute is a nonprofit organization whose mission is "...to improve human health through regenerative farming and organic gardening." The Institute's Research Center, founded in 1972 by Robert Rodale, focuses most of its research on regenerative land use practices that can lead to a more sustainable agriculture and society.

[2] The term "co-composting" was used to signify the combining of materials that are traditionally treated separately, in this case, municipal yard wastes and agricultural manures. Major funding for this project has been received from the Rockefeller Brothers Fund, The Pew Charitable Trusts and the Pennsylvania Department of Environmental Resources.

Benefits associated with composting and the use of compost have the potential to address all of these issues simultaneously.

WASTE MANAGEMENT ISSUES

Municipal and rural waste management practices have produced a national crisis marked by closing landfills, environmental degradation, and escalating costs. The USEPA estimates that more than 80% of existing landfills will be closed within the next 10 yr (USEPA, 1990a). The northeastern states have become the most vulnerable to increasing waste disposal costs and closing landfills (Glenn & Riggle, 1991a), and are, therefore, among the most likely to seek out alternative approaches to waste management (Inst. for Local Self Reliance, 1990).

Of the seven types of solid waste management practices identified in a recent survey of solid waste professionals, composting was seen as both most likely to increase and to experience the most growth (Roper, 1992). The same survey noted that collection of source-separated compostables also will increase in the future. Source-separated compostables are cleaner than mixed solid waste, and depending on the source, can go straight to composting with relatively little preprocessing. Examples of source-separated recyclables that could go to an agricultural composting operation include food processing wastes, grocery store wastes, fishery by-products, and yard wastes.

Yard waste represents one of the largest single components of the municipal solid waste (MSW) stream in the USA (approximately 18% by USEPA estimates), second in weight percentage only to paper and paperboard products. Twenty-three states have banned the disposal of yard waste in landfills, and legislation is currently pending in others (Kashmanian, 1993). The annual survey by *Biocycle* magazine highlights bans on yard waste disposal in landfills as the "ban which can have the greatest effect on reducing the amount of waste being disposed of" (Glenn & Riggle, 1991b).

The amount of agricultural waste is greater than municipal solid waste. The USDA estimates that over 1600 million wet tons of manure are produced annually by cattle, swine, sheep, and poultry (USDA, 1972). The USEPA estimates that nonpoint sources (NPS) contribute 45% of pollutants to estuaries, 76% in lakes, and 65% in rivers (USEPA, 1990b). Nitrate from animal manure is one of the major pollutants in these systems (Kashmanian et al., 1990). Efforts to clean up marine and fresh water environments, such as the Chesapeake Bay Program, have focused on reducing the impacts of agriculture on water quality, particularly from manure handling and disposal practices.

Off-site impacts of farming are not only in the form of water pollution. Many farmers find themselves facing a growing nonfarm population, as suburbs and developments push further into rural areas. Dust and odors that were never an issue now become serious problems to contend with. While right-to-farm legislation and agricultural security zones can afford some protection, neighbor relations is a reality many farmers can not ignore.

COMPOSTING AND SUSTAINABLE AGRICULTURE

A growing movement within agriculture is called "sustainable agriculture." Simply put, it is the movement towards an agricultural system that is both economically viable and environmentally responsible. It is founded on practices that conserve and rebuild the soil. Composting is a key technology for recycling of nutrients and building soil organic matter that is a part of a sustainable agriculture. Composting offers a number of other benefits to farmers, including flexible manure management, potential added income, increased odor control, weed control, and reduced disease pressures (NRAES, 1992).

RURAL/URBAN LINKAGES

Composting has the potential to serve as a new major link between the agricultural and nonagricultural sectors (Fig. 4-1). Compostable materials include yard wastes, food processing wastes, and/or nonrecyclable fiber products (i.e., paper) from municipalities or industries, and agricultural wastes, such as manures, from farms. Cooperative arrangements among municipalities and farmers to compost these organic resources can benefit all parties. The compost can either be produced on the farm or at a municipal

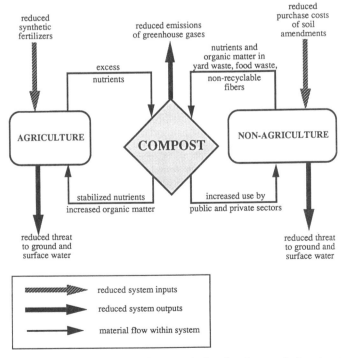

Fig. 4-1. Potential relationship of composting to agricultural and nonagricultural sectors.

or commercial composting facility. The compost that is produced can then get used on the farm, or in public or private applications, as a source of soil conditioning and fertility. A municipality could benefit from reduced solid waste management costs, and reduced costs of purchased soil amendments for their public works projects. A farm could benefit from increased revenue, increased organic matter and reduced input costs.

The stabilized nutrients in compost are less likely to cause environmental problems than the raw materials, and there is a reduced need for mineral fertilizers as well (Reider et al., 1992; Maynard, 1993). Increasing soil organic matter content by repeatedly applying compost reduces the risk of surface and groundwater pollution through increased water infiltration and improved water- and nutrient-holding capacity (Mays et al., 1973; Dick & McCoy, 1993). Furthermore, the release of certain greenhouse gases, such as methane and ammonia, could be reduced through the incorporation of N into the microbial biomass, though this aspect of the composting process would have to be optimized for significant benefits to be realized (S. Mathur, personal communication, 1993).

Given the range of potential on- and off-farm benefits of agricultural composting, why are more farmers not composting municipally generated wastes on their farms? To answer this key question, we must understand the barriers to on-farm composting and develop strategies for overcoming them. A systems approach, where the system is the regional flow of organic wastes and the major comopnents are the municipal and agricultural organizations within the system, can highlight opportunities to affect structural and functional change (Smith & Oshins, 1993).

FIRST- AND SECOND-ORDER CHANGE

The most commonly understood type of change is *first-order* change. First-order change corresponds to a change in the *structural* properties of a situation, without changing the *functional* properties of the relationships among components of the system. The focus is on changing the efficiency within a component, not on the overall effectiveness of the system. The "game" in which the change is taking place remains the same (Smith, 1989).

First-order change at the urban scale means improved collection and processing of yard waste through municipal composting. Rural examples include the composting of animal manure to produce soil amendment for use on the farm. The focus of change in these situations is intraorganizational, a farm or municipality. The measure of performance is the change in efficiency of operations.

First-order changes tend to treat rural and urban systems as separate components. Manure is perceived as too expensive to transport to municipalities, and yard waste compost has too little nutrient value to market to farms. Regulatory barriers often compound the difficulties of mixing urban and rural waste streams. The economic, regulatory, and social hurdles create sufficient resistance to change to thwart most attempts to resolve the inherent limitations of the present system.

A second-order change represents a change in the functional relationships among actors in a situation. These actors, who include farmers, waste haulers, municipal planners and recyclers, environmentalists, and state agency employees, play a different game, with different rules for action. An example of second-order change is collection of yard wastes from municipalities and then paying farmers tipping fees to receive these materials. Farmers may directly apply material to their fields or use the high C content of leaves by mixing them with manure to produce compost. The compost may be used in crop production or sold to a landscaper who uses it on urban lawns.

Farmers benefit from a new "cash crop" which requires little or no new capital expenditure. Municipalities benefit by avoiding capital and operating expenditures for municipal compost facilities. The more proactive, second-order approach to reframing the rural/urban waste situation calls for fundamental changes in the relationships among urban and rural actors. The focus of second-order change is thus interorganizational, and the measure of performance is the effectiveness in pursuing individual and shared objectives.

FARM CO-COMPOSTING PROJECT

Rodale adopted a stakeholder-driven, empirical research approach toward overcoming barriers to farm cocomposting. This includes building a network of commitments across organizational and political boundaries with a well-defined regional and technology focus to foster the kind of second-order change described in the previous section. The Project is being conducted in four broad, overlapping phases:

1. Conduct Stakeholder-Driven Research. Identify and evaluate the barriers and opportunities facing regional practitioners in waste management and agriculture. Build consensus among diverse interests on the shared opportunities in regional approaches to transforming rural/urban wastes into agricultural resources.

2. Support Local Initiatives. Identify economic and other incentives for rural/urban partnerships between farmers and waste management practitioners. Broker new relationships between regional farmers and municipalities and counties.

3. Create Regional Institutions. Cosponsor public/private projects to support on-going research and development of regional farm cocomposting practices. Participate in the creation of regional institutions to foster such practices.

4. Act Regionally, Disseminate Nationally. Serve as a resource to similar initiatives in other regions through dissemination of a "tool-kit" for farm co-composting. Augment these efforts by publicizing results in print and visual media, through cooperative projects with national associations, and in various ways shape national policy debate and practices.

Identifying Barriers and Opportunities

The key to Rodale's systems approach is stakeholder-driven planning. The social science research involved key stakeholders of the rural/urban system in identifying barriers and opportunities to farm cocomposting at the regional scale. The research provides a results-oriented focus to understanding the web of transactional relationships among diverse public and private interests who comprise the whole system.

The stakeholders include regional public and private sector interests with a stake in waste management and agricultural practices. These include county solid waste management authorities, farmers, farm support agencies, private hauling companies, financial institutions, municipal recycling and community development offices, urban and rural residents, community-based civic groups, and local environmental groups. The research focused on four counties in southeastern Pennsylvania: Northampton, Lehigh, Berks, and Lancaster. These counties represent a full spectrum of urban, suburban, and agricultural communities.

The stakeholder-driven research agenda has been developed in consultation with these diverse public and private interests. Methods included interviews with numerous stakeholders to map the divergence of views, interests, and practices, and to develop a database of contacts. These interviews were followed by a surveys of more than 60 Pennsylvania municipalities on their yard waste management practices. Focus sessions in the four counties were held with representatives from different interest groups to facilitate the convergence of views, interests, and practices.

This participatory approach served to enhance stakeholder understanding and acceptance of the research findings and recommendations. The research identified an array of interrelated barriers and opportunities, including technical, economic, and institutional constraints (Table 4-1). Some of these constraints are faced by any potential agricultural composter, while others are particularly pertinent to those farmers using off-farm inputs in their composting operation.

Technical/Educational Issues

From a technical standpoint, the issues reflect a lack of familiarity of farmers with composting more than a problem with the technology itself. From an "innovation adoption" perspective, composting is at an early stage in the diffusion process (Rogers, 1983). This is complicated by the fact that composting is not simply a piece of machinery that may or may not be *adopted*, but a process that must be *adapted* to the specific farming situations. Manure composition and production, site layout, availability of off-farm inputs, and resource availability will all vary from one farm to the next. Thus, the adoption of composting requires a relatively high level of understanding on the part of the adopter.

Sources of information that a farmer might access to gain the necessary understanding are currently inadequate. These sources include print media,

Table 4-1. Barriers to working with farms to manage municipal yard wastes.

Barrier	Factors	Strategies
Getting the materials to the farm	Coordination of collectors/haulers and farmers	Have a broker (public or private) serve several municipalities
	Distance to the sites	Compare to other alternatives
	Differences among materials	Different farms will need/want different materials at different times
		Woody materials should get used for mulch by a landscaper or grounds department
Reliability	Availability of fields during inclement weather, during growing season	All-weather drop-off/storage site
		Multiple sites
	Security of arrangements	Provide compensation; Negotiate a contract
PA-DER† regulations	Hard to understand	Rewrite in "farmer friendly" format
	Ambiguous and inconsistent	Coordinate with manure management guidelines
	Antiregulation attitude	Rewrite in "farmer friendly" format
	Future change could create a liability	Err to caution, work with PA-DER and local agencies
Contamination	Trash	Design collection system to minimize trash, educate residents on where materials are going; Provide compensation
		Provide for disposal of trash in yard waste at no cost to farmer
	Pesticides	Preliminary data shows these are not a problem, more research needs to be done to fully allay fears
		Reduce potential through a residential program that promotes minimizing pesticide use, grass recycling, and not cutting sprayed lawns for several days
Costs (other than for collection and transportation)	On-farm: Equipment, operations, site improvements, guidelines compliance	Direct compensation: tipping fees (per ton or cubic yard), flat fee for season, hourly wage
		Cost-share: PA-DER recycling grants, USDA's ASCS, and other, e.g.,: USEPA's Chesapeake Bay Program
	Municipal: administration, personnel	Compare to avoided costs
Education	Awareness	Included in DER yard waste manual
		Increase visibility of programs
	Knowledge	Expand programming of Cooperative Extension
		Increase training of farm advisors
	Participation	Host or attend field days, workshops

† PA-DER = Pennsylvania Department of Environmental Resources.

service organizations, and direct contact. Educational materials on farm-based composting have included bulletins (e.g., Small Farm Energy Project, 1980), conference proceedings (e.g., Rynk, 1987), reports (e.g., Zabriskie, 1991) and periodicals (e.g., *Biocycle* magazine). The Northeast Regional Agricultural Engineering Service (NRAES) recently published a manual on on-farm composting (NRAES, 1992) that has begun to fill this niche.

Service organiztions are another source of information for many farmers. These include cooperative extension (ES), the Soil Conservation Service (SCS), county conservation districts, and industry trade associations. Feed, fertilizer, and machinery salespeople also function as information sources for many farmers. Of all these, cooperative extension has the clearest educational mission, but suffers from the same lack of educational materials as farmers. In addition, extension agents generally have received no formal training in composting. Those agents that are able to provide advice to farmers do so because of their own initiative. As more farmers become interested in composting, cooperative extension and other groups must make a bigger commitment to providing their clients with the needed information. Some states, such as Maine, Pennsylvania, and Michigan, are moving forward faster than others.

The third source of knowledge about composting available to a farmer is the exchange of information with other farmers who have tried composting. In fact, farmers often turn first to other farmers as sources of information. However, because very few farmers are composting, a farmer interested in starting a composting operation cannot readily go and see an established operation. Moreover, those farmers that are composting are "innovators" or "early adopters" (Rogers, 1983), which means they are likely atypical in operation, resources, or motivation so their experiences are of limited relevance to most other farmers.

While agricultural composting in general faces these educational challenges, they are especially acute for farm cocomposting. In a recent survey of farm composters in Pennsylvania, only a small percentage (12 out of 54, or 22%) were receiving municipal materials (Oshins & Fiorina, 1993). With the addition of off-farm inputs, concern over contaminants became important. Farmers tended to be more worried about trash that could come with loads of leaves, such as plastic, glass and metal. Municipal people on the other hand tended to be more focused on toxic chemicals, such as pesticide residues on grass clippings. Their concerns were in fact related to the fear of liability that is aroused whenever toxic or hazardous substances are handled.

Economic Issues

Composting requires resources in the form of labor (to process and use the materials), land (for staging, processing and curing areas) and capital (for processing machinery such as windrow turners and material grinders and for purchased raw materials). These costs must be balanced against the benefits that may accrue in terms of fertilizer value and other soil enhancing properties of the compost, tipping fees that may come with certain amendments, and avoided costs of other manure handling systems. The benefits of compost in other-than-nutrient terms has not been quantified, yet are generally agreed to be present. Quantifying these benefits, such as reduced erosion or improved tilth, would allow additional economic benefits to be assigned to compost.

For farm cocomposting, transactional and transportation costs can be overwhelming. Transaction costs refer to the time and expense in forming and maintaining relationships. A municipality may need to deal with a number of farms, in order to assure adequate field availability during changing seasons. For municipalities mandated to manage their wastes, reliability is a key issue. One recycling coordinator suggested that this type of relationship "creates a dependence on farmers, leaving the municipality vulnerable if the farm should stop accepting leaves or be sold to developers." On the other hand, a large livestock producer may need to work with a number of municipalities and industries to obtain sufficient material for composting large volumes of manure. The seasonality issue also is important, because leaves come only during the fall, but manure is generated every day of the year.

Transporting materials is expensive, so a farm must be reasonably close to the source to make it worthwhile, depending on the distance and cost of the next closest alternative. Regulations often drive these economics, especially near state borders, where differing regulations can greatly affect the cost of doing business. Hence, many Pennsylvania farmers receive grass clippings from New Jersey municipalities, because the cost of regulatory compliance in that state is much higher.

Institutional/Cultural Issues

Issues identified by farmers and municipalities focus strongly on regulatory barriers to farm cocomposting. Regulations are often cited as being hard to understand and ambiguous. Composting regulations are written by solid waste agencies and aimed for a municipal audience, and are difficult for farmers and other nonregulators to decipher. Regulations examined in this study did not directly address the possibility of farm co-composting, thus creating a large amount of uncertainty and increasing the risk for investment. In addition, the regulatory jurisdiction for farm cocomposting remained unclear, because animal manure and yard waste are regulated by different state agencies.

Farmers do not want to be perceived as "solid waste facilities" for cocomposting clean organic materials. They see composting as sound resource and farm management. Many farmers are very reluctant to seek regulatory approval for on-farm practices. On the municipal side, an interesting but pervasive attitude exists that if farmers are interested in their materials, such as leaves, then they shouldn't need to be paid to take it. They seem to fail to recognize that the value of the material to the farmer does not outweigh the farmer's costs of handling and regulatory compliance.

Targeted Strategies: Advocacy, Education and Research

The barriers to adoption identified through the stakeholder-driven research formed the foundation for much of the planning and activities that followed. A mix of advocacy, education and research activities are being used to overcome the institutional, economic, technical and educational barriers that were identified.

Efforts to overcome institutional barriers began with the creation of a Project Advisory Committee (PAC). The PAC is composed of key leaders representing diverse public and private interests, local and statewide, with a stake in the project. The PAC helps keep the project focused on stakeholder issues, serves as a sounding board for Rodale project staff, and taps the political, economic and technical resources of a large network of stakeholders. This greatly enhances the credibility and legitimacy of the project, and magnifies its impact.

Institutional barriers can be addressed through both legislative and regulatory channels. Rodale has helped form a Compost Working Group, with representatives from key regulatory and technical agencies and organizations. This group is advising the Pennsylvania Department of Environmental Resources (PA-DER) on regulatory revision or alternatives.

Cultural and educational barriers are being addressed through a number of projects, in cooperation with Pennsylvania State Cooperative Extension, the PA-DER, and the Pennsylvania Association for Sustainable Agriculture (PASA). Getting farmers and municipal officials out to see working examples of farm cocomposting is both the best way to teach about the technology and to change people's misconceptions and uninformed atittudes. To achieve this, farm "field days" or "open houses" have been a core activity. A range of technologies and ingredients have been highlighted, and attendance has included farmers, farm adivsors, regulators, and municipal officials. Field days also offer opportunities for equipment demonstrations, evaluating composting sites and technologies, and examining field plots researching or demonstrating the use of compost.

Other educational activities include development and/or distribution of such materials as brochures, fact sheets, manuals, slide shows, and a directory of on-farm composters. This latter item is a tool to help connect farmers to farmers. It gives them enough information to know they are connecting with someone with a similar condition. Presentations have been prepared for various groups, such as municipal committess, regional workshops, and state and national conferences.

Research is another key strategy for overcoming barriers. Rodale is documenting several case studies, representing a spectrum of technologies, systems and arrangements by working with farmer–collaborators. The farmer–collaborators are addressing a variety of questions, including costs and benefits of a number of composting scenarios, direct use of grass clippings and leaves, and impact of yard residue utilization on odor generation and nutrient conservation in different settings. These studies will be useful in making a stronger case for fair compensation of their services.

One of the more innovative cases involved a swine producer who was searching for a way to reduce odors from his manure lagoon. He experimented with using grass clippings from a nearby city. He pushed the grass into the lagoon, which spread out into an 8-cm mat over the surface. This mat significantly reduced odors, according to a number of extension agents he worked with as well as other visitors to the farm. In addition to the odor control, the grass clippings increased the nutrient and organic matter levels of the manure.

Finally, a number of activites are directed towards the technical and economic barriers to farm cocomposting commercializations. Limited technical assistance is provided free of charge to farmers trying or planning to compost and to small companies seeking to offer composting services and equipment for farms. In collaboration with Pennsylvania State University, we are documenting the economics of agricultural composting as a manure management method and we are doing a market analysis of farm-based compost and compost services in Pennsylvania. Rodale is working to link haulers, municipalities, and farmers by providing contacts, clarifying regulations, facilitating arrangements, and advocating for a fair fee for the farmer.

RECOMMENDATIONS

Based on experiences and information gathered during the Farm Co-Composting Project so far, there are four key recommendations.

Compensate Farmers

First is the importance of compensating the farmer. The value of the service that the farmer is providing to the municipality must be recognized, and paid for adequately. The level of compensation is something that must be worked out on a case-by-case basis, but typically falls in the $1 to $5 per cubic yard range. Aside from tipping fees, other arrangements include payment on an hourly basis for landspreading, or simply a flat fee for a season.

Paying the farmer not only compensates for the direct costs of utilizing the materials, but also addresses the reliability issue. Payment gives the farmer an incentive to stay with the program and provides a sense of responsibility, obligation, and value. It also compensates for dealing with the inevitable trash that shows up in yard waste. At a winter meeting of over 20 farmers participating in Lancaster County's aggressive farm application program, all the farmers had stories of "contaminants" (everything from brush to fast food wrappers to dirty diapers), yet none were dropping out of the progrma.

The effectiveness of farmer compensation is evident where such a system is the norm. In Lancaster County, the demand for materials, as represented by approved farm capacity, exceeds the supply of leaves and grass clippings by three times (personal communication, T. Breneisen, 1993). One farmer stated that getting paid to take leaves made the difference between profit and loss during 1992.

Coordinate Resources

The second conclusion is the value an intermediary can play in coordinating the farmer-municipal connection. For many smaller municipalities, it would make more sense for one person or organization to act as "yard waste broker." A broker could coordinate delivery of materials from several municipalities to several farms more efficiently than could each borough or township by itself.

There are several good examples where such linkage systems exist in both public and private sectors. The highest rate of diversion of materials to farms in this study was in Lancaster County, where the Lancaster County Solid Waste Management Authority works to facilitate linking haulers and farmers. At the state level, New Jersey runs a state hotline for linking municipalities and farmers for direct land application of leaves in the fall. Finally, a private company, Compost Connections in Pittston, Maine, is brokering materials to a network of composting farms (Jones, 1992).

Streamline Regulations

Third, regulatory barriers need to be addressed. State leaf composting regulations are written for a municipal audience, and though usually less onerous than full permitting, can be hard for a farmer to decipher. In addition, the cocomposting of yard wastes with manures is generally not addressed, in either yard waste regulations or manure management guidelines. Leaves represent an attractive C source for farmers wishing to compost high N manures, but many are reluctant to enter this regulatory arena, believing it will involve a lot of red tape, and be time consuming, expensive and intrusive.

Massachusetts has an innovative program worked out with the cooperation of the state's Departments of Environmental Protection and Food and Agriculture. A farmer wishing to compost yard or food waste (up to specified limits) can simply obtain a copy of the state *Guide to Agricultural Composting* (Watson & Tierney, 1992) and use the attached registration form to register with the Massachusetts Department of Food and Agriculture. This user-friendly model is allowing significant numbers of farmers to work within the system (van de Kamp, 1992).

Continue Research and Education

Research and education need to be continued and expanded. While there are many questions, key areas are economic feasibility, environmental impacts, and recommendations for use. The actual costs of establishing and operating an on-farm composting operation need to be better detailed under a variety of scenarios. Only then can truly fair compensation be achieved. The environmental impacts of agricultural composting, both pro and con, must be documented and compared to alternatives. Finally, we must better understand how compost works in the field. This means building up a database documenting a range of composts, crops, and conditions. This must include a detailed enough description of the compost(s) so that meaningful comparisons can be made. The ultimate goal should be that a farmer could send in a sample of his or her compost, along with a soil sample and intended use, to a soils laboratory and get back reliable recommendations on how to use the compost.

Equally important is continued education of farmers, haulers, recycling coordinators, and policy makers on the advantages and possibilities of working with farms. This can not only expand their options for managing certain "wastes," but also can help maintain a productive local agriculture, improving the relationships among farmers, their neighbors, and their municipalites.

SUMMARY

Surveys and focus groups conducted by the Rodale Institute Research Center identified a variety of barriers to the cocomposting of agricultural and nonagricultural wastes on farms. Institutional, technical, cultural, and economic barriers were all distilled. Based on this research, strategies were developed to reduce or overcome these barriers. Strategies include advocating regulatory change, documenting a diversity of case studies, supporting technical and economic research, supporting local and regional initiatives, and developing an educational strategy with a variety of educational tools. Key recommendations for increasing farm cocomposting are compensating farmers, coordinating resources, and streamlining regulations. Increased farm cocomposting can help protect the environment from farm and municipal pollution while helping to sustain agriculture and be cost effective for municipalities.

REFERENCES

Dick, W.A., and E.L. McCoy. 1993. Enhancing soil fertility by addition of compost. p. 622–641. *In* H.A. Hoitink and H.M. Keener (ed.) Science and engineering of composting: Design, environmental microbiological and utilization aspects. Ohio State Univ. Press, Wooster, OH.

Glenn, J., and D. Riggle. 1991a. The state of garbage in America. BioCycle 32:34–38.

Glenn, J., and D. Riggle. 1991b. The state of garbage in America, part 2. BioCycle 32:30–35.

Institute for Local Self Reliance. 1990. Beyond 40%: Record setting recyling and composting programs. Ind. Local Self Reliance, Washington, DC.

Jones, B.J. 1992. Composting food and vegetative waste. Biocycle 33:69–71.

Kashmanian, R. 1993. Predicting the tonnage of yard trimmings to be composted in 1996. Biocycle 34:51–53.

Kashmanian, R., C. Gregory, and S. Dressing. 1990. Where will all the compost go? Biocycle 31:38, 39, 80–83.

Maynard, A.A. 1993. Nitrate leaching from compost-amended soil. Compost Sci. Utiliz. Spring, p. 65–72.

Mays, D.A., G.L. Terman, and J.C. Duggan. 1973. Municipal compost: Effects on crop yield and soil properties. J. Environ. Qual. 2:89–92.

Northeast Regional Agricultural Engineering Service. 1992. On-farm composting handbook. R. Rynk (ed.) NRAES-54. Riley Robb Hall, Coop. Ext., Ithaca, NY.

Oshins, C., and L. Fiorina. 1993. Challenges of on-farm composting. Biocycle 34:72–73.

Reider, C., C. Oshins, A. Pronk, R. Janke, and J. Moyer. 1992. Compost utilization for field crop production: Part 2. Rodale Inst. Res. Center Tech. Rep. Rodale Inst., Kutztown, PA.

Roper Organization. 1992. What is the future of composting: Perspective of experts. Novon Products Group/Warner Lambert Co., Rockford, IL.

Rynk, R. (ed.) 1987. Proc. On-Farm Composting Conf. Amhearst, MA. 15 January. Publ. no. AG196-2/88-300. Univ. Massachusetts Coop. Ext., Amhearst, MA.

Small Farm Energy Project. 1980. Composting of farm manure. Proj. Focus no. 8, Center for Rural Affairs, Hartington, NE.

Smith, A.E., and C.S. Oshins. 1993. Composting wastes in to resources: a rural/urban framework. J. Resour. Manage. Technol. 21:62-68.

Smith, A.E. 1989. Enterpreneurial development of organizational communities: An events-focused planning approach. Ph.D. diss. Univ. Pennsylvania, Philadelphia (UMI Order no. 9015167).

U.S. Department of Agriculture. 1972. From agricultural waste to feed or fuel. Chemical Engineering News. 29 May, p. 14.

U.S. Environmental Protection Agency. 1990a. Characterization of municipal solid waste in the united states: 1990 Update. U.S. Gov. Print. Office, Washington, DC.

U.S. Environmental Protection Agency. 1990b. Selected non-point source problems and solutions. U.S. Gov. Print. Office, Washington, DC.

van de Kamp, M. 1992. Farm composters play significant management role. BioCycle 33:67-69.

Watson, G., and S. Tierney. 1992. Guide to agricultural composting. Publ. No. 17163-38-599-7/92. Dept. Food and Agric., Boston, MA.

Zabriskie, P. (ed.). 1991. Composting organic wastes on the farm. Gardener's Supply Co., Burlington, VT.

5 Mineralogy of High Calcium/ Sulfur-Containing Coal Combustion By-Products and Their Effect on Soil Surface Sealing

L. Darrell Norton

Purdue University
West Lafayette, Indiana

Coal combustion by-products consist of many different types with different physical and chemical properties. The type of coal and burning technology dictates the kind and amount of by-product (ashes or residues) produced. The Clean Air Act (CAA) of 1976 as amended in 1990 mandates that coal burning facilities reduce the amount of SO_2 delivered to the atmosphere. This may be accomplished by burning low-S coal or using Clean Coal Technologies (CCT). Because of the CCA and the use of the CCT, a growing tonnage of ashes and residuals are being produced. The types of by-products produced include: fly ashes, bottom ash, cyclone slag, spent bed material from fluidized bed combustion (FBC), and flue gas desulfurization sludges (wet and dry). Some advanced processes (Torrens & Radcliff, 1990) produce ammonium sulfate and ammonium nitrate, as well as, elemental S, sulfuric acid and wallboard-quality gypsum. The most widely studied by-products are fly ashes, which are generally very fine sand or silt-sized and capable of being carried by the flue gases from combustors. Fly ash is divided into highly alkaline (Class C) and acid-neutral (Class F). Class F fly ash is generally derived from eastern U.S. high-S coals while Class C fly ashes are derived from western low-S coals and lignite. Class F fly ashes may have a considerable amount of sulfates as well as heavy metals present (Brieger et al., 1992).

Clean coal technologies (CCT) also produce a large amount of residuals other than fly ashes. They are produced as a result of coal conversion (e.g., elemental S and sulfuric acid), precombustion cleaning (i.e., pyrite), combustion cleaning [fluidized bed combustion (FBC) residuals from atmospheric and pressurized systems] or post–combustion cleaning such as flue gas desulfurization (FGD). Flue gas desulfurization systems consist of wet

Copyright © 1995 American Society of Agronomy, Crop Science Society of America, Soil Science Society of America, 677 S. Segoe Rd., Madison, WI 53711, USA. *Agricultural Utilization of Urban and Industrial By-Products*. ASA Special Publication no. 58.

or dry scrubbing of the flue gases and produce materials high in calcium sulfite and/or calcium sulfate. Advanced FGD system may produce high-quality gypsum suitable for the manufacture of wallboard. Estimates are as high as 50 million tons of FGD production per year by the year 2000 (USEPA, 1988). As of the end of 1988, 218 units in the USA utilized or planned to use FGD to lower S emissions. This represented a total scrubbing capacity of almost 90 000 MW of electrical production (Hance et al., 1990). The wet limestone FGD systems (with and without forced oxidation) are the most widely used today and except for small amounts used in making dry wall, most of the material is landfilled.

While FGD is the most common method chosen for retrofitting existing combustors (Torrens & Radcliff, 1990), newer plants often choose fluidized bed combustion (FBC) due to lower costs. Fluidized bed combustion fly ashes and bottom ashes contain a considerable amount of calcium sulfate and have a high pH due to the presence of calcium oxide. Because of the high pH and the presence of beneficial alkaline oxides, FBC has been evaluated for its potential for agricultural land application (Stout et al., 1988). This work focused mainly on the beneficial effects of liming, as well as the loading of heavy metals (Bennett et al., 1985). Fluidized bed combustion also has been evaluated as a material for the preparation of concrete (Rose et al., 1986), but the quality is generally poor because of expansion during ettringite formation (Solem & McCarthy, 1992). Some work has been conducted to look at the effect of adding these materials on the physical properties of the soil, particularly those related to infiltration and soil loss (Norton et al., 1993). Similar to FGD, FBC residues are generally landfilled and their use in agriculture at this point is limited. Presently, landfill sites are becoming more rare and the cost of landfilling is rapidly increasing, therefore, it is highly desirable to find alternative uses for these materials such as a soil amendment.

The objective of this paper is to examine the different mineralogies and properties of some commonly produced high Ca and S coal combustion by-products and summarize several studies to evaluate their potential for use as a soil amendment for reduction of surface sealing and erosion. This paper will not deal with fly ashes but will focus on the emerging by-products that are more chemically soluble including FGD and FBC.

COAL COMBUSTION BY-PRODUCTS

Following a review of various CCT's, samples of high-Ca by-products were collected to represent a range of materials commonly produced by burning high-S coal. Samples were collected from three plants in Indiana and one plant in Florida for study (Table 5-1). All samples were obtained in the waste stream prior to mixing with fly ash in order to study the high Ca/S materials. The mineralogical composition of the materials was determined by x-ray diffraction and thermal gravimetric analysis. Calcite was determined gasometrically (Dreimanis, 1962) and total S content with a Leco sulfur

Table 5-1. Locatons of S containing coal combustion by-products studied.

Plant	Process	Location	pH	CCE†
Big Bend	Forced oxidation FGD‡	Tampa, FL	7.3	4.5
Gibson	Wet limestone FGD	Owensville, IN	8.1	27.4
Merom	Wet limestone FGD	Merom, IN	8.3	30.3
Purdue	Atmospheric FBC§	West Lafayette, IN	12.5	45.1

† By acid neutralization.
‡ FGD = Flue gas desulfurization residue (provided by Dr. Dale Ritchie, Beckley, West Virginia).
§ FBC = Fluidized bed combustion ash.

analyzer (Leco Corp., St. Joseph, MI). The FBC was further analyzed for total selected trace elements in the bottom ash and leachates from it using an induced coupled plasma spectrometer (ICP). Toxic characteristic leachate procedure (TCLP) was performed for the eight Resource Conservation and Recovery Act (RCRA) elements following standard procedures (USEPA, 1991).

MINERALS CONTAINED IN COAL COMBUSTION BY-PRODUCTS

Depending on the method of removing SO_2 gas during coal combustion, the minerals contained in the by-products may vary considerably (Fowler et al., 1992). Generally, the S-containing minerals produced are quite soluble, but depending on the method of combustion and the coal composition considerable amounts of insoluble residues may be produced which are relatively nonreactive compared to the soluble minerals. Table 5-2 is a list of the commonly produced soluble residues in FGD and FBC and their solubil-

Table 5-2. Common soluble minerals that occur in FGD sludges and FBC bottom ash.†

Soluble residues		Solubility in cold water	Thermic activity
		$g\ L^{-1}$	
$CaSO_4$	Anhydrite	2.1	triclinic-rhombohedral > 200
			$-SO_3 > 950$
$CaCO_3$	Calcite	0.014	$-CO_2 > 710$
$CaSO_4 \cdot 2H_2O$	Gypsum	2.4	$-3/2\ H_2O > 128$
			$-2\ H_2O > 163$
			$-SO_3 > 950$
$CaSO_3 \cdot 1/2H_2O$	Hannebachite	0.043	$-1/2H_2O > 350$
CaO	Lime	1.31	
CaS	Oldhamite	0.21	
$CaSO_4 \cdot 1/2H_2O$	Bassanite	3.00	$-1/2H_2O < 165$
			$-SO_3 < 950$
$Ca(OH)_2$	Portlandite	1.85	$-H_2O > 580$

† Flue gas desulfurization, FBC = fluidized bed combustion.

Table 5-3. Common insoluble components in FGD and FBC.†

Mineral		Solubility in cold water	Melting point
FeS_2	Pyrite	0.0049	1171
Fe_2O_3	Hematite	Insoluble	1565
SiO_2	Quartz	Insoluble	1610
SiO_2	Glass	Insoluble	1600
Al_2O_3	Corundum	Insoluble	2015
MgO	Periclase	Insoluble	2852

† FGD = flue gas desulfurization, FBC = fluidized bed combustion.

ity and thermal activities and Table 5–3 lists some of the insoluble residues that may be present.

The high Ca/S compounds are chemically active due to their solubilities, whereas, the insoluble materials are relatively inert. All of the materials listed in Tables 5–2 and 5–3 are crystalline and their presence can be identified with x-ray diffraction techniques except for the amorphous silica glassy phases. The three most intense d-spacings for the common minerals produced are given in Table 5–4.

It should be noted that since many of these materials produced are chemically active, that the primary minerals may easily convert to secondary minerals. For example, lime may quickly hydrate even in atmospheric conditions to portlandite which may further carbonate at atmospheric conditions to form calcite. Lime may further hydrate at high pH with sufficient water present to ettringite, which is a secondary mineral that may quickly form but is unstable as the pH becomes lowered in the soil environment. Additional transformations of the minerals also may occur in the soil environment. Especially, the bio-oxidation of the sulfite minerals to sulfates. Therefore, the mineralogy of the coal combustion by-products may vary considerably over time due to the method of production, treatment and storage.

Table 5-4. X-ray powder diffraction data for minerals found in FGD and FBC by-products from coal-fired power plants.†

Mineral name	Nominal formula	PDF card no.	d-spacings (intensity)		
Anhydrite	$CaSO_4$	37-1496	3.50	2.85	2.33
Calcite	$CaCO_3$	5-586	3.04	2.29	2.10
Ettringite	$Ca_6Al_2(SO_4)_3(OH)_{12}26H_2O$	41-1451	9.72	5.61	3.87
Gypsum	$CaSO_4 2H_2O$	33-311	2.87	4.28	2.68
Hannebachite	$CaSO_3 \cdot 1/2H_2O$	39-725	3.16	2.63	5.56
Hematite	Fe_2O_3	33-664	2.70	2.52	1.69
Lime	CaO	37-1497	2.41	1.70	2.78
Oldhamite	CaS	8-464	2.85	2.01	1.64
Periclase	MgO	4.829	2.11	1.49	1.22
Portlandite	$Ca(OH)_2$	4.733	2.63	4.90	1.93
Quartz	SiO_2	33-1161	3.34	4.26	1.82

† FGD = flue gas desulfurization, FBC = fluidized bed combustion.
‡ Joint Committee on Powder Diffraction Files reference powder diffraction card number.

Table 5-5. Mineralogy of S containing coal combustion by-products studied.

Plant	Process	Location	Mineralogical composition (formula)
Big Bend	FGD†	Tampa, FL	100% Gypsum ($CaSO_4 - 2H_2O$)
Gibson	FGD	Owensville, IN	60% Hannebachite ($CaSO_3 - 1/2H_2O$)
			18% Calcite ($caCO_3$)
			15% Gypsum ($caSO_4 - 2H_2O$)
Merom	FGD	Merom, IN	97% Hannebachite ($CaSO_3 - 1/2H_2O$)
			2% Calcite ($CaCO_3$)
			1% Gypsum ($CaSO_4 - 2H_2O$)
Purdue	FBC‡	West Lafayette, IN	73% Anhydrite ($CaSO_4$)
			23% Lime (CaO)
			trace Quartz (SiO_2)
			1% Calcite ($CaCO_3$)
			3% Portlandite ($Ca(OH_2)$)

† FGD = flue gas desulfurization residue.
‡ FBC = fluidized bed combustion ash seived to removed material other than FBC.

MINERALOGY OF THE HIGH CALCIUM/SULFUR
BY-PRODUCTS STUDIED

The semiquantitative mineralogy of the specific materials sampled was determined using x-ray diffraction, thermal analyses and supportive chemical analyses (Table 5-5). The Big Bend plant is an advanced FGD system with forced oxidation to produce wallboard-quality gypsum which was found to contain only gypsum by x-ray powder diffraction (Fig. 5-1a) and thermal gravimetric analysis. The Gibson plant was a wet limestone FGD system which contained predominantly hannebachite with a substantial amount of unreacted calcite and some gypsum (Fig. 5-1b). The Merom plant (Fig. 5-1c) was a wet limestone FGD system with a retrofitted forced oxidation system to prevent scaling problems commonly associated with this system (Reynolds, 1990). At the time of sampling the forced oxidizer was not functioning and the material produced was found to be almost pure hannebachite with minor amounts of calcite and gypsum by TGA (Table 5-5).

The widest variety of chemically active minerals were found in the Purdue FBC bottom ash (Fig. 5-1d). The predominant mineral was anhydrite with a significant amount of lime and a minor amount of unreacted calcite with a trace amount of quartz (Table 5-5). Since the material had been exposed to moisture in the air, some of the lime had hydrated to portlandite. This mineral was not detected by x-ray diffraction but was found using TGA. Ettringite was not found since a sufficient amount of water had not been present for its formation. The anhydrite was probably the less reactive rhombohedral form since the temperature of production was greater than 200 °C although this could not be confirmed by x-ray diffraction.

The solubility of these materials was compared to the more soluble minerals portlandite and bassanite and the more weakly soluble calcite and phosphogypsum in Fig. 5-2. The traces in Fig. 5-2 contain no y-axis since each trace represents the change in conductivity of pure water with time after immersion of the material. The y-scale for each trace ranges from zero to the final number labeled as saturated EC. The Merom FGD had the slowest

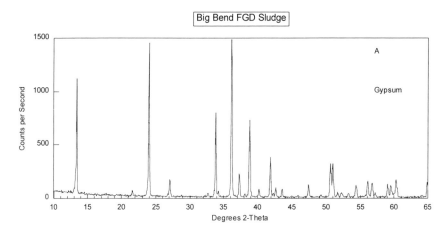

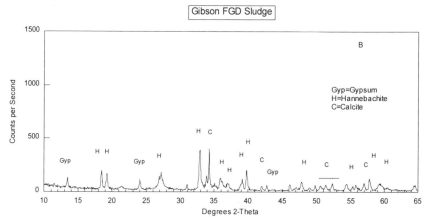

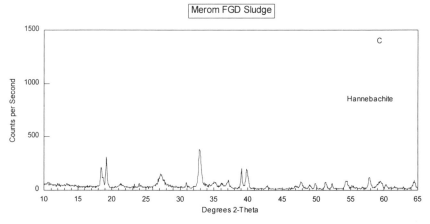

Fig. 5-1. X-ray powder diffractograms for the by-products studied (*A*, Big Bend; *B*, Gibson; *C*, Merom).

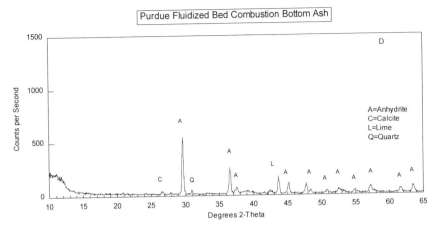

Fig. 5-1. Cont'd. *D*, Purdue.

release of electrolytes and the lowest saturated electrical conductivity (EC) of the materials studied because of the predominance of hannebachite. The Gibson FGD-contained more gypsum than Merom and therefore had a greater saturated EC which was equal to that of Big Bend. All of these materials released significantly greater amounts of electrolytes than did calcite and were similar to phosphogypsym. The Purdue FBC, however, released considerably more electrolytes than the above FGD materials and the total was slightly less than that of standard portlandite or bassanite. Both portlandite and bassanite released electrolytes very quickly when immersed in water, whereas, the Purdue FBC released electrolytes more slowly (Fig. 5-2). This is an im-

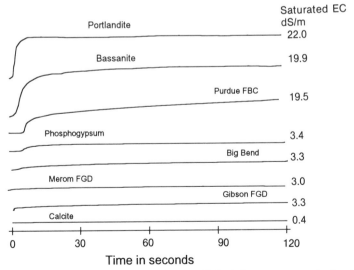

Fig. 5-2. Electrolyte release from the by-products studied relative to some standard materials. The ending number represents the final saturated electrical conductivity. The *y*-axis (unlabeled) is the same magnitude for all materials.

portant phenomena for erosion since the release of electrolytes has been used
to explain the increased infiltration and reduced soil loss with addition of
phosphogypsum (Miller et al., 1990; Norton et al., 1993). This slower release
of electrolytes would allow for the longer beneficial effect on erosion during
a rainfall event.

EFFECT OF DIFFERENT MATERIALS ON SURFACE SEALING

The FBC and the two FGD sludges were compared to phosphogypsum
in an earlier study (Norton et al., 1993) and the results are reviewed here.
Addition of gypsum to sodic soils at a rate of 5MT ha^{-1} was previously
found to be effective on reducing surface sealing by Agassi et al. (1981). Similar results were found for phosphogypsum on nonsodic soils from the
southeastern USA (Miller & Bharuddin, 1986). Norton et al. (1993) compared three by-product materials and phosphogypsum for their ability to
reduce surface sealing on a nonsodic Miami soil typical of the Cornbelt. Figure
5-3 shows the comparison of the effectiveness of these materials on increasing infiltration rates for this soil. Both the phosphogypsum and FBC significantly improved the final infiltration rate compared to the FGD sludges or
the control. The increased effectiveness of the phosphogypsum relative to
the FGD sludges was probably due to the greater release of electrolytes. As
can be seen in Fig. 5-2, the phosphogypsum released electrolytes more quickly
than the FGD sludges although the saturated EC was only slightly greater.
This was probably due to the greater homogeneity and crystallinity of the
FGD sludges. The Merom sludge released less electrolytes than did Gibson
yet it was much more effective in improving infiltration than Gibson. The
reason for this was that the Gibson material was much finer (<20 μm) and
was observed to physically plug the pores. The release of electrolytes nearly

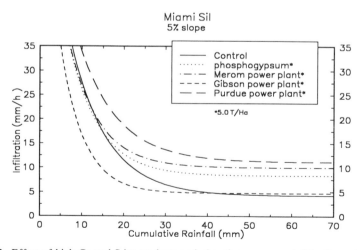

Fig. 5-3. Effect of high Ca and S by-products and phosphogypsum on infiltration rate with
cumulative rainfall for the Miami soil (Norton et al., 1993).

equaled this negative effect and was not significantly different from the control. This observation points to the importance of the physical size of these materials as a variable in their effectiveness for erosion control as well as their ease of handling and application.

The infiltration curves in Fig. 5-3 show the difference in surface sealing processes. All curves have a high initial infiltration rate and decline with cumulative rainfall to a low steady-state rate. In the control, this is due to the combination of physical and chemical processes which include dispersion, slaking and compaction by raindrops (LeBissonaise et al., 1989). However, for the FBC curve, the effect of chemical processes have been largely eliminated by the release of electrolytes and the lowering of the infiltration rate is due to mechanical processes during the rainstorm. Therefore, the differences in the infiltration rates between the control and FBC or phosphogypsum represent the magnitude of chemical processes. The difference from the initial infiltration rate to final rate of the FBC curve is due to the mechanical processes. The magnitude of the mechanical processes are considerably greater than chemical processes for this soil even at the low rainfall intensity of 37 mm h^{-1}.

Total soil loss also was compared with addition of these materials. Fluidized bed combustion and phosphogypsum had a greater effect in lowering soil loss than the two FGD's (Fig. 5-4). This again appears to be a direct result of the release of electrolytes, with the FBC having the greatest release followed by phosphogypsum and the FGD sludges. However, the effect of FBC is not quite so simple since its pH is high. The phosphogypsum, FGD sludges and the soil all have pH's near neutrality, whereas, the FBC is around 12.5. At this high pH, the Miami soil (fine-silty, mesic Typic Hapludalf) should be dispersed and have greater soil loss than the control. However,

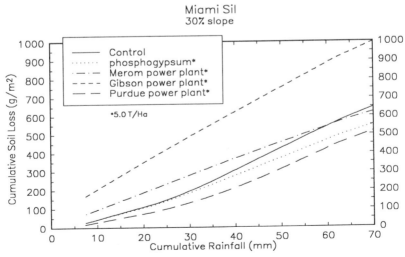

Fig. 5-4. Effect of high Ca and S by-products and phosphogypsum on cumulative soil loss with cumulative rainfall for the Miami soil (Norton et al., 1993).

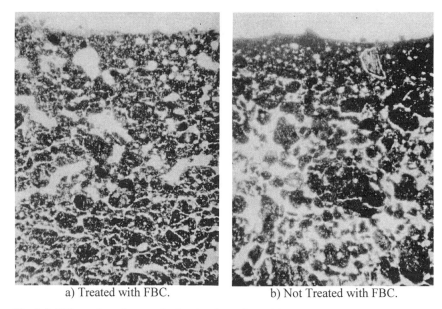

| a) Treated with FBC. | b) Not Treated with FBC. |

Fig. 5–5. Thin-section micrographs showing the surface of the Miami soil with (a) and without (b) addition of 5 MT ha^{-1} FBC. Each surface received a total of 70 mm rainfall. The frame width of each micrograph is 7 mm.

the high pH and the electrolytes released are mainly from the CaO, therefore, the system remains highly flocculated since the CFC in FBC was found to be around 1 mmol L^{-1} and the EC of the runoff was greater than this value. If dispersion were to occur, the clay content in the runoff would have been greater, however, this was not found. Further evidence of the flocculation of the clays was confirmed by observations from thin sections. For the untreated Miami soil (Fig. 5–5b) the surface seal contained a thin continuous layer of fine material which covered compressed and coalesced aggregates. The thin sections from the treated soil (Fig. 5–5a) did not show this fine material only a thin layer of mechanically disrupted aggreates and no evidence of clay dispersion in the fabric.

EFFECT OF FLUIDIZED BED COMBUSTION ON SURFACE ROUGHNESS

In the earlier section, the importance of the physical effects of surface sealing in addition to the chemical effects were discussed related to infiltration. Surface roughness has an additional importance in erosion since it imparts a surface that has a high depressional storage to prevent runoff from occurring and has a high hydraulic friction to slow flowing water and therefore decrease its transport capacity. In order to determine if addition of FBC to the surface could promote greater surface roughness maintenance a field study was conducted on a Miami silt loam soil by F.F. Eltz (1993). Surface microtopography was studied following incremental rainfall using a modified laser microtopographer similar to that described by Huang and Bradford (1992). Fluidized bed combustion was compared to control plots with different initial surface roughness in order to determine if FBC was effective in stabilizing the physical structure as measured by fractal and other surface roughness indices.

Starting with different initial tillage induced roughness in the field, with and without addition of 5 MT ha^{-1} FBC, and a rainfall simulator Eltz (1993) found that there was no difference in the physical form of the surfaces with addition of FBC at all rainfall intensities studied (Fig. 5-6a, b, c, d) as measured by a number of roughness indices. The overall conclusion was that addition of FBC at this rate had no effect on the physical process of surface sealing or the maintenance of surface roughness. Eltz (1993) also found that addition of FBC lowered the total soil loss and increased infiltration to a similar extent as found in laboratory studies, therefore, the effect of adding FBC was mainly chemical. This conclusion points out that management systems with a combination of residue cover to prevent the physical aspects of surface sealing from occurring and FBC to reduce the chemical processes may be a viable option for problematic soils.

EFFECT OF FLUIDIZED BED COMBUSTION ON SOIL SURFACE SEALING FOR SOILS WITH A WIDE RANGE OF PROPERTIES

Surface sealing occurs as the result of a combination of processes. For the Miami soil, the mechanical effects dominated over the chemical processes in surface sealing. However, for other soils this may not be the case. The main chemical processes in surface sealing are dispersion and slaking of aggregates due to the low electrolyte content of the rainwater and the mechanical impact of raindrops. The mechanical processes can be controlled easily with surface residue cover; however, the chemical effects are controlled by many soil factors which are not well understood. In order to sort out the relative importance of soil properties on the chemical processes of surface sealing, FBC was added to soils with a wide range of chemical, physical and mineralogical properties (Reichert & Norton, 1994a, b; Reichert et al., 1994).

Fluidized bed combustion addition was found to be effective for the Miami soil by Norton et al. (1993); however, its effect on soils with a wide range of properties was unclear. In order to evaluate its usefulness on other soils, J.M. Reichert (Reichert & Norton, 1994a, b; Reichert et al., 1994), conducted a similar study of soils from Australia, Brazil and the USA. The soils studied (Table 5-6) were sampled at field moist conditions, air dried and ground to pass a 4-mm sieve. The sieved soils were then packed into small 0.14-m^2 erosion pans over a sand base acting as a tension table (Bradford & Ferris, 1987). Prior to rainfall, the pans were equilibrated to 5-cm tension and placed at 5% slope. Two pans each of a control and 5 MT ha^{-1} surface-applied FBC were subjected to variable intensity rainfall using the Purdue Programmable Rainfall Simulator (Neibling et al., 1981). During the

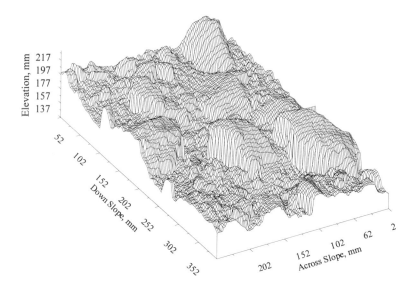

a) 0 mm simulated rainfall, 0 MJ mm/ha/h EI Control

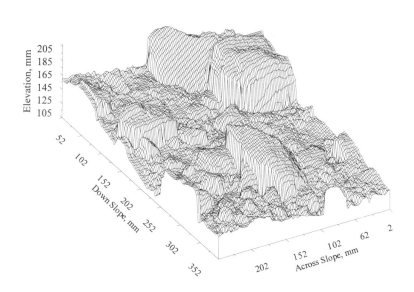

b) 0 mm simulated rainfall, 0 MJ mm/ha/h EI FBC

Fig. 5-6. Fish net microtopographic representations of surfaces receiving no FBC (a and c) and those receiving 5 MT ha^{-1} FBC (Eltz, unpublished data, 1993).

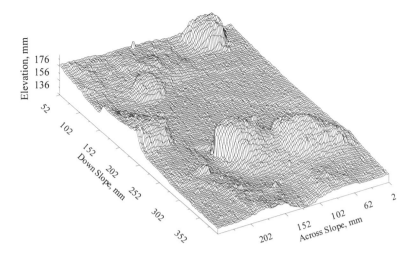

c) 162 mm simulated rainfall, 1627MJ mm/ha/h EI Control

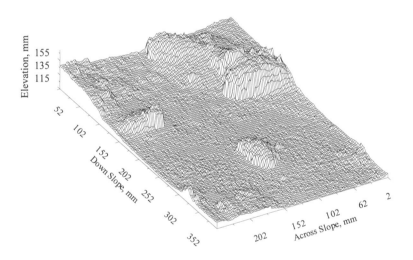

d) 162 mm simulated rainfall, 1627 MJ mm/ha/h EI FBC

Fig. 5-6. Continued.

rainfall, runoff, sediment concentration and infiltration were monitored at 5-min intervals. Runoff and infiltration samples were weighed on an electronic balance connected to a personal computer. Sediment concentration was determined following flocculation of the runoff with alum, decanting the supernatant and oven drying to determine the weight of the soil. Sediment concentration was taken as the weight of sediment divided by the weight of the runoff including sediment.

Table 5-6. Soils studied and some of their properties.

Soil	Location	CMS†	Sand	Silt	Clay	WD‡	OC§	CFC¶
			g kg⁻¹					mmol$_c$L⁻¹
Bela Vista Sandy loam	Parana, Brazil	Oxidic	778	56	166	77	6	2.5
Grey Clay	Queensland, Australia	Smectitic	278	262	460	22	32	0.5
Irving clay	Queensland, Australia	Smectitic	28	370	602	202	23	1.5
Heiden Clay	Texas	Smectitic	86	396	518	23	24	3.5
Hoytville Silty clay	Ohio	Illitic	28	370	566	313	24	3.5
Miami Silt loam	Indiana	Mixed	42	727	231	75	26	2.0
Middleridge Clay	Queensland, Australia	Oxidic	147	389	464	197	27	1.0
Pierre Clay loam	North Dakota	Smectitic	422	276	302	208	17	1.5

† CMS = clay mineral system.
‡ WD = water dispersible clay.
§ OC = organic C.
¶ CFC = critical flocculation concentration.

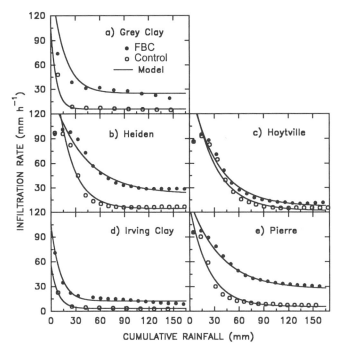

Fig. 5-7. Effect of FBC on infiltration rate for the smectitic soils studied (Reichert, unpublished data, 1993).

Water-dispersible clay was determined by shaking the untreated soil overnight in ultrapure deionized water on an oscillating shaker and decanting after the appropriate settling time. The clay fraction was then placed in a series of FBC solutions to determine the critical flocculation concentration (Miller et al., 1990).

In this study, FBC was added to soils ranging from coarse to fine textures and included oxidic, siliceous, kaolinitic, illitic, and smectitic clay mineral systems. For the fine-textured soils, addition of FBC to the surface generally improved infiltration and reduced soil loss. Smectitic soils (Fig. 5-7) had the greatest increase in infiltration rates with FBC even to a greater extent than the Miami soil which had mixed mineralogy. However, for two oxidic soils there was actually a negative effect of adding the FBC on infiltration (Fig. 5-8). For this soil, which contained highly dispersible clay in a matrix of sand, adding FBC flocculated the clay by releasing electrolytes and actual plugged the pores which otherwise would drain and allow the dispersed clay to percolate with the water. Evidence for this included dispersed clay in infiltrated water in the control and the lack of it in the treated plots.

Figure 5-8 also shows the transient nature of surface sealing. First, the surface seal forms and as runoff increases it is eroded causing infiltration to increase until it reforms and erodes again. In Reichert and Norton (1994a, b), soil clay mineralogy was found to be an important property to predict

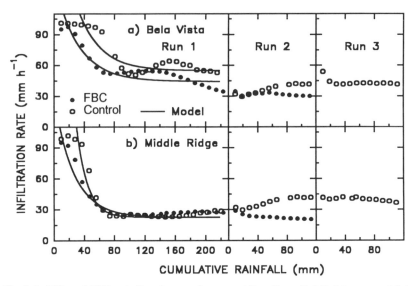

Fig. 5-8. Effect of FBC on infiltration rate for two oxidic soils studied (Reichert, unpublished data, 1993).

the behavior of adding FBC on infiltration and soil loss. He found that the beneficial effect of adding FBC on infiltration and soil loss was greater in the order of smectitic > illitic > kaolinitic > oxidic mineralogies. Water-dispersible clay appeared to be an important property, but only when dispersion of the clay caused a decrease in infiltration rates.

ENVIRONMENTAL CONSIDERATIONS FOR LAND APPLICATION OF FLUIDIZED BED COMBUSTION

Since FBC is quite soluble, its potential for causing environmental harm should be considered. The high pH is a definite factor to be considered when applying it to the soil in order to avoid fertility problems. Standard liming practices should be followed when determining the amount that may be applied to the soil without incurring micronutrient deficiencies in plants. Several other factors should also be considered, particularly, the B content since B has a narrow concentration range from which it is a nutrient or a toxin (Bennett et al., 1985).

The Purdue FBC bottom ash was analyzed for a variety of environmentally important properties and those data are presented in Table 5-7. The data indicate that the FBC has little value for supplying the macronutrients N, P, K and has a low NO_3 content. The total Ca content is very high as would be expected with the potential to be an excellent liming material. The median diameter (D_{50}) is also a good size for land application by broadcasting for use as a liming agent.

Table 5-7. Environmentally important data for Purdue FBC Bottom Ash.

Property	Value
pH	12.5
D_{50} [†]	0.35 mm
Total dissolved solids	4.7 g L^{-1}
Saturated electrical conductivity	19.5 dS m^{-1}
Soluble B[‡]	0.74 mg L^{-1}
Total C[§]	400 g kg^{-1}
Soluble Chloride	65 mg L^{-1}
Soluble Fluoride	0.39 mg L^{-1}
Soluble Sulfate	1.6 g L^{-1}
Total Sulfide	12 mg kg^{-1}
Total N	0.49 g kg^{-1}
Total Nitrate N	2.6 mg kg^{-1}
Total P	13 mg kg^{-1}
Total K	0.24 g kg^{-1}

† D_{50} = diameter at 50% passing.
‡ Soluble components by American Society for Testing Materials leaching procedure 3987-85 (ASTM, 1990).
§ Total components by USEPA 200.7 (USEPA, 1989) and ICP.

The more soluble components probably warrant the most concern for land application. The sulfate content in the leachates is quite high amounting to over one-third of the total dissolved solids. The chloride content in the leachate is also quite high and could contribute to salinity problems in some environments. The most important component of the leachate is the B content. Although it amounts to only 0.75 mg L^{-1}, this amount may be harmful to sensitive plants. In fact some B toxicity was noticed in the early growth stages of corn when only applying 5 MT ha^{-1} in a field experiment at Beckley, West Virginia, in the spring of 1993[1]. The corn recovered from the toxic effects after the first several rainfall events, but application at higher rates could have had a devastating effect.

In addition to the elements listed in Table 5-7, the eight RCRA elements need to be considered. Table 5-8 shows the Toxic Characteristic Leachate Procedure values for the Purdue FBC and the amount for the national drinking water standard (USEPA, 1989). Most of the elements were below the detectable limit. Of those that were detectable only Cr was near the drinking water standard. This was probably due to the nature of the coal that was being burned at the time of production which was northern Indiana high-S coal. The other two detectable elements Se and Ar were well below the drinking water standard. With this particular FBC land application would appear relatively safe considering only the RCRA elements. More research is needed to address the environmental aspects of land application of FBC bottom ash.

---

[1] Personal communication with Dr. Dale Ritchey, USDA-ARS Beckley, West Virginia.

Table 5-8. Toxic characteristics leachate procedure (TCLP) and resource conservation and recovery act RCRA limits data for Purdue FBC bottom ash.

Element	TCLP	RCRA drinking water limit	Detection limit
		————— mg kg^{-1} —————	
As	0.01	0.05	0.005
Ba	BDL†,‡	1.0	2.0
Cd	BDL	0.01	0.0020
Cr	0.04	0.05	0.05
Pb	BDL	0.05	0.010
Hg	BDL^2	0.002	0.0050
Se	0.002	0.01	0.010
Ag	BDL	0.05	0.04

† BDL = below detectable limit.
‡ Detection limit is greater than drinking water standard.

SUMMARY

The production of by-products from desulfurization using CCT is rapidly increasing. A growing tonnage of by-products are presently being landfilled which represents an undesirable situation. These materials are mostly high in Ca and S, two elements with beneficial uses in agriculture. Depending on the process of CCT used, these elements may be present in different minerals. The commonly produced minerals from CCT are reviewed in this paper. Flue gas desulfurization processes produce a variable mixture of hannebachite and gypsum with unreacted starter reagents. Fludized Bed Combustion produces bottom ash that contains anhydrite, lime, portlandite and calcite. The reactivity and solubility of the FBC was found to be much greater than several FGD's studied.

Addition of FGD and FBC reduced surface sealing by release of electrolytes to the eroding rainwater. However, the FGD was considerably less effective than the FBC in improving infiltration or reducing soil loss. The considerable effect of increasing infiltration and reducing soil loss by addition of FBC was due to its chemical action. Fluidized bed combustion was not effective in maintaining surface roughness or changing the physical processes of surface sealing.

Adding FBC to the surface of soils with different properties demonstrated that it is more effective on improving infiltration and reducing soil loss on soils containing smectite. A negative effect on infiltration was found for soils with low activity clays and caution is warranted for addition of FBC to these soils for erosion control. Although the FBC appears to be relatively environmentally safe for land application based on the eight RCRA elements, the high alkalinity and B content must be considered when applying it to an agricultural system. Additional information on the effect of adding FBC to land is needed for environmental safe use as a soil amendment.

ACKNOWLEDGMENTS

Thanks to Dr. Dale Ritchey of the USDA-ARS, Beckley, West Virginia for providing the sample of the advanced FGD from the Big Bend Power Plant. Many thanks also go to Dr. Miguel Reichert and Dr. Flavio Eltz for their hard work in evaluating the effect of FBC on erosion. Also thanks go to Jody Tishmack for assembling information on the Purdue FBC and to Dr. Jerry Bigham, Dep. of Agronomy, Ohio State University for assistance with thermal analyses and the interpretation of the mineralogical components in the by-products studied.

REFERENCES

American Society of Testing Materials. 1990. Annual book of ASTM standards. ASTM, Philadelphia, PA.

Agassi, M., I. Shainberg, and J. Morin. 1981. Effect of electrolyte concentration and soil sodicity on infiltration rate and crust formation. Soil Sci. Soc. Am. J. 45:848–851.

Bennett, O.L., J.L. Hern, H.D. Perry, R.L. Reid, W.L. Stout, J.H. Edwards, and K.O. Smedley. 1985. p. 558–576. In Agricultural uses of atmospheric fluidized bed combustion residue (AFBCR)—A seven year study. Proc. 2nd Annu. Ptitsburgh Coal Conf., Pittsburgh, PA. 11–20 September.

Bradford, J.M., and J.E. Ferris. 1987. Effect of surface sealing on infiltration, runoff, and rainsplash. p. 417–428. In Y. Fok (ed.) Proc. Int. Conf. Infiltration Development and Application, Honolulu, HI. 12–15 January. Water Resources Res. Center, Univ. Hawaii, Manoa.

Brieger, G., J.R. Wells, and R.D. Hunter. 1992. Plant and animal species composition and heavy metal content in fly ash ecosystems. Water, Air Soil Pollut. 63:87–103.

Dreimanis, A. 1962. Quantitative gasometric determination ofcalcite and dolomite by using the Chittick apparatus. J. Sediment. Petrol. 32:520–529.

Eltz, F.L. 1993. Surface roughness changes as affected by tillage and rainfall erosivitiy. Ph.d. diss. Purdue Univ., West Lafayette, IN (Diss. Abstracts 94-40793).

Fowler, R.K., J.M. Bigham, U.I. Soto, and S.J. Traina. 1992. Mineralogy of clean coal technology by-products. p. 241–246. In Proc. 9th Annu. Int. Pittsburgh Coal Conf., Pittsburgh, PA. 24–29 August.

Hance, S.L., R.S. McKibben, and F.M. Jones. 1990. Utility flue gas desulfurization survey, January–December 1988. Proj. Summary DOE Contract no. DE-AC05-84OR21400. U.S. Gov. Print. Office, Washington, DC.

Huang, C., and J.M. Bradford. 1992. Applications of a laser scanner to quantify soil microtopography. Soil Sci. Soc. Am. J. 56:14–21.

LeBissonaise, Y., A. Bruand, and M. Jamagne. 1989. Laboratory experimental study of soil crusting: Relations between aggregates breakdown and crust structure. Catena 16:377–392.

Miller, W.P., and M.K. Bharuddin. 1986. Relationship of soil dispersibility to infiltration and erosion of southeastern soils. Soil Sci. 142:235–240.

Miller, W.P., H. Frenkel, and K.D. Newman. 1990. Flocculation concentration and sodium/calcium exchange of kaolinitic soil clays. Soil Sci. Soc. Am. J. 54:346–351.

Neibling, W.H., G.R. Foster, R.A. Nattermann, J.D. Nowlin, and P.V. Holbert. 1981. Laboratory and field testing of a programmable plot-sized simulator. p. 405–414. In Erosion and sediment transport measurement, Florence, Italy. June. Int. Assoc. Hydrol. Sci. Publ. no. 133. Int. Assoc. of Hydrol. Sci. Press, Inst. Hydrology, Wallingford, Oxfordshire, England.

Norton, L.D., I. Shainberg, and K.W. King. 1993. Utilization of gypsiferous amendments to reduce surface sealing in some humid soils of the eastern United States of America. Catena Suppl. 24:77–92.

Reichert, J.M., and C.D. Norton. 1994a. Fluidized bed bottom-ash effects on infiltration and erosion of swelling soils. Soil Sci. Soc. Am. J. 58:1483–1488.

Reichert, J.M., and L.D. Norton. 1994b. Aggregate stability and rain-impacted sheet erosion of air-dried and prewetted clayey surface soils under intense rain. Soil Sci. 158:159–169.

Reichert, J.M., L.D. Norton, and C. Huang. 1994. Sealing, amendment, and rain intensity effects on erosion of high-clay soils. Soil Sci. Soc. Am. J. 58:1199–1205.

Reynolds, P. 1990. In-situ forced oxidation retrofit on 125 Megawatt module at the Merom generating station an update. *In* Proc. Flue Gas Desulfurization Users Conf., Jacksonville, FL. 27–29 November. Hoosier Energy, Bloomington, IN.

Rose, J.G., A.E. Bland, and C.E. Jones. 1986. Production of concrete using fluidized bed combustion waste and power plant fly ash. Tennessee Valley Authority Contract no. TV-60443A. U.S. Govt. Print. Office, Washington, DC.

Solem, J.K., and G.J. McCarthy. 1992. Hydration reactions and ettringite formatoin in selected cementitious coal conversion by-products. Advanced cementitious systems: Mechanisms and properties. Materials Res. Soc. Symp. Ser. 245-71-79.

Stout, W.L., J.L. Hern, R.F. Korcak, and C.W. Carlson. 1988. Manual for applying fluidized bed combustion residue to agricultural lands. USDA-ARS ARS-74, 15pp.

Torrens, I.M., and P.T. Radcliffe. 1990. SO$_2$ control in the 90's an EPRI perspective. 1990. p. 5–19. SO$_2$ Control Symp, 8–11 May 1990, New Orleans, LA.

U.S. Environmental Protection Agency. 1988. Wastes from the combustion of coal by electric utility power plants. EPA/530-SW-88-002. U.S. Govt. Print. Office, Washington, DC.

U.S. Environmental Protection Agency. 1989. Stabilizing/Solidification of CERCLA and RCRA wastes, physicla tests, chemical testing procedures, technology, and field activities. EPA 600/2-83-001. U.S. Govt. Print. Office, Washington, DC.

U.S. Environmental Protection Agency. 1991. Code of Federal Regulations. Vol. 40, Part 261, App. II, pp 66–81.

6 Utilization of Coal Combustion By-Products in Agriculture and Horticulture

R. F. Korcak

USDA-ARS, Fruit Laboratory
Beltsville, Maryland

Coal ash produced from the burning of coal has become a generic term for all types of coal combustion by-products. Specifically, fly ash is that portion of the ash stream that has a sufficiently small size (0.001–0.1 mm) to be carried from the boiler in the flue gas. These particles are either mechanically or electrostatically captured or emitted via the stack.

The furnace residues from the combustion process (bottom ash and boiler slag) are common to all types of coal combustion. Both materials have a particle size generally within the 0.1- to 10-mm range. The amounts of boiler slag produced are projected to decrease as newer boiler technologies become on-line. Currently, the ash stream consists of 5% boiler slag and about 25% bottom ash.

Total ash production varies considerably with the type of coal consumed as well as the source. Anthracite coal produces the highest ash content (about 30%) while bituminous coal can range from 6 to 12% ash (USEPA, 1988). Subbituminous and lignite coals have a slightly wider range of ash contents, 5 to 19%. Currently coal combustion produces, on average, about 10% ash (USEPA, 1988).

There exist a number of other by-products which are dependent on the type of combustion process and/or methodology used to reduce gaseous emissions, usually sulfur oxides. These include flue gas desulfurization (FGD) by-products (either wet or dry), fluidized bed combustion (FBC) by-products, and coal gasification ash. Coal gasification ash results from the conversion of coal into a synthetic gas or liquid fuel. Conventional ash produced is similar to fly ash, therefore coal gasification ash will not be dealt with separately.

Fluidized gas desulfurization by-products result from postcombustion treatment (scrubbing) of the flue gas with an absorbent [usually lime (calcium oxide), limestone or dolomite] to reduce S emissions. Such treatment may be performed under dry or wet conditions which affects the moisture status of the end-product. In the wet method, flue gases pass through a slurry of absorbent in a contact chamber. In the dry method, a fine spray of absorbent is injected into the flue gas stream as it passes through the contact chamber. The water in the fine spray largely evaporates in the gas stream, leaving a dry powder end-product. The wet method tends to be more efficient (about 90%) than the dry method (about 70%) in the removal of S from the flue gases. Thus, dry scrubbing is usually performed when low S coal is consumed.

The major types of FGD systems currently in use are presented in Table 6-1. These types are classified as recovery or nonrecovery systems based on the production of a salable (recovery) by-product such as S, sulfuric acid or liquid sulfur dioxide. This classification scheme (recovery and nonrecovery) is not indicative of the potential for agricultural utilization. However, recovery and industrial utilization of these S-based by-products is limited. Therefore, the agricultural utilization of these S-based by-products is warranted.

Of the FGD systems listed in Table 6-1, direct lime and direct limestone are the wet, nonrecovery methods most used in the industry (USEPA, 1988). The alkaline fly ash scrubber system is used primarily with highly alkaline western coals for S removal. The dual-alkali process uses a mixture of lime and Na salts for S removal.

Dry scrubbing methods such as spray drying and dry sorbent injection have been under study by the industry since 1988. A newer FGD system currently under study is the Pircon-Peck process. In this system, calcium phosphate (rock phosphate) is used as an absorbent rather than calcium carbonate (limestone). The by-product of this process contains both gypsum (calcium sulfate) and acidic P. The initial by-product is then ammoniated producing a mixture of gypsum and ammonium phosphate which provides four of the five nutrients needed in largest quantities by crops. However, even if the Pircon-Peck process does prove to be economically feasible and used by the entire fertilizer market for phosphate, it would use less than 15% of the S by-product expected from coal combustion in 2000 A.D.

The simultaneous combustion of coal and an absorbent (usually limestone or dolomite) in FBC results in end-products very different from those produced by combustion of coal alone per se. The Ca reacts in the furnace as an absorber of S, thereby reducing flue emissions of S and producing large amounts of a dry by-product. The bottom and fly ashes display an alkaline pH (usually about pH 10) and contain substantially higher concentrations of Ca including calcium sulfate and some calcium oxide than by-products

Table 6-1. Some examples of nonrecovery and recovery types of FGD systems. Recovery systems produce recyclable end-products such as elemental S.[†]

Nonrecovery systems		Recovery systems	
Wet	Dry	Wet	Dy
Direct lime (CaO)	Spray drying	Wellman-Lord	Alumina/Cu Surbent
Direct limestone	Dry sorbent injection	Magnesium oxide	Activated C sorbent
Alkaline fly ash			
Dual-alkali			

† USEPA, 1988.

from conventional power plants. This technology is used mainly in new boilers.

A similar new technology is the limestone injection multistage burner (LIMB). A Ca-based sorbent is injected into the burner to achieve S removal as with FBC. The LIMB technology is used to retrofit existing boilers. The dry by-product obtained is easier to handle than traditional wet scrubber wastes.

A diverse range of coal combustion by-products exists ranging from those produced in conventional facilities to an ever-growing list of FGD by-products (either calcium sulfate or calcium sulfite) plus those from newer combustion technologies such as FBC (calcium oxide and calcium sulfate). The potential exists to utilize many of these by-products in agriculture rather than place them in ever shrinking landfill space.

AMOUNTS PRODUCED

Fly and Bottom Ash

The utilization of coal to produce electric power or heat consumes large quantities of coal annually. For example, coal consumption in Georgia alone was 24.5 million megagrams annually resulting in 2.0 million megagrams of fly ash and 0.5 million megagrams of bottom ash (McIntosh et al., 1992). Therefore, fly and bottom ashes are produced at about 10% of the coal consumed. These figures do not reflect amounts due to new clean air technologies such as FGD scrubber by-products. Installation of scrubbers, at selected power plants in Georgia alone, will produce an additional 1 million megagrams of FGD materials annually (USEPA, 1988).

About 63.6 million megagrams of fly and bottom ash were produced nationally in 1984 (USEPA, 1988) (Table 6–2). Projected future use of coal will increase this figure to about 109 million megagrams annually by the year 2000 (USEPA, 1988). These figures do not include the amounts of FGD and FBC material generated that may be produced.

Table 6–2. Past, present and projected amounts of by-products produced by the coal combustion industry.

By-product	Type	Past production (1984)	Present production (1991)	Projected production (2000)
			million Mg	
Ash	Total†	64	65.2	111
	Fly	--	47.3	79
	Bottom	--	12.2	27
FGD		15	16	46

† Includes boiler slag (USEPA, 1988), Am. Coal Ash Assoc. (1991, unpublished data).

Flue Gas Desulfurization By-Products

Approximately 95% of current FGD technologies produce nonrecovery by-products (USEPA, 1988). Both technologies are further classified as wet or dry systems to indicate the moisture status of the end-product. From an agricultural utilization viewpoint, the end-products produced are quite variable and will require different strategies for their effective use.

The FGD production in 1985 was about 14.5 million megagrams which includes all types of by-products (USEPA, 1988). It is estimated that this figure will increase to about 45.4 million megagrams annually by the year 2000.

Fluidized Bed Combustion

Because both FBC and other Ca-based dry by-product technologies such as LIMB are just beginning to come on-line in significant capacity, annual production figures for these types of by-products are not available. It has been estimated that a 1000-MW FBC plant would generate about 1800 Mg of dry waste per day or about 0.64 million megagrams annually (Ruth, 1975). About 110 FBC plants are currently in operation with an additional 13 plants under construction (J. Tishmack, 1992, personal communication). Residue production is about 18.2 million megagrams per year. Because of their coherent, dry condition and easily workable particle-size and density, these gypsiferous materials may be among the most suited for agricultural utilization.

CHEMICAL COMPOSITION OF ASH

Conventional Fly Ash and Bottom Ash

The chemical constituents of ash can vary greatly depending on the coal type, source and plant operating parameters. Major constituents include Al, Ca, Fe, Mg, K, Si, Na, and Ti. Of these eight primary constituents which make up 95% of the mass of the ash, five are essential plant nutrients. The concentrations of these primary constituents are listed in Table 6–3.

Trace element concentrations in ash also are variable and can have a direct impact on the potential agricultural utilization of these materials. Ranges of trace element concentrations present in a broad sample of ash and types of coal are presented in Table 6–4, and the agriculturally important trace elements are identified. Significant variations in trace element concentrations are also known for various coal sources (eastern, midwestern, and western). Midwestern coal ashes are usually highest in Cd, Zn, and Pb while Ba and Sr are highest in western coal. Selenium tends to be greater in eastern and midwestern coals.

Coal cleaning, prior to combustion, can significantly reduce elemental concentrations of S, Se, and other trace elements. The cleaning is performed using physical (usually density differences separating out pyritic-sulfur), chem-

Table 6-3. Concentration ranges for major constituents of ash.[†]

Element	Fly ash	Bottom ash
	g kg^{-1}	
Essential plant nutrient		
Ca	0.1–177	8.0–51
Fe	8.0–289	27–203
Mg	5.0–61	4.0–32
K	2.0–35	7.0–16
Si	196–271	180–273
Other elements		
Al	11–144	88–135
Na	1.0–20	20–13
Ti	<0.1–16	30–7

[†] Utility Solid Waste Activities Group (1982, unpublished data).

ical or biological pre-combustion cleaning. The latter two methods are newer and not extensively used at present. Coal cleaning also can have a significant impact on the amount of ash generated. In Virginia, raw coal produced an average 9.7% ash while cleaned coal produced 5.7% ash (Randolph et al., 1990). Precombustion cleaning of coal is one of several categories of clean coal technology currently being funded and developed under the Department of Energy (U.S. Dep. Energy, 1992).

Elemental concentrations also vary with the particular portion of the ash stream sampled. Fly ash contains significantly greater quantities of As,

Table 6-4. Trace element concentration ranges in ash averaged over all ash and coal types.[†]

Element	Overall range
	mg kg^{-1}
Essential plant nutrient	
B	10–1300
Cu	3.7–349
Mn	56.7–767
Mo	0.84–100
Zn	4.0–2300
Other elements	
As	0.50–279
Ba	52–5790
Cd	0.10–18
Cr	3.4–437
Co	4.9–79
Fl	0.4–320
Pb	0.4–252
Hg	0.005–4.2
Ni	1.8–258
Se	0.08–19
Ag	0.040–8
Sr	30–3855
Tl	0.10–42
Vn	11.9–570

[†] Tetra Tech, Inc. (1983).

Cu, and Se than bottom ash. Distribution of elements in the ash stream is highly dependent on boiler temperature and therefore can vary greatly. Some components such as elemental S and Hg are essentially completely volatilized, thereby reducing their concentrations in bottom ash in conventional coal-burning plants.

Fly ash types are typically classified on the basis of major components. Class-C fly ash contain less than 70% but greater than 50% of silica, alumina, and iron oxides and are usually denoted as high-lime, western ashes. If the concentration of silica, alumina, and iron oxides exceeds 70%, ashes are classified as Class F. These are usually generated from eastern coals (Environ. Manage. Serv., 1992). A modification of this fly ash classification system has been proposed. This modification is based on a more detailed chemical composition (Roy et al., 1981). The three basic groupings proposed are silica (Si-Al-Ti-oxides), calcic (Ca-Ma-K-Na oxides), and ferric (Fe-Mn-S-P oxides). Use of such a classification system would be an initial step in helping to identify potentially useful by-products for specific purposes. Expansion of the list to include agriculturally related parameters such as plant nutrient availability indices; potential elemental phytotoxicity indices, and total alkalinity would facilitate communication between different reserach groups and development of cost effective and environmentally beneficial uses of these ashes.

Flue Gas Desulfurization By-Products

Independent of the type of process used (see Table 6–1) to scrub the flue gas, all FGD products include spent reagent in combination with sulfites or sulfates plus unreacted reagent. Additionally, the FGD material may contain water (in wet processes) and coprecipitated fly ash. The quantity of reagent used is usually proportional to the S concentration of the coal burned but is also a function of the percentage SO_x recovery desired as well as system operating parameters.

Wet scrubbers generally produce slightly smaller particle sized material (0.001–0.05 mm) than dry scrubbers (0.002–0.074 mm). Wet scrubber sludge can vary from 16 to 43% or more moisture.

The chemical composition of FGD sludges will by necessity vary depending on the process employed as well as quantity of reagent used, amount of fly ash present, the S content of the burned coal and whether or not forced oxidation is used in the treatment process. The degree of forced oxidation has a significant impact on the potential use of the material in agriculture since oxidation increases the amount of sulfates compared to the amount of sulfites present in the end-product. A comparison of FGD by-products from different sources is illustrated in Table 6–5.

Dual alkali and spray drying systems that utilize a Na absorbent produce FGD sludges containing $NaSO_4$ (oxidized) or $NaSO_3$ sodium sulfite (reduced). Since Na deteriorates soil structure these Na-containing FGD sludges are probably not useful for agriculture.

Table 6-5 Effect of FGD process and coal source (eastern and western) on the chemical components of the end-products produced.[†]

| Component | Direct lime | | Direct limestone | | Alkaline fly ash |
	East	West	East	West	West
	% of FGD sludge (dry wt basis)				
Ca-sulfate	15–19	17–95	5–23	85	20
CaSO$_3$ 1/2H$_2$O	13–69	2–11	17–50	8	15
Ca-sulfite	1–22	0–3	15–74	6	--
Fly ash	16–60	3–59	1–45	3	65

† USEPA, 1988.

A comparison of the liquid fraction of a direct lime (calcium oxide) FGD vs. a dual-alkali FGD is shown in Table 6-6. The differences in Ca and Na contents of the two FGD scrubber liquors highlights the need for an awareness of the process used in relation to agricultural utilization. The ratio of sulfate to sulfite also is important in relation to the solubility and/or toxicity of the end-product. Sulfites are much lower in solubility. Installation of an oxidizing step in the FGD process, although an additional expense, aids not only in increasing the solubility of the end-product but also increases the potential for agricultural utilization due to greater gypsum concentration. On the other hand, there is preliminary evidence that sulfite sludges applied to soils several weeks prior to planting crops are oxidized to sulfates before the crops begin to grow (K.D. Ritchey, 1993, personal communication). Consequently there is a lack of an initial negative effect on plant growth that is observed when they are applied at planting time.

Investigations on the behavior of sulfite materials in the soil/plant system are currently being performed at the Beckley, West Virginia, USDA-ARS Laboratory (R.B. Clark, 1993, personal communication).

The pronounced difference in trace element concentrations in the solid and liquid component of wet scrubber sludges is shown in Table 6-7. With the exception of B, most trace elements remain associated with the solid FGD material. The percentage of soluble B present in these materials may be of

Table 6-6. Chemical composition of the liquors emanating from FGD processes that use a C-based absorbent (direct lime) or a Na-based sorbent (dual alkali). Both determinations are based on burning eastern coal.[†]

Component	Direct lime	Dual alkali
	mg L^{-1}	
pH	8–9.4	12.1
K	11–28	320–380
Na	36–137	53 600–55 300
Mg	24–420	0.1
Sulfate	800–4500	80 000–84 000
Sulfite	0.9–2.7	--

† USEPA, 1988.

Table 6-7. Trace element concentration ranges in wet FGD solids and liquors.[†]

Element	Solids	Liquors
Essential plant nutrient		
B	42–530	2–76
Cu	6–340	<0.01–0.5
Other elements		
As	0.8–52	<0.01–0.1
Cd	1.6–180	<0.01–0.1
Cr	1.6–180	<0.01–0.3
Fluoride	266–1017	0.2–63
Hg	0.01–6	<0.01–0.1
Pb	0.2–290	<0.01–0.5
Se	2–60	<0.01–1.9

[†] USEPA, 1988.

significance in the soil/plant system. Many plants are sensitive to B in a narrow range of concentrations. High available B levels could induce a toxicity. However, where B is deficient, additions of B-bearing sludges on deficient soils may be beneficial. Such potential benefits from careful management are explored later in this report (see Trace Element Dilemma section).

Fluidized Bed Combustion By-Products

Fluidized bed combustion by-products also vary in elemental composition. Range of elemental concentrations from a representative FBCa plant utilizing eastern coals are presented in Table 6-8. The large amounts of Ca present in the by-product are primarily in the form of anhydride, unreacted

Table 6-8. Major and trace element concentration ranges in FBC by-products compared to ranges normally found in soils.[†]

Element	FBC	Soil
	$g\ kg^{-1}$	
Ca	240–460	7.0–500
AL	3–20	40–300
S	72–140	0.1–20
Fe	<1–16	7.0–550
Mg	5–12	0.6–6
K	<1–8	0.4–30
	$mg\ kg^{-1}$	
P	380–500	50–2000
Mn	210–685	200–3000
B	95–170	2–100
Mo	0.12–0.28	0.2–5
Cu	12–19	2–100
Zn	12–19	10–300
Ni	13–29	5–500
Pb	1.5–7.5	2–200
Cd	0.5	0.01–0.7
Cr	9–23	5–1000
Se	0.16–0.58	0.1–2

[†] Stout et al. (1988), and Page et al. (1979).

sorbent, and calcium oxide. A typical analysis indicates that a spent bed material has an aqueous pH of about 12 and contains (in % dry weight) 52 $CaSO_4$, 33 CaO, 0.6 $CaSO_3$, 0.8 MgO, 0.3 NaCl, 0.02 P_2O_5, 4.5 R_2O_3 (R = Fe and Al), and 7 SiO_2 (Korcak, 1988). This combination of Ca by-product components is especially useful for agriculture being high in both acid neutralizers (calcium oxide) and the relatively soluble and mobile Ca and sulfate in the gypsum. Generally, trace element concentrations are similar to those of other coal combustion by-products and these levels vary depending primarily on the constituents of the coal and sorbent utilized in the combustion process.

ORGANIC COMPOSITION OF COAL COMBUSTION BY-PRODUCTS

There are many incompletely oxidized organic compounds in fly ash. Roy et al. (1981) list a number of carcinogens and/or mutagens found in ash. However, it is difficult to track organics in flue gases exiting power plants due to climatic and atmospheric effects on composition of air entering the burners. Additionally, compositional changes which occur in the stack prior to sample capture may not accurately reflect potential component toxicity estimates.

Organics have received little attention in studies on agricultural utilization of coal combustion by-products. Research on the transformation and fate of organics in the soil/plant system is difficult. Additionally, stock-piled, weathered ash may present a different organic composition than fresh materials. Based on past studies in related areas it appears that the primary hazards to human health would be via direct inhalation by operators applying these materials rather than via plant uptake and food consumption. However, the potential hazards for contamination by organics needs to be documented.

MINERALOGY OF COAL COMBUSTION BY-PRODUCTS

There is considerable information on the mineralogy of coal combustion by-products. Most of this work has been performed on fly ash and has examined particulates. This work has shown a strong association between fly ash particle size and trace element concentration (Davidson et al., 1974). Concentrations of Se, Ca, As, Pb, Ni, Cr and Sn increased with decreasing particle size. Similarly, Phung et al. (1979) found enhanced levels of B, Cr, Mo, Ni, As, and Se associated with fly ash particle sizes < 53 μm. This association may prove beneficial for the further utilization of fly ash in agriculture, since mechanical removal and separation of small fly ash particles at the power plant could reduce the risk of contamination from some trace elements. However, this additional operation may not be economically feasible.

Many trace elements, which can induce phytotoxicities and nutrient imbalances in animals (i.e., B, Se, and Mo), tend to be associated with the finer particle-sized fly ash. Therefore, these finer materials should be carefully analyzed and applied on a prescription basis according to need. Utilization of bottom ash and FGD materials not mixed with fly ash could probably be applied in larger amounts in agriculture. The same would hold true for bottom ashes from FBC and newer technologies such as LIMB. However, further documentation is warranted on the full range of by-products considered for agricultural use.

Only recently have studies been initiated on the mineralogy of coal combustion by-products applied to agricultural soils. The short- and long-term fate of mineral forms in the soil system needs to be examined.

The pozzolanic nature of FBC materials was used to benefit apple orchards (Korcak, 1988). Rates of spent bed material up to 112 Mg ha^{-1} (50 tons acre^{-1}) were applied as a within row cap in an established apple (*Malus domestica* Borkh.) orchard. The surface applied material formed a porous cement which prevented weed growth for up to 4 yr after application. Over a period of 6 yr, cumulative yields were increased in three of four cultivar-rootstock combinations by the applied surface mulch of FBC material. Foliar Mg levels decreased with time due to high FBC material application, reflecting the greatly increased soil Ca status and the decrease in Mg levels in the surface soil horizons caused by leaching.

These apple plots were re-examined 12 yr after the initial application and 5 yr after the plots were plowed (R.F. Korcak, 1993, unpublished data). X-ray diffraction patterns of remnant cemented pieces of the applied spent bed ash showed that most of the original calcium oxide had converted to calcium carbonate (calcite). Besides calcite the other dominant mineral present was quartz. Secondary minerals present were gypsum and ettringite (Fig. 6–1). The formation of calcium carbonate with time is expected and leads to the maintenance of a relatively high pH (not exceeding pH 8.3, which is the equilibrium pH for calcium carbonate). Surface pH values from these plots after 12 yr were about 7.6. However, the mineral ettringite is unstable at pH levels less than 10. Therefore, the presence of even trace ettringite indicates the existence of microenvironments with a pH of at least 10 within the soil matrix. This further indicates that some unreacted calcium oxide is still present albeit the amount is small. Therefore, application of these materials at relatively high rates, 112 Mg ha^{-1}, can have long-lasting effects on the soil environment and soil mineralogy.

Studies are needed on the mineralogy of coal combustion by-products not only for fresh materials but also for weathered by-products and materials that have been exposed to the soil environment for various periods of time when used as a surface mulch or other high volume applications. Such studies will provide information on the eventual fate of trace elements included in these minerals as well as on changes in soil chemistry and soil mineralogy.

CURRENT NON-AGRICULTURAL DISPOSITION

Coal combustion by-products are generally regulated by individual states under solid waste regulations. These regulations vary greatly from state to

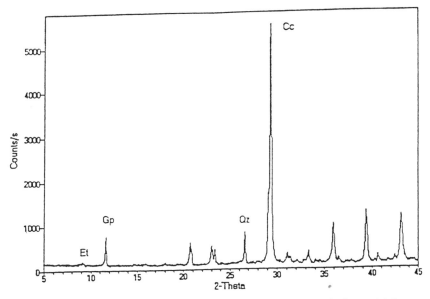

Fig. 6-1. X-ray diffraction pattern from a 12-yr-old piece of FBC spent bed material that was surface applied and later soil incorporated (Cc = calcium carbonate, Qz = quartz, Gp = gypsum, Et = ettringite).

state ranging from very stringent to total exemption from regulation for on-site disposal. Approximately 80% of coal combustion by-products are treated, stored, and/or disposed by means of land management with the remaining 20% recycled (USEPA, 1988). Land management techniques include surface impoundements, landfills, and placement in mines and quarries. Impoundments and landfills are the two most widely used land management methods, with about 77% of facilities using one of these methods.

The overall cost incurred in the management of coal combustion wastes ranged from $2.20 to $34.14 per megagram in 1988 (USEPA, 1988). This cost is generally rising rapidly as are costs for landfill disposal of other wastes. The range in cost is based on the type and size of facility and the characteristics of the waste generated. Generally, fly ash is more costly to manage than bottom ash or FGD wastes. Environmentally sound recovery/recycling of coal combustion by-products can have a significant effect in reducing the costs incurred by the industry in dealing with waste streams. This affects all consumers of such products as electricity.

Degree of recovery and/or recycling of coal combustion by-products varies with the particular end-product. Coal ash utilization increased from 18% for the 1970 to 1980 period to 27% in 1985 (USEPA, 1988). However, less than 1% of all FGD products were recovered and utilized. This situation may change as industry develops more efficient recovery/utilization processes. A summary of recovery/utilization processes is shown in Table 6-9. Although about 20% of ash products are utilized or recycled the current expectation is that this percentage will not increase in the foreseeable future.

Table 6-9. Current nonagricultural utilization of various coal combustion by-products.

Byproduct	Recovery use	Used
		%
Bottom ash	Blasting grit, road and construction fill, roofing granules	33
Fly ash	Concrete admixture, cement additives, grouting, road and construction fill, stabilization of hazardous wastes, clay liner additive, magnetite production, asphalt amendment	17
FGD products	Sulfuric acid, S, other S products (currently listed in scope), gypsum	<1
FBC products	Cementation of hazardous wastes, cement additive	?

Conventional Bottom and Fly Ash

Fly ash and bottom ash may exhibit pozzolanic properties whereby the dried material forms a hard cementlike material. Carefully selected ashes are used as pozzolans in the manufacture of cement. However, high concentrations of sulfates or nitrates reduces suitability for this purpose. As noted previously, the pozzolanic nature of some coal combustion by-product materials can be beneficial in certain surface application techniques in agriculture.

Flue Gas Desulfurization By-Products

Recovery/recycling of FGD by-products is limited. Many of the construction related uses of fly ash are not appropriate for FGD materials since they generally lack adequate cementing properties. Some FGD processes such as forced oxidation wet scrubbing result in gypsum which can be used as a replacement for mined gypsum in wallboard production. However, this use could account for only a few percentage points of the FGD by-products expected in response to the 1990 Clean Air Act since there is competition from other gypsum sources. However, FGD by-product gypsum is considered to be equal or superior to mined gypsum for this use. Newer technologies are under study to increase the production of S-related chemicals from FGD sludges.

Fluidized Bed Combustion By-Products

Due to the relatively recent increase in the number of FBC plants, data is lacking on recovery/recycling uses for FBC by-products (see Table 6-9). Some of the advantages for utilization of these materials for construction purposes include a dry nature which reduces hauling costs, and some degree of pozzolanic properties. However, the initial changes in volume of the minerals as they absorb large amounts of water and generate heat make them difficult to use for engineering purposes. The same would hold true for similar by-products from the LIMB FGD materials.

Land Utilization

Coal combustion by-products have been used as amendments for disturbed lands. Work has been performed on mined land reclamation and renovation of coal refuse piles around the USA (Jastrow et al., 1981; Fail, 1987; Taylor & Schuman, 1988; Stehouwer & Sutton, 1992). Haering and Daniels (1991) reviewed the use of fly ash in mined land reclamation. The extremely acidic nature of these lands (i.e., resulting from oxidation of S and sulfides) often requires basic material additions to bring pH into the range where plants grow and trace element concentrations are at desired concentrations in soil solution. Consequently the use of power plant by-products which are generally alkaline can assist in maintaining pH at the desired levels to reduce trace element availability. Projects looking at the coutilization of coal combustion by-products and organic amendments such as sewage sludge in disturbed land reclamation are, and have been, ongoing. The addition of sewage sludge provides a N source for plant establishment and growth. As with coal combustion materials, the addition of sewage sludge requires the maintenance of a suitable pH to keep trace elements in the desired concentration ranges.

The utilization of fly ash and other coal combustion by-products with or without the addition of an organic material may allow revegetation without application of a topsoil cap. Addition of a topsoil cap is generally the major expense in reclamation of disturbed lands. Successful revegetation with trees on abandoned ash basins has been reported without the need for a topsoil capping (Carlson & Adriano, 1991).

AGRICULTURAL UTILIZATION OF COAL COMBUSTION BY-PRODUCTS

Overview

Any amendment to the soil/plant system must exhibit clearly defined benefits to the environment (soil, water and/or air) or to the quality of the crops produced to justify utilization. These benefits must exceed the costs and hazards whether one is applying a fertilizer, organic mulch, irrigation, or an industrial by-product. Potential benefits and hazards associated with the agricultural utilization of various coal combustion by-product materials are noted below followed by a review of past and ongoing research on the effects of coal combustion by-products on soil chemical, physical, and microbiological parameters as well as plant growth.

Potential Benefits

There are a number of potential benefits of applying coal combustion by-products to agricultural soils. These can be classified as either chemical or physical. Chemical benefits result from supplying essential plant nutrients for crop production (e.g., supplying B to a B-deficient soil) or by modifying the soil to create a more favorable medium for plant growth (e.g., increasing soil pH, decreasing Al toxicity, enhancing root penetration, etc.).

Physical benefits include increased water infiltration and aggregation of the soil which under certain conditions result from gypsum applications.

As noted, gypsum is a major constituent of most FGD by-products and residues from FBC and LIMB. Sulfate-containing coal combustion by-product materials are the most likely candidates for agricultural utilization.

Definition of potential benefits from coal combustion by-product utilization may be complex in that separation of a chemical or a physical benefit is not feasible. For example, application of high-gypsum FGD material may improve soil chemistry and increase water availability and crop yield as a result of reduction of subsoil chemical restrictions on rooting depth, while at the same time increasing water infiltration into the surface soil.

In addition to providing a clearly definable benefit, the applicaton of coal combustion by-products must not create hazardous conditions in the soil, groundwater, plants, or the food chain. Prevention of adverse conditions will, in most cases, be attainable by selecting appropriate by-products and utilizing them at appropriate rates. A distinction must be made between utilization benefits and disposal application rates.

Potential Hazards

The primary potential hazards from agricultural utilization of coal combustion by-products are excessive trace element loadings resulting in increased food chain metals, highly soluble salt loadings which may reduce initial plant growth, sodicity resulting from high Na-containing by-products which reduce water infiltration; negative effect of sulfites on crop growth and leaching of toxic substances into the groundwater. Although the potential for these hazards exists, all can be controlled in agricultural situations by logical application of selected coal combustion by-products. For instance, careful limitation in the use of fly ashes known to be rich in trace elements can control the loading of these elements to the soil and keep their concentrations in the beneficial or benign range in terms of leaching and/or plant uptake.

Many coal combustion by-product materials are highly alkaline or high in soluble salts which can inhibit plant establishment. One method to alleviate this potential hazard is to surface apply coal combustion by-products and then plow to incorporate the material essentially as a layer below the germinating seeds (K.D. Ritchey, 1993, personal communication). Sulfite by-products applied at planting have reduced rates of crop growth, but oxidation to sulfate may be sufficiently rapid in some soils that application of sulfite-bearing by-products a few months before planting will avoid harmful effects.

Induction of plant nutrient deficiencies of P and Mg are secondary potential problems in certain situations. Application of FGD by-products or FBC materials originating from facilities using a Ca-based sorbent can create an imbalance in the soil Ca/Mg ratio. This may result in an induced Mg deficiency. Although Mg deficiency is usually easily corrected by the soil application of $MgSO_4$ (Epsom salts), this is an extra expense and care should be taken to monitor the Ca/Mg ratio of the material applied and in the soil at the application site.

The high level of Ca and/or Fe and Al in some coal combustion by-products can result in the formation of insoluble complexes with P. These complexes reduce the availability of P to plants which may result in an induced P deficiency.

However, there may be situations where the formation of these insoluble P complexes is desirable. The potential coutilization of FBC ash and poultry manure is currently being examined (R.F. Korcak, 1993, unpublished data). One of the limitations on land utilization of poultry manure in intensive poultry producing states is the potential for P pollution of surface water supplies. Coutilization of the high Ca FBC material with poultry litter may form insoluble Ca-P complexes which will reduce potential P pollution problems.

PHYSICAL-CHEMICAL INTERACTIONS OF COAL COMBUSTION BY-PRODUCTS IN THE SOIL/PLANT SYSTEM

The Trace Element Dilemma

Most reports on the utilization of coal combustion by-products in agriculture conclude that the most serious potential hazards stem from B, Se, As, and Mo accumulation in soils and plants. However, coal combustion by-products can act as a supplementary source of Ca, S, B, Mo, Se, and other trace elements when soil contents are deficient for adequate plant/animal growth. Rates of ash application which achieve sufficient or excessive concentrations of these trace elements are often site-specific and, therefore, need to be more thoroughly examined before coal combustion by-products are utilized on a large scale.

Selenium is not an essential element for higher plant growth, although it has been shown to be a required element for some lower plant species. However, Se is an essential element for animal growth. The problem is accentuated since Se in animal nutrition is needed only in very low concentrations, slightly greater concentrations cause Se toxicity. Recommended food and feed concentrations to provide adequate animal Se range from 0.1 to 1 mg kg^{-1} dry weight. Food and feed Se concentrations above 5 mg kg^{-1} are detrimental, if these plants make up 100% of the animal ration. On the other hand, low Se also is detrimental. It is estimated that one-third of the forage and grain crops in the USA are below optimal in Se (Mengel & Kirby, 1987). Welch et al. (1991) provide an excellent discussion of micronutrient needs and availability in soils and maps showing areas where Se, Cu, and Mo are high in crops, areas where they are sufficient and others where additions of these elements are needed to optimize crop production and animal health.

A clearly defined benefit of coal combustion by-product use in agriculture would be to correct suboptimal concentrations of plant Se. Molybdenum, Cu, and B deficiencies may also be similarly ameliorated.

Trace element deficiencies of B, Mo, Cu, and Zn have been corrected by the application of coal combustion by-products (Page et al., 1979; Adriano et al., 1980; Aitken et al., 1984; El-Mogazi et al., 1988; Brieger et al., 1992; Environ. Manage. Serv., 1992). The application of by-products should be based on crop needs and current soil levels of the particular nutrient. In some ashes, B and Se appear to have sufficiently high concentrations to warrant limitations on the use of these ashes on agricultural soils (Ransome & Dowdy, 1987).

Soluble Salts

The other major concern, besides trace elements, with the agricultural utilization of coal combustion by-products is the high soluble salt content of many materials. At high application rates, salt injury can occur to germinating seeds or established plants. The problem of high soluble salts can be alleviated in a number of ways. As noted earlier, surface application of coal combustion by-products followed by plowing allows seeds to germinate without contacting the high salt zone. A similar technique was used by Jacobs et al. (1991) where ash was banded into the soil at a 45° angle to the surface. These two methods isolate the applied material from initial root contact. Most application methods homogenize the applied ash into the surface soil and maximize seed contact. Additionally, the timing of application can have a significant impact on avoiding initial soluble salt-related problems.

Another method to avoid soluble salt problems would be the use of weathered or stockpiled material from which a substantial portion of the soluble salts has been removed by percolation and some of the oxides and hydroxides have been stabilized by carbonation from air. Weathered vs. fresh fly ash was compared on field trials with maize (*Zea mays* L.) (Martens & Beahm, 1976). Weathered ash could be used at rates up to 131 Mg ha^{-1} while salt-related problems occurred at 87.2 Mg ha^{-1} using fresh ash. Also of interest was a decrease in the incidence of B toxicity with weathered ash. As previously noted (Table 6–7) a relatively high percentage of the B in ash is soluble. Therefore, lower amounts of water soluble B will be applied to soils when weathered ash is utilized. The use of weathered materials also decreases the dust hazard associated with applying fresh dry coal combustion by-products because bonding and recrystalization during moist weathering reduces the proportion of small-size particles.

A third method to reduce the potential for soluble salt problems has been the successful use of FBC residues as a soil "cap" where in a thick (5-cm) layer of FBC residue is surface applied and not plowed or mixed with the soil (Korcak, 1988). Used with horticultural crops, this method provides either sufficient soil mass for the roots to avoid contact with the initial flush of soluble salt or reduces this initial flush to levels that can be tolerated by crops. The "cap" of coal combustion by-product remains porous, thus allowing water to infiltrate. An associated benefit of the "cap" method for utilization is that the "cap" acts as a one-way valve: it allows water to infiltrate but decreases evaporation from the surface since large pores let water in, but the "cap" provides little air exchange.

Effect on Soil Chemical Properties

Because of the alkaline nature of many coal combustion by-products, a number of studies have examined their effect on modifying soil chemistry, primarily pH. The basic property of coal combustion by-products measured to quantify the effect on soil pH is the $CaCO_3$ equivalence of the materials. The neutralizing effect of pure $CaCO_3$ is 100% and that of coal combustion by-products usually ranges from 20 to 60%. Therefore, if a coal combustion by-product has a $CaCO_3$ equivalence of 50%, twice as much coal combustion by-product as $CaCO_3$ is needed to neutralize the same amount of soil acidity. Successful modification of soil pH has been demonstrated with a wide range of coal combustion by-products. Agricultural applications in most situations will probably be based on soil pH modification.

The FBC residues and oxidized FGD materials also contain significant amounts of gypsum and/or its anhydride. The potential benefits derived from gypsum applications in certain soils make those coal combustion by-products materials enriched with gypsum strong candidates for agricultural utilization.

A majority of the S currently being deposited in FGD processs is in the form of Ca–sulfites. Seedlings of some crop species grown in the presence of significant amounts of sulfites are not benefitted as they are with sulfates, and actual growth reductions have been observed (R.B. Clark, 1993, personal communication). Increased oxidation in the FGD process can result in the production of sulfates rather than sulfites. However, pilot power plant estimates indicate that this will add about $5.50 per metric ton (about $6 per ton) to the cost of the S by-product.

Another avenue would be to wait until natural processes oxidize the sulfites to sulfates. The somewhat gelatinous nature of the sulfite by-product hampers drying and invasion of the stored by-product by the air phase. Consequently the rate of oxidation of sulfites stored in large impoundments is generally extremely slow and often practically negligible. On the other hand, there are indications that the rate of oxidation increases rapidly when the sulfites are applied to soils. Whether this is due to better access to O_2 or the inoculation of the sulfite by oxidizing organisms from the soil is not known, but there are indications that the sulfite can oxidize to sulfate within a few weeks. Timing the soil application to allow oxidation to occur prior to plant growth may facilitate conversion of FGD sulfite-bearing materials to sulfates. Properly managed oxidation in the soil might then change the hundreds of millions of tons of FGD sulfite-bearing materials which are currently impounded as a hazardous waste into a sulfate resource with significant value.

Effect on Soil Physical Properties

A number of soil physical and related properties have been positively affected by the use of coal combustion by-products. Improved soil texture

(Chang et al., 1988) with concomitant increase in aeration and reduced bulk density result from application of silt-sized coal combustion by-products. Although increases in water-holding capacity in some soils have been reported from some ash applications, it is unclear whether this effect translates directly into increased available water for plant growth. The existing literature is not clear on this point. However, an interesting study on water relations and ash application was performed by Jacobs et al. (1991). They banded ash into the soil at a 45° angle to the surface. Corn roots were concentrated at the ash-band which was water saturated following a rain event. Corn yields increased in the ash-banded plots.

The pozzolanic activity of some coal combustion by-products can be viewed as either a positive or negative attribute. Ash materials which exhibit pozzolanic activity have been shown to reduce soil hydraulic conductivity as well as root growth. These effects can be lessened by using weathered materials or lower application rates. As noted above, banding ash into the soil can avoid some of these problems. No reports were noted on trenching of coal combustion by-products in agriculture. Current studies are underway to examine trenching (15 cm wide by 120 cm deep) of FBC materials alongside tree rows in establishing apple orchards (R.F. Korcak, 1993, unpublished data). The purpose of trenching is to prevent lateral root growth to initiate early fruit bearing and to reduce soil volume exploited by the root systems to facilitate management of tree nutrition by fertigation. Additionally, trenching will allow tree roots the alternative of growing into the fringes of the FBC trench to pick up needed Ca, S, and micronutrients or staying away from the relatively high concentrations of these elements if they are deleterious to root growth.

The soil "cap" technique also has a positive effect on precipitation use efficiency. A "cap" of by-product increases sustained infiltration rates, reduces transpiration by weeds (R.F., Korcak, 1993, unpublished data), reduces evaporation losses from the soil surface (K.D. Ritchey, 1993, personal communication) increases rooting depth in acid soils (Sumner, 1990), to increase plant water use efficiency. The resulting improvement in water use efficiency and consequent reduction in water stress on crops would probably be beneficial in many of the crop producing areas of the USA.

High Na-containing ash or FGD by-products may present a potential sodicity hazard with accompanying soil dispersion and reduction in infiltration rates. Application of by-products high in Na in dry climates, even if mixed in the soil, could create sodicity as the Na is carried to the surface and deposited. This also could be a potential hazard in humid areas particularly over longer time periods. Consequently, highly sodic materials should generally be identified and their application to agricultural soils should be avoided.

Overall, effects of coal combustion by-products application in agricultural soils should be beneficial to soil physical properties if the type of materials are well characterized before use and highly sodic materials are avoided. In fact, some of the major advantages of coal combustion by-products may be in the area of enhanced soil water availability for plant growth. This concept needs additional evaluation.

Effect on Soil Microbiological Properties

The microbiology of the soil/plant system as affected by ash application has received the least emphasis by researchers. Most of the research performed to date has examined either soil microbial activity or soil respiration activity (Cervelli et al., 1987; Pichtel & Hayes, 1990). Results of these and other studies are generally inconclusive, although a tendency for reduced soil respiration and microbial number following ash application usually occurs. The exact cause of this response has yet to be firmly elucidated.

Amelioration of reduced soil microbial activity may be made by simultaneous addition of an organic amendment such as sewage sludge (Pichtel & Hayes, 1990). The ratio of organic C to N in soils has a significant effect on soil microbiology. Little or no N is supplied by ash materials, and the C content varies depending on the particular ash by-product. Normally, most C in these materials is inorganic and would have little direct effect on microbial activity in any case. The effect of applied ash on the equilibrium soil C/N ratio requires more research. Additionally, the effect of higher C levels in some coal combustion by-products as well as the effects of coutilization with an organic source (e.g., sewage sludge, manures, newspaper, etc.) should be evaluated.

Utilization of Bottom Ash

It is worth singling out conventional power plant bottom ash as a potential soil amendment since this material represents one of the more useful coal combustion by-products for agriculture. These granular materials are generally applied at rates at or near the lime requirement for particular soils. They have a positive effect on soil texture and a modifying influence on soil pH and increased infiltration.

A management plan for the agricultural utilization of FBC bottom ash was recently proposed (Sell et al., 1989). This plan incorporates an economic analysis for agricultural utilization which showed that a 62% savings for land spreading vs. conventinoal landfill disposal could be achieved. Additional land management plans need to be developed, perhaps on a state-by-state basis, for the agricultural utilization of bottom ash. This should take into consideration soil type, crops grown, and climatic factors.

Utilization of Fluidized Bed Combustion and Fluidized Gas Desulfurization Residues

Research has been conducted on the agricultural utilization of FBC by-products but, no reviews are available. It is difficult to discern whether FBC materials used in many studies was bottom ash or a combination of bottom ash and captured fly ash. The research has generally involved rates equal to the lime requirement of the soil or multiples thereof. Crops studied include corn and peanuts (*Arachus hypogae*) (Terman, 1978), peach (*Prunus persica*) (Korcak et al., 1984; Edwards et al., 1985), forages (Stout et al., 1979), and apple (Korcak, 1979, 1980, 1982, 1984, 1985; Wrubel et al., 1982). Fluidized bed combustion residue also was used as an amendment for acid mine spoils (Sidle et al., 1979).

Utilization of high application rates of FBC materials is limited by the high alkalinity produced when the material is mixed with the soil (Terman, 1978). Mayes et al. (1991) incorporated FBC by-products at rates of 0, 20, 102, and 508 Mg ha^{-1} for corn, soybean (*Glycine max* L.), tall fescue (*Festuca arundinacea* Schreb.), and alfalfa (*Medicago sativa* L.). Annual application rates up to 20 Mg ha^{-1} or a single applicaton of 102 Mg ha^{-1} had no adverse effects on the yield of any of the crop plants tested. The highest rate led to crop failure primarily due to high soil pH and very high levels of soil Ca and S. The pozzolanic nature of the by-product created large chunks of the material in the field.

The data base on agricultural utilization of FGD materials is sparse, particularly with unoxidized materials. Compared to FBC materials, unoxidized scrubber sludges will probably require more careful monitoring and application rates and beneficial agricultural use will be at lower rates. Most scrubber sludges contain some fly ash and fly ash is often added at the end of the waste stream to aid in stabilization of the slurry (Terman, 1978). Scrubber sludges also must be kept from reaching an anaerobic stage due to the potential for generating hydrogen sulfide gas (Raiswell & Bottrell, 1991).

The FGD sludges oxidized at the coal combustion plant result in material that is high in $CaSO_42H_2O$ (gypsum); and if they are not oxidized, $CaSO_3$ (calcium sulfite) predominates (Terman, 1978). Calcium sulfate is an agriculturally valuable product, widely used to supply Ca to peanut in a soluble form. It also has potential for decreasing subsurface soil acidity and increasing plant rooting depth and drought tolerance. The dissolution of several gypsum-containing FGD materials was compared to phosphogypsum and mined gypsum (Bolan et al., 1991). The FGD materials were 99+% pure gypsum while the phosphogypsum was 97.5% gypsum compared to 82.5% in the mined material. The overriding difference was the higher (12.4%) content of $CaCO_3$ in the mined gypsum. All of the FGD materials and the phosphogypsum had higher dissolution rates than the mined gypsum. Dissolution of all samples was three to eight times faster in the presence of soil than in water.

Gissel-Nielsen and Bertelsen (1988) evaluated a number of FGD products in trials with barley (*Hordeum vulgare* L.). One of these contained 10% SO_3^{2-}, 24% SO_4^{2-}, 8% fly ash, and 0.5% NO_3^-. Although not noted, the high amount of sulfate present apparenlty indicated some oxidation of the material. They noted that plant Se concentrations were increased from 0.05 mg kg^{-1} in the control to 0.18 mg ka^{-1} at the highest application rate (5 g kg^{-1}) of coal-derived FGD. At these concentrations, Se in plants is considered an adequate source for animal nutrition.

Scrubber sludge containing 4.1 g B per kilogram was utilized as a B source to correct a B deficiency on a loamy sand soil (Ransome & Dowdy, 1987). Soybean yields were decreased during the first application year with the 10, 20, 40 Mg scrubber sludge per hectare applications due to elevated salt content. Yields were enhanced by scrubber sludge at all rates by the 3rd yr. Adequate soil B for soybean growth was achieved with the 20 Mg ha^{-1} application rate. The type of scrubber sludge, whether oxidized or not, was not indicated. The authors also expressed a need to determine the location of B that was apparently leached out of the root zone or otherwise inactivated.

There is a continuing need to examine the potential for the utilization of FGD by-products in agriculture. The FGD materials currently being produced, particularly dry, oxidized materials, are among those coal combustion by-products best suited for agricultural utilization. Research is currently underway to examine some of the wet FGD by-products which are high in calcium sulfite (K.D. Ritchey and R.B. Clark, 1993, personal communication). These studies as well as studies involving new by-products coming on-line are needed.

SUMMARY

Total production of coal combustion by-products will reach nearly 154 million megagrams (170 million short tons) annually in the USA by the year 2000. Besides conventional combustion of coal for electric power which generates bottom and fly ashes, a number of newer by-products are generated by this industry. These newer by-products emanate from the need to reduce S emissions. Typically, the desired desulfurization of the flue gases is accomplished by precipitating the sulfur oxides with Ca in the flues, or in the fire boxes with newer combustion technologies such as fluidized bed combustion systems. The diversity of products is further increased by differences in power plant design, operating parameters, sources and types of coal consumed, and, in the case of FGD, the types of reactive reagents utilized.

The lack of current utilization of most of these by-products, their diversity, and potentials for benefiting agriculture create the need for a database to facilitate agricultural utilization. The majority of the available data base information has been geared towards engineering properties of landfilled ash. An agricultural-oriented data base will facilitate the selection of those by-products exhibiting clearly definable benefits to the soil/plant system and identify components such as B, Se, and heavy metals whose presence in soil should be maintained within certain limits.

Potential agricultural benefits from coal combustion by-products include alleviating soil trace elemental deficiencies, modification of soil pH, and increasing levels of needed Ca and S, infiltration rates, depth of rooting, and drought tolerance. Flue gas desulfurization products and residues from FBC which contain appreciable amounts of gypsum appear to have particularly high potentials for improving water use efficiency, product quality and productivity of soil–crop systems.

The existing literature on agricultural utilization of coal combustion by-products needs to be expanded to include data from long-term exposure of these materials in the soil environment. Potential sites for examination exist. Additionally, cooperative research should be initiated with the Department of Energy and private industry to evaluate the potential agricultural utilization of by-products resulting from new clean air technologies as they are developed. These studies should address not only new by-products but also incorporate innovative strategies for application and clear documentation of benefits derived.

Documentation of hazards involved and benefits derived, especially from field studies, will be required to reduce present regulatory barriers to agricultural utilization of coal combustion by-products. Current inexpensive disposal costs, usually on-site, discourage land application of coal combustion by-products. However, on-site disposal may result in environmentally hazardous concentrations of certain elements in water supplies and the food chain.

RESEARCH NEEDS

A number of broad research areas need to be approached to discern which materials should be utilized in agriculture and what data needs to be forthcoming to evaluate these materials. Among the needs are: (i) a coal combustion by-product data base should be developed to incorporate agriculturally important parameters since existing engineering data bases are not readily applicable. Such a data base would assist in the selection of the most appropriate coal combustion by-products for agricultural utilization. (ii)Cooperative work should be utilized with the appropriate agencies and industry to evaluate new clean air technology-based by-products as these technologies are being developed. This research needs to be conducted in several climatic zones with different soil types. (iii) Chemical data is needed on the fate of coal combustion by-products in the soil environment. Studies should be initiated at the laboratory and field level to ascertain fate of potential contaminants. Old agricultural sites previously treated with coal combustion residue should be identified and evaluated. (iv) Assay techniques to identify potentially hazardous by-products should be developed. Such assays should be plant oriented, simple to perform and short term. Parameters to be assayed should include soluble salts, trace element phytotoxicities, and excessive alkalinity. (v) Application methods should be evaluated including surface incorporation, banding, trenching, and surface "capping." (vi) Coal combustion by-products which contain significant amounts of gypsum should be examined as potential soil amendments. Research should complement reported and ongoing work on mined gypsum and phosphogypsum. (vii) Studies should be initiated to examine the potential benefits of mixtures and/or composts of coal combustion by-products and other waste streams. In many cases, it appears that these mixtures would enhance the agronomic value of the by-products, and (viii) A better understanding of the chemical behavior of sulfite in the soil environment is needed to manage agricultural use of wet scrubber-type FGD by-products which contain significant amounts of sulfite.

REFERENCES

Adriano, D.C., A.L. Page, A.A. Elseewi, A.C. Chang, and L. Straughan. 1980. Utilization and disposal of fly ash and other coal residues in terrestrial ecosystems: A review. J Environ. Qual. 9:333–344.

Aitken, R.L., D.J. Campbell, and L.C. Bell. 1984. Properties of Australian fly ashes relevant to their agronomic utilization. Aust. J. Soil Res. 22:443–453.

Bolan, N.S., J.K. Syers, and M.E. Sumner. 1991. Dissolution of various sources of gypsum in aqueous solutions and in soil. J. Sci. Food Agric. 57:527–541.

Brieger, G., J.R. Wells, and R.D. Hunter. 1992. Plant and animal species composition and heavy metal content in fly ash ecosystems. Water Air Soil Pollut. 63:87–103.

Carlson, C.L., and D.C. Adriano. 1991. Growth and elemental content of two tree species growing on abandoned coal fly ash basins. J. Environ. Qual. 20:581–587.

Cervelli, S., G. Petruzelli, and A. Perna. 1987. Fly ashes as an amendment in cultivated soils. I. Effect on mineralization and nitrification. Water Air Soil Pollut. 33:331–338.

Chang, A.C., A.L. Page, L.J. Lund, J.E. Warneke, and C.O. Nelson. 1989. Municipal sludges and utility ashes in California and their effects on soils. p. 125–139. In B. Bar-Yosef et al. (ed.) Inorganic contaminants in the vadose zone. Ecol. Stud. Vol. 74. Springer-Verlag, Berlin.

Davidson, R.L., D.F.S. Natusch, and J.R. Wallace. 1974. Trace elements in fly ash. Dependence of concentration on particle size. Environ. Sci. Technol. 8:1107–1113.

Edwards, J.H., B.D. Horton, A.W. White, Jr., and O.L. Bennett. 1985. Fluidized bed combustion residue as an alternative liming material and Ca source. Commun. Soil Sci. Plant Anal. 16:621–637.

El-Mogazi, D., D.J. Lisk, and L.H. Weinstein. 1988. A review of physical, chemical, and biological properties of fly ash and effects on agricultural ecosystems. Sci. Total Environ. 74:1–37.

Environmental Management Services. 1992. Land application of coal combustion by-products: Utilization in agriculture and land reclamation. Electric Power Res. Inst., Proj. no. RP 3270-00. EPRI, Waupaca, WI.

Fail, J.L., Jr. 1987. Growth response of two grasses and a legume on coal fly ash amended strip mine spoils. Plant Soil 101:149–150.

Gissel-Nielsen, G., and F. Bertelsen. 1988. Inorganic element uptake by barley from soil supplemented with flue gas desulfurization waste and fly ash. Environ. Geochem. Health 10;21–25.

Haering, K.C., and W.L. Daniels. 1991. Fly ash: Characteristics and use in mined land reclamation—A literature review. Virginia Coal Energy J. 3(Summer):33–46.

Jacobs, L.W., A.E. Erickson, W.R. Berti, and B.M. NacKellar. 1991. Improving crop yield potentials of coarse textured soils with coarse flyl ash amendments. p. 59-1 to 59-16. In L.W Jacobs et al. (ed.) Proc. 9th Int. Ash Use Symp. Vol. 3. EPRI GS-7162, Orlando, FL. 10–15 January. Am. Coal Ash Assoc., Washington, DC.

Jastrow, J.D., C.A. Zimmerman, A.J. Dvorak, and R.R. Hinchman. 1991. Plant growth and trace-element uptake on acidic coal refuse amended with liem or fly ash. J. Environ. Qual. 10:154–160.

Korcak, R.F. 1979. Fluidized bed material as a calcium source for apples. HortSci. 14:163–164.

Korcak, R.F. 1980. Effects of applied sewage sludge compost and fluidized bed material on apple seedling growth. Commun. Soil Sci. Plant Anal. 11:571–576.

Korcak, R.F. 1982. Effectiveness of fluidized bed material as a calcium source for apples. J. Am. Soc. Hort. Sci. 107:1138–1142.

Korcak, R.F. 1984. Utilization of fluidized bed material as a calcium and sulfur source for apples. Commun. Soil Sci. Plant Anal. 15:879–891.

Korcak, R.F. 1985. Effect of coal combustion waste used as lime substitutes on nutrition of apples on three soils. Plant Soil 85:437–441.

Korcak, R.F. 1988. Fluidized bed material applied at disposal levels: Effects on an apple orchard. J. Am. Soc. Hort. Sci. 113:189–193.

Korcak, R.F., J.J. Wrubel, and N.F. Childers. 1984. Peach orchard studies utilizing fluidized bed material. J. Plant Nutr. 7:1597–1604.

Martens, D.C., and B.R. Beahm. 1976. Growth of plants in fly ash amended soils. p. 657664. *In* J.H. Faber et al. (ed.) Proc. 4th Int. Ash Utilization Symp., St. Louis, MO. 12–15 May. MERC SP-76/4. ERDA Morgantown Energy Res. Ctr., Morgantown, WV.

Mays, D.A., P.M. Giodano, and A.D. Behel, Jr. 1991. Impact of fluidized bed combustion waste on metal content of crops. Water Air Soil Pollut. 57–58:307–317.

McIntosh, C.S., W. Kriesel, W.P. Miller, and M.E. Sumner. 1992. Utilization of coal combustion by-products in agriculture and land reclamation: Market analysis for southeast region. Res. Proj. 3270. Electric Power Res. Inst., Palo Alto, CA.

Mengel, K., and E.A. Kirby. 1987 Principles of plant nutrition, Int. Potash Inst., Bern, Switzerland.

Page, A.L., A.A. Elseewi, and I.R. Straughan. 1979. Physical and chemical properties of fly ash from coal-fired power plants with reference to environmental impacts. Residue Rev. 71:83–120.

Phung, H.T., L.J. Lund, A.L. Page, and G.R. Bradford. 1979. Trace elements in fly ash and their release in water and treated soils. J. Environ. Qual. 8:171–175.

Pichtel, J.R., and J.M. Hayes. 1990. Influence of fly ash on soil microbial activity and populations. J. Environ. Qual. 19:593–597.

Raiswell, R., and S.H. Bottrell. 1991. The disposal of flue gas desulfurization waste: Sulphur gas emissions and their control. Environ. Geochem. Health 13:119–126.

Randolph, J., J.R. Jones, and L.J. Prelaz. 1990. Virginia coal. Virginia Center for Coal and Energy Res., Virginia Polytechnic Inst. State Univ., Blacksburg, VA.

Ransome, L.S., and R.H. Dowdy. 1987. Soybean growth and boron distribution in a sandy soil amended with scrubber sludge. J. Environ. Qual. 16:171–175.

Roy, W.R., R.G. Theiry, R.M. Schuller, and J.J. Suloway. 1981. Coal fly ash: A review of the literature and proposed classification system with emphasis on environmental impacts. Illinois State Geol. Surv., Environ. Geol. Notes 96.

Ruth, L.A. 1975. Regeneration of $CaSO_4$ in FBC. p. 425–438. *In* Proc. 4th Int. Conf. on Fluidized-bed Combustion, McLean, VA. 5–8 April. The MITRE Corp., McLean, VA.

Sell, N., T. McIntosh, C. Severance, and A. Peterson. 1989. The agronomic land spreading of coal bottom ash: Using a regulated solid waste as a resource. Resour. Conserv. Recycl. 2:118–129.

Sidle, R.C., W.L. Stout, J.L. Hern, and O.L. Bennett. 1979. Solute movement from fluidized bed combustion waste in acid soil and mine spoil columns. J. Environ. Qual. 8:236–241.

Stehouwer, R., and P. Sutton. 1992. Treatment of acid mine spoil with dry FGD by-products: Leachate quality and plant growth. p. 211–225. *In* Proc. Abandoned Mine Lands Conf., Chicago, IL. 23–26 Aug. Illinois Abandoned Mine Lands Reclam. Council, Springfield, IL.

Stout, W.L., J.L. Hen, R.F. Korcak, and C.W. Carlson. 1988. Manual for applying fluidized bed combustion residue to agricultural lands. USDA-ARS no. 74. Natl. Tech. Inform. Serv., Springfield, VA.

Stout, W.L., R.C. Sidle, J.L. Hern, and O.L. Bennett. 1979. Effects of fluidized bed combustion waste on the Ca, Mg, S, and Zn levels in red clover, tall fescue, oat, and buckwheat. Agron. J. 71:662–665

Sumner, M.E. 1990. Gypsum as an amendment for the subsoil acidity syndrome. Final Rep., Proj. no. 83-01-024R. Florida Inst. Phosphate Res., Bartow, FL.

Taylor, E.M., and G.E. Schuman. 1988. Fly ash and lime amendment of acidic coal spoil to aid re vegetation. J. Environ. Qual. 7:120–124.

Terman, G.L. 1978. Solid wastes from coal-fired power plants: Use or disposal on agricultural lands. Bull. Y-129. Tennessee Valley Authority, Muscle Shoals, AL.

Tetra Tech, Inc. 1983. Physical-chemical characteristics of utility solid wastes. Electric Power Res. Inst., no. EA-3226. EPRI, Palo Alto, CA.

U.S. Environmental Protection Agency. 1988. Wastes from the combustion of coal by electric utility power plants. EPA/530-SW-88-002. USEPA, Washington, DC.

U.S. Department of Energy. 1992. Clean coal technology demonstratino program. Program update 1991. U.S. Dep. Energy, Washington, DC.

Welch, R.M., W.H. Allaway, W.A. House, and J. Kubota. 1991. Geographic distribution of trace element problem. p. 31–57. *In* J.J. Mortvedt et al. (ed.) Micronutrients in agriculture. SSSA Book Ser. 4. SSSA, Madison, WI.

Wrubel, J.J., Jr., R.F. Korcak, and N.F. Childers. 1982. Orchard studies utilizing fluidized bed material. Commun. Soil Sci. Plant Anal. 13:1071–1080.

7 Coal Combustion By-Product Use on Acid Soil: Effects on Maize Growth and Soil pH and Electrical Conductivity

R. B. Clark, S. K. Zeto, K. Dale Ritchey, R. R. Wendell, and V. C. Baligar

USDA-ARS
Appalachian Soil and Water Conservation Research
  Laboratory
Beckley, West Virginia

Many coal combustion by-products (CCBP's) such as fly ashes (FA's), fluidized bed combustion BP's (FBC-BP's), and flue gas desulfurization BP's (FGD-BP's) are produced when coal is burned for generation of electricity. It has been estimated that in the year 2000, 110 million megagrams of fly and bottom ashes and 45 megagrams of FGD-BP's will be produced annually (USEPA, 1988). Most of these CCBP's (estimate about 80%) are disposed in landfills/surface impoundments (USEPA, 1988), and interest has arisen in using these BP's on agricultural land, especially as an acid soil amendment.

Information on physical and chemical properties, use on land/soil, and potential environmental effects of fly and bottom ashes is extensive and has recently been reviewed (Carlson & Adriano, 1993). Application of FBC-BP's also have been reported (Korcak & Kemper, 1993; Stout et al., 1988). Only limited information is available on use of FGD-BP's on acid soil.

High-sulfate gypsum BP's can enhance plant growth (Alcordo & Rechcigl, 1993; Shainberg et al., 1989). Several studies showed no effect of gypsum the 1st yr of application in field experiments even when it was applied at relatively high levels. Although some studies showed enhanced growth due to Ca and S, enhanced gypsum effects often occurred during droughty years which might indicate improved subsoil physical/chemical properties to improve plant access to subsurface water and nutrients. Plants grown on soils with especially low or marginal soil Ca might respond to Ca as a nutrient, and gypsum might be expected to enhance plant growth at any level of applicaton (Ritchey et al., 1982). Sources of gypsum used in most studies have been mined gypsum or phosphogypsum BP's derived from the phosphate fertilizer industry, and not CCBP's. Gypsum has high S (and some P in phosphogypsum BP's) so S (and P) as nutrients should not be ruled out as reasons for plant enhanced responses to gypsum.

Growth of weeping lovegrass [*Eragrostis curvula* (Schrader) Nees] and lespedeza [*Lespedeza cuneata* (Dum.-Cours.) G. Don] was poor when plants were grown with 100% FGD-BP's that were unleached (Shahandeh & Sumner, 1993). Growth reductions were attriuted to ion toxicity (probably B and/or Cl), salinity [high electrical conductivity (EC)], and poor hydraulic properties (poor texture from high silt particles) of the FGD-BP's used. Growth was reduced unless the FGD-BP's were leached and mineral nutrients added, especially N, P, and K; N was the most limiting of the nutrients. Differences among FGD-BP's also were noted for root penetration and distribution. In other studies, alfalfa (*Medicago sativa* L.) grown in a greenhouse with acid soil (pH 4.6) containing two FBC-BP's and one FGD-BP mixed at varied levels (0 to 2.8% in soil mix) had enhanced growth over unamended soil in each of six cuttings except the first (Sutton & Stehouwer, 1992). In the first cutting, growth was lower when plants were grown at the highest level of FGD-BP compared to the lower levels. Tall fescue (*Festuca arundinacea* Schreber) grown with similar soil mixes as alfalfa had similar or only slightly higher growth at each level of CCBP in each of six cuttings as the unamended soil. The first cutting of tall fescue was reduced when plants were grown with mine spoil underclay (subsoil) mixed with varied levels (0–24% in soil mix) of a FBC-BP and a FGD-BP compared to unamended soil (Stehouwer et al., 1993). However, growth reductions at each level of BP were overcome in later cuttings. When these BP's were mixed at similar levels with acid coal mine spoil overburden, the growth of six cuttings of tall fescue was enhanced at the lower levels of BP and growth inhibitions at the highest level were overcome in later cuttings.

A FGD scrubber sludge impoundement was successfully vegetated with four of seven herbaceous [tall wheatgrass (*Agropyron elongatum* (Host) Beauv.), tall fescue, yellow sweet clover (*Melilotus officinalis* Lam.), and Japanese millet (*Echinochloa crusgalli* (L.) Beauv.)] and two of six tree [eastern cottonwood (*Populus deltoides* Marsh.) and eastern red cedar (*Juniperus Virginiana* L.] species tested when fertilizer (N, P, and K) and other amendments (cow manure for herbaceous plants and wood chips for trees) were added (Mulhern et al., 1989). The treated plots had greater productivity than untreated plots 7 yr after the initial application of the products (Wilson et al., 1991). The greater productivity on the treated plots was attributed, in part, to mycorrhizal associations with roots.

Objectives of our studies were to obtain information on effects of different levels of CCBP's on maize (*Zea mays* L.) growth and on some soil chemical properties.

MATERIALS AND METHODS

Experiments were conducted in a greenhouse (25 ± 3 °C) using natural and artificial light to extend short days to 14-h d lengths and to provide ex-

Table 7-1. Coal combustion by-products added to acid soil to determine effects on maize growth and soil pH and EC.

---

1. Fly ashes (FA's)
    BP-#12† Class F
    BP-#18   Class C
2. Fluidized bed combustion by-products (FBC-BP's)
    BP-#15   Atmospheric FBC residue containing mixed bottom and fly ash
    BP-#21   Pressurized FBC residue containing FA and Mg + Ca sulfate and oxide
    BP-#26   Atmospheric FBC residue
3. Fluidized gas desulfurization by-products (FGD-BP's)
    High-sulfite FGD-BP's
    BP-#1‡   Scrubber FGD residue before addition of Calcilox (Ca + Mg alumino-silicate material)
    BP-#2‡   Scrubber FGD sludge after addition of Calcilox
    BP-#4†   Scrubber FGD sludge after centrifugation and contains Ca + sulfite
    BP-#5†   Stabilized scrubber FGD sludge containing portions of BP-#4, BP-#10, and BP-#12
    BP-#6§   Vacuum filter scrubber FGD sludge containing sulfite/sulfate
    BP-#8§   Stabilized vacuum filter scrubber FGD sludge containing FA, sulfite/sulfate, and CaO
    High-sulfate (gypsum quality) FGD-BP's
    BP-#16   Ex situ forced oxidation FGD residue (wallboard quality gypsum)
    BP-#22   Limestone based in situ forced oxidation FGD residue
    BP-#27   In situ forced oxidation FGD residue [Mg enhanced lime residue or high purity gypsum containing about 6% Mg(OH)$_2$]
4. Miscellaneous
    BP-#10† CaO material

---

† Represents BP's from the same location.
‡ Represents BP's from the same location.
§ Represents BP's from the same location.

tra light during cloudy days (400–500 μmol m^{-2} s^{-1} at plant height). Artificial light was provided by high-pressure sodium lamps. Maize (PA329 × PA353P) was grown with 15 CCBP's (Table 7–1) at different levels (level of CCBP varied because levels were sought that would reduce plant growth) with an acid Porters soil (coarse-loamy, mixed, mesic, Umbric Dystrochrepts) which had been fertilized with 50 mg N kg^{-1} soil as NH$_4$NO$_3$ (100 kg N ha^{-1}) and 400 mg P kg^{-1} soil as KH$_2$PO$_4$ (400 kg P ha^{-1}). See legend of figures for amount of each CCBP used.

The CCBP's were thoroughly mixed with soil by hand to assure homogeneity of the soil mix. Soil–CCBP mixes had deionized water added to provide 0.033 MPa, enclosed in plastic bags, and equilibrated 7 d at ambient temperature before being placed in pots (1.0 kg soil mix pot^{-1}) for plant growth. A small amount (20–25 g) of the initial soil mix from each pot was saved for pH and EC measurements. Some chemical properties of the original Porters soil were: 13.2% organic matter; 4.22 (1 soil/1 water) and 3.88 (1 soil/1 0.01 M CaCl$_2$) pH; 2.70 P (Bray-1-extractable) and 3.86 Mn (0.005 M DTPA-extractable) in micrograms per gram; 6.11 total acidity and 5.38 Al (1 M KCl-extractable) in molar charge per kilogram (81% Al saturation); and 0.18 K, 0.10 Mg, 0.24 Ca, and 0.03 Na (1 M ammonium acetate-extractable) in molar charge per kilogram.

Maize seeds were surface sterilized with 0.5 M NaOCl for 5 min, rinsed thoroughly with deionized water, and germinated between wrapped germination papers moistened with deionized water containing dilute $CaSO_4$ to assure good root development. Three 3-d-old maize seedlings were placed in each pot of soil mix, and irrigated with deionized water. Deionized water was added every other day initially and daily after plants became established to provide sufficient water for plant growth. Water was added manually and leaching from soil was prevented.

Different experiments were conducted over time to test the many CCBP's. Control treatments in each experiment consisted of plants grown in unamended soil with added fertilizer only. The experimental design for each experiment was completely randomized blocks with four replications. After 21 d growth in pots, shoots were severed about 0.5 cm above soil surface, leaves were separated from each other (sheaths not included), leaf areas were measured with a Li-Cor 3100 leaf area meter (Li-Cor Co., Lincoln, NE[1]), and leaves and sheaths were combined, dried in a forced-air oven at 60 °C, and weighed. Soil was shaken from roots and representative soil samples were saved for pH-water (1 soil/1 water), pH-$CaCl_2$ (1 soil/1 0.01 M $CaCl_2$), and EC (1 soil/1 water) measurements. Roots were thoroughly washed from soil, blotted dry, weighed fresh, cut into 1- to 2-cm segments, and mixed thoroughly. A representative sample was used to determine total root length (RL) using a Comair RL scanner (Commonwealth Aircraft Corp., Ltd., Melbourne, Australia[1]). Remaining roots were dried and weighed. Fresh root samples used for RL measurements were dried, weighed, and weight added to that of the other roots to give total root dry matter yield (DMY). Specific RL was calculated: specific RL = total RL/total root DMY (m g^{-1} root DMY).

Some chemical properties [pH, EC, calcium carbonate (lime) equivalency (CCE), and total S, sulfite S, and sulfate S] were determined on samples of each CCBP (Table 7-2). The pH and EC values were determined on 1:1 and 1:2 (CCBP/water) pastes. Even though 1:1 (CCBP/water) pastes are commonly reported for pH determinations, some of the CCBP's solidified (cemented) extensively in 1:1 pastes. Thus, 1:2 pastes also were used to overcome solidification problems. The CCE values were determined using a boiling HCl (0.5 M) digestion and back titrating with NaOH (0.05 M) against bromocresol green indicator (ASTM, 1985). The S in the CCBP's was assumed to be sulfite S or sulfate S. Total S and sulfite S were determined using a gravimetric (Method L4) for total S and an I-thiosulfate titration (Method M2) for sulfite S from Electric Power Research Institute (EPRI, 1988a, b). Sulfate S was determined by subtracting sulfite S from total S.

Representative samples of each CCBP (0.5 g) were digested for determination of mineral element concentrations using USEPA Method 3050A (USEPA, 1992). Digested samples were poured through funnels containing

---

[1]Mention of company names or prroprietary products does not indicate endorsement by the USDA, and does not imply their approval to the exclusion of other companies or products that may also be suitable.

Table 7-2. pH, EC, CCE, total S, sulfate S, and sulfite S of CCBP's used to grow maize in acid soil.[†]

CCBP no.	pH (1:1)	pH (1:2)	EC (1:1)	EC (1:2)	CCE	Total S	SO_4-S	SO_3-S
			dS m^{-1}		%	g kg^{-1}		
FA's								
BP-#12 (Class F)	11.60	12.89	2.96	3.54	5.5	6.4	4.0	2.4
BP-#18 (Class C)	11.60	12.68	1.78	1.82	44.8	14.6	14.5	0.1
FBC-BP's								
BP-#15	11.50	13.59	3.96	7.77	75.4	31.0	29.5	1.5
BP-#21	11.80	13.17	6.68	5.56	88.9	91.2	90.6	0.6
BP-#26	12.40	12.80	6.18	8.45	55.2	173.8	170.6	3.2
FGD-BP's (high-sulfite)								
BP-#1	8.75	9.76	2.41	3.38	7.6	216.1	47.6	168.5
BP-#2	9.90	10.76	4.74	4.65	66.6	167.3		
BP-#6	8.51	9.43	2.24	3.54	38.1	187.9	35.1	152.8
BP-#8	10.60	11.27	2.69	2.94	38.8	103.7	13.7	90.0
Selected CCBP's from same location. BP-#4 + BP-#10 + BP-#12 = BP-#5								
BP-#4	8.44	9.50	3.46	4.04	63.0	211.0	64.5	146.5
BP-#10	11.60	13.67	4.27	6.72	130.5	3.1		3.6
BP-#12	11.60	12.89	2.96	3.54	5.5	6.4	4.0	2.4
BP-#5	9.57	10.43	2.13	2.77	38.8	110.5	16.5	94.0
FGD-BP's (high-sulfate)								
BP-#16	8.63	9.31	1.80	1.73	<1.0	219.4	217.3	2.1
BP-#22	8.91	8.96	1.67	1.92	5.0	216.9	216.1	0.8
BP-#27	9.53	9.65	3.35	3.29	13.1	177.2	176.2	1.0

† EC = electrical conductivity, CCE = calcium carbonate (lime) equivalent, (see Materials and Methods for procedures used for determination of each property).

dried and preweighed filter paper (Whatman no. 42), rinsed thoroughly, and collected in preweighed plastic bottles. They (filter papers + residues) were dried at 70 °C for 48 h and weighed to calculate percentage insoluble residue for each digested CCBP. Bottles containing the digested solution had internal standard elements added [20 μg yttrium (Y) and 10 μg rhodium (Rh) per 50 mL] before bottles were brought to near 50 mL and weighed for determination of final volume. Samples were analyzed for concentrations of 34 mineral elements using inductively coupled plasma (ICP) spectrometry (Model 46P Jobin-Ivon, Lonjumeau Cedex, France[1]). Representative samples also were sent to a commercial laboratory for determination of concentrations of leachable mineral elements using the TCLP (toxicity characteristic leaching procedure) Method 1311 (Fed. Reg., 1990). Insoluble residue after digestion and mineral element concentrations in digests for each CCBP are given in Tables 7-3 to 7-5, and TCLP mineral element concentrations of the CCBP's are given in Tables 7-6 and 7-7.

[1] Mention of company names or proprietary products does not indicate endorsement by the USDA, and does not imply their approval to the exclusion of other companies or products that may also be suitable.

Table 7–3. Digest residue percentages and S, Ca, Mg, K, P, Fe, Mn, Zn, Cu, and B concentrations for CCBP's used to grow maize in acid soil (HNO_3 + HCl + H_2O_2 digest).†

CCBP no.	Residue‡	S	Ca	Mg	K	P	Fe	Mn	Zn	Cu	B
	%					— mg kg^{-1} —					
FA's											
BP-#12 (Class F)	89.4	5 443	25 480	3 689	6 138	1 970	13 410	89.5	20.25	19.0	431
BP-#18 (Class C)	51.8	10 250	174 500	24 240	4 595	4 854	33 560	380.2	22.21	90.8	358
FBC-BP's											
BP-#15	31.7	34 630	333 400	4 287	6 855	213.0	5 423	125.2	4.16	1.84	7.8
BP-#21	19.8	95 090	232 700	14 020	1 214	67.9	11 610	163.6	18.74	3.51	75.4
BP-#26	21.5	159 600	413 800	3 650	24	117.4	8 689	192.2	14.47	0.67	170.8
FGD-BP's (high-sulfite)											
BP-#1	6.3	217 900	287 400	11 930	554	3.5	1 024	89.0	2.43	<0.01	46.5
BP-#2	13.9	165 000	271 600	22 760	4 394	28.6	1 721	403.0	2.53	<0.01	98.2
BP-#6	17.4	168 600	220 500	8 789	2 604	77.9	5 276	84.7	6.12	2.07	144.6
BP-#8	46.4	98 540	165 700	6 996	6 456	224.7	14 860	96.8	14.28	7.56	174.9
Selected CCBP's from same location. BP-#4 + BP-#10 + BP-#12 = BP-#5											
BP-#4	6.2	214 500	298 900	9 303	1 536	79.1	1 687	107.4	3.98	<.01	52.8
BP-#10	11.0	6 394	384 300	14 200	5 806	12.2	1 077	136.5	1.98	<.01	<0.02
BP-#12	89.4	5 443	25 480	3 689	6 138	1 970	13 410	89.5	20.25	19.00	430.9
BP-#5	43.4	100 300	162 600	6 513	6 002	660.5	8 716	89.8	7.71	6.90	170.9
FGD-BP's (high-sulfate)											
BP-#16	5.4	189 700	240 500	468	368	7.6	754	195.7	<0.01	<0.01	<0.02
BP-#22	4.9	177 300	238 500	230	32	60.7	441	58.4	<0.01	<0.01	<0.02
BP-#27	7.4	162 600	208 600	22 740	165	<0.03	1 050	85.5	2.52	0.12	99.0

† Means of five representative samples for each CCBP.
‡ Percent residue after digestion.

Table 7–4. Concentrations of Mo, Co, Ni, Al, Si, Ag, As, Cd, Cr, Pb, and Se for CCBP's used to grow maize in acid soil (HNO₂ + HCl + H₂O₂ digest).†

CCBP no.	MO	Co	Ni	Na	Al	Si	Ag	As	Cd	Cr	Pb	Se
						mg kg^{-1}						
FA's												
BP-#12 (Class F)	7.28	7.87	19.00	1 175	26 790	1 497	<0.01	<0.06	<0.01	55.3	175	<0.10
BP-#18 (Class C)	8.17	31.26	54.23	<0.05	72 390	740	<0.01	<0.06	<0.01	118.1	1 130	<0.10
FBC-BP's												
BP-#15	1.21	3.93	8.51	1 218	11 460	570	<0.01	<0.06	<0.01	110.2	27.3	<0.10
BP-#21	4.66	2.78	7.01	1 005	5 231	675	<0.01	<0.06	<0.01	74.9	127.7	7.82
BP-#26	13.40	3.27	15.65	157	1 598	346	<0.01	<0.06	0.21	154.4	80.6	11.41
FGD-BP's (high-sulfite)												
BP-#1	0.61	1.40	9.53	425	856	156	<0.01	<0.06	<0.01	84.8	8.5	<0.10
BP-#2	0.50	1.28	1.49	1 200	5 514	643	<0.01	<0.06	<0.01	90.7	3.1	<0.10
BP-#6	5.31	2.63	5.94	1 246	4 075	448	<0.01	<0.06	<0.01	72.3	28.4	<0.10
BP-#8	14.73	6.34	17.33	1 305	9 885	585	<0.01	<0.06	<0.01	65.2	218.0	2.23
Selected CCBP's from same location. BP-#4 + BP-#10 + BP-#12 = BP-#5												
BP-#4	0.54	1.47	4.63	1 068	2 217	364	<0.01	<0.06	<0.01	104.1	9.0	0.42
BP-#10	0.41	0.95	3.26	1 151	3 527	447	<0.01	<0.06	<0.01	148.4	0.3	<0.10
BP-#12	7.28	7.87	19.00	1 175	26 790	1 497	<0.01	<0.06	<0.01	55.3	175	<0.10
BP-#5	3.19	4.62	9.23	1 191	12 310	642	<0.01	<0.06	<0.01	72.0	75.7	<0.10
FGD-BP's (high-sulfate)												
BP-#16	<0.01	0.37	0.72	796	360	49.9	<0.01	<0.06	<0.01	72.9	2.4	<0.10
BP-#22	<1.61	2.08	2.11	483	37	43.1	<0.01	<0.06	0.14	86.0	17.1	16.51
BP-#27	0.63	1.45	6.10	424	1 224	306	<0.01	<0.06	<0.01	74.2	10.8	6.62

† Means of five representative samples for each CCBP.

Table 7-5. Concentrations of Hg, Ba, Be, La, Li, Sb, Sc, Sr, Ti, Tl, and V for CCBP's used to grow maize in acid soil (HNO_3 + HCl + H_2O_2 digest).†

CCBP no.	Hg	Ba	Be	La	Li	Sb	Sc	Sn	Sr	Ti	Tl	V
							mg kg^{-1}					
FA's												
BP-#12 (Class F)	<0.05	714	12.4	13.76	57.2	<0.07	6.70	<0.08	265	857	<0.06	98.6
BP-#18 (Class C)	<0.05	32 010	72.3	57.16	71.8	<0.07	20.95	22.02	2 763	6 043	0.39	270.0
FBC-BP's												
BP-#15	<0.05	388	129.3	5.70	46.3	<0.07	2.94	<0.08	328	515	<0.06	25.6
BP-#21	<0.05	117	80.8	2.55	47.0	<0.07	0.92	<0.08	116	278	<0.06	13.8
BP-#26	<0.05	195	178.4	4.30	0.01	<0.07	2.86	<0.08	312	261	1.09	42.6
FGD-BP's (high-sulfite)												
BP-#1	<0.05	106	105.7	<0.01	0.4	<0.07	0.01	<0.08	205	70	<0.06	0.2
BP-#2	<0.05	244	113.7	2.59	30.6	<0.07	1.42	<0.08	253	300	<0.06	5.1
BP-#6	<0.05	174	85.1	1.25	17.2	<0.07	1.36	<0.08	208	199	<0.06	13.7
BP-#8	<0.05	345	63.5	4.69	60.1	<0.07	3.73	<0.08	217	507	6.39	40.7
Selected CCBP's from same location. BP-#4 + BP-#10 + BP-#12 = BP-#5												
BP-#4	<0.05	153	127.3	<0.01	5.2	<0.07	0.37	<0.08	232	123	<0.06	4.6
BP-#10	<0.05	286	184.5	1.83	11.5	<0.07	0.50	<0.08	329	198	<0.06	11.9
BP-#12	<0.05	714	12.5	13.76	57.2	<0.07	6.70	<0.08	265	857	<0.06	98.6
BP-#5	<0.05	353	62.1	6.44	45.6	<0.07	3.30	<0.08	397	526	<0.06	46.4
FGD-BP's (high-sulfate)												
BP-#16	0.59	79	94.7	<0.01	<0.01	<0.07	<0.01	<0.08	175	18	<0.06	<0.01
BP-#22	<0.05	75	97.2	5.51	<0.01	<0.07	0.32	<0.08	230	0.10	2.73	9.48
BP-#27	<0.05	85	88.1	0.98	1.57	<0.07	0.17	<0.08	175	87.8	<0.06	0.44

† Means of five representative samples for each CCBP.

Table 7-6. Concentrations of TCLP mineral elements S, K, Ca, Mg, P, Fe, Mn, Al, Si, and Na for CCBP's used to grow maize in acid soil.[†]

CCBP no.	S	K	Ca	Mg	P	B	Fe	Mn	Al	Si	Na
						mg kg^{-1}					
FA's											
BP-#12 (Class F)	20.7	18.3	817	88.6	2.2	19.0	0.27	0.53	0.4	1.8	
BP-#18 (Class C)	20.6	39.5	1730	192	0.3	7.80	2.00	0.07	0.8	5.0	45.3
FBC-BP's											
BP-#15	61.6	47.2	355	0.5	0.1	0.48	0.25	0.03	0.5	0.8	4.2
BP-#21	40.7	27.9	1410	324	0.1	1.44	0.33	0.04	0.7	1.2	1.76
BP-#26	755	11.8	2420	<0.1	0.6	4.60	0.2	<0.1	1.2	0.4	
FGD-BP's (high-sulfite)											
BP-#1	38.9	39.6	1320	200	0.1	3.4	5.3	0.29	15.5	3.4	5.7
BP-#2	27.5	30.2	1720	360	0.1	4.1	0.4	0.54	0.6	3.8	35.3
BP-#6	36.6	17.2	951	133	0.1	4.8	0.5	0.21	3.6	3.0	
BP-#8	52.6	20.6	1440	152	0.1	9.6	18.2	1.00	10.3	6.0	11.7
Selected CCBP's from same location. BP-#4 + BP-#10 + BP-#12 = BP-#5											
BP-#4	38.0	8.5	1480	155	0.1	2.3	7.3	0.32	19.2	2.8	2.5
BP-#10	8.1	18.4	295	1.0	0.06	0.4	0.43	0.03	0.6	0.7	4.5
BP-#12	20.7	18.3	817	88.6	0.2	19.0	0.27	0.53	0.4	1.8	
BP-#5	35.6	15.8	1750	139	0.3	9.6	33.0	0.96	24.0	11.5	10.0
FGD-BP's (high-sulfate)											
BP-#16	36.1	49.3	848	12.2	0.08	21.1	1.00	1.20	4.3	1.5	
BP-#22	72.7	46.6	930	3.7	0.4	0.28	0.23	0.10	0.8	1.1	
BP-#27	824	9.7	1120	250	0.7	10.2	0.3	<0.1	13.6	24.3	

† Values of one representative sample for each CCBP.

RESULTS

Shoot and Root Dry Matter Yields

BP-#12 (Class-F FA) and BP-#18 (Class-C FA) did not reduce shoot and root DMY below that of plants grown in unamended acid soil at any level of FA used (Fig. 7-1). However, slight reductions in shoot and root DMY below the maximum obtained at 2% FA in soil mixes occurred in plants grown with BP-#12 as level increased to 5%. The DMY of plants grown with BP-#18 was about threefold higher for shoots and twofold higher for roots than plants grown in unamended soil at levels as high as 25% BP in the soil mix. Shoot and root DMY remained relatively constant over the wide range of BP-#18 added to the soil (0.5–25%).

Plants produced about twofold higher shoot and root DMY when grown in soil mixes containing up to 1% BP-#15 and up to 3% BP-#21 (FBC-BP's) compared to unamended acid soil (Fig. 7-2). Above these levels of BP-#15 and BP-#21, both shoot and root DMY decreased as level of BP increased. Shoot DMY at the highest level of BP-#15 and BP-#21 used was similar to shoot DMY of plants grown in unamended soil. Root DMY showed more dramatic decreases at the highest levels of these two FBC-BP's than shoot DMY. At the highest levels of these two FBC-BP's, root DMY was 71% for BP-#15 and 30% for BP-#21 that of plants grown in unamended soil. In contrast, shoot and root DMY increased only slightly above that of plants

Table 7–7. Concentrations of TCLP mineral elements Zn, Cu, Mo, Co, As, Ba, Cd, Cr, and Pb for CCBP's used to grow maize in acid salt.†

CCBP no.	Zn	Cu	Mo	Co	As	Ba	Cd	Cr	Pb	Se	Hg
						mg kg^{-1}					
FA's											
BP-#12 (Class F)	0.04	0.04	0.25	0.12	0.50	<0.01	<0.02	0.40	<0.2	<0.2	0.0003
BP-#18 (Class C)	0.04	0.05	0.20	0.14	<0.2	<0.01	<0.02	0.77	<0.2	<0.3	0.0006
FBC-BP's											
BP-#15	0.02	0.05	0.04	0.12	<0.2	<0.01	<0.02	0.09	<0.2	<0.2	<0.0002
BP-#21	0.03	0.06	0.08	0.15	<0.2	<0.01	<0.02	0.12	<0.2	<0.2	0.0004
BP-#26	<0.1	0.3	<0.1	0.1	0.01	<0.01	<0.02	0.04	<0.2	<0.2	
FGD-BP's (high-sulfite)											
BP-#1	0.14	0.10	0.04	0.08	<0.2	<0.01	<0.02	0.08	<0.2	<0.2	0.0002
BP-#2	0.06	0.06	0.09	0.09	<0.2	<0.01	<0.02	0.09	<0.2	<0.2	0.0004
BP-#6	0.04	0.04	0.09	0.11	<0.2	0.16	<0.02	0.06	<0.2	<0.2	0.0003
BP-#8	0.48	0.10	0.21	0.16	<0.2	<0.01	<0.02	0.13	<0.2	<0.2	0.0043
Selected CCBP's from same location. BP-#4 + BP-#10 + BP-#12 = BP-#5											
BP-#4	0.11	0.07	0.42	0.11	<0.2	<0.01	<0.02	0.15	<0.2	<0.2	0.0007
BP-#10	0.02	0.06	0.09	0.10	<0.2	<2.3	<0.02	0.07	<0.2	<0.2	0.0003
BP-#12	0.04	0.04	0.25	0.12	0.50	<0.01	<0.02	0.40	<0.2	<0.2	0.0003
BP-#5	0.25	0.08	0.11	0.15	<0.2	<0.01	<0.02	0.12	<0.2	<0.2	0.0031
FGD-BP's (high-sulfate)											
BP-#16	0.08	0.06	<0.03	0.12	<0.2	<0.01	<0.02	0.06	<0.2	<0.2	0.0002
BP-#22	0.13	0.05	0.03	0.15	<0.2	<0.01	<0.02	0.06	<0.2	<0.2	0.0005
BP-#27	0.1	0.3	<0.1	0.2	0.5	<0.01	<0.02	0.4	<0.2	0.5	

† Values of one representative sample for each CCBP.

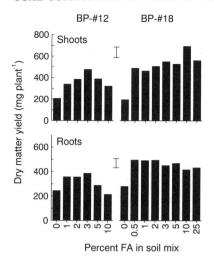

Fig. 7-1. Shoot and root DMY of maize grown with the FA's BP-#12 and BP-#18 on acid soil. Vertical bars are LSD (0.05) values.

grown in unamended soil when 0.5% BP-#26 (FBC-BP) was mixed with soil, and decreased markedly when 1% or higher levels of this BP was mixed with soil (Fig. 7-2). BP-#26 gave the poorest growth of the unleached FBC-BP's tested.

    Three of the four high-sulfite FGD-BP's (BP-#1, BP-#2, and BP-#6) showed relatively little beneficial effect on shoot and root DMY at levels up to 1%, and DMY decreased below that of plants grown in unamended soil as level of BP increased from 1 to 5% (Fig. 7-3). One of these BP's (BP-#8) enhanced shoot and root DMY when added up to 2% in soil mix before DMY decreased to near that of plants grown in unamended soil at the highest level (5%) added. The effects of these FGD-BP's on enhancing shoot and root

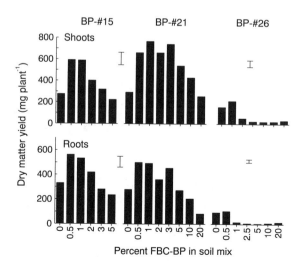

Fig. 7-2. Shoot and root DMY of maize grown with the FBC-BP's BP-#15, BP-#21, and BP-#26 on acid soil. Vertical bars are LSD (0.05) values.

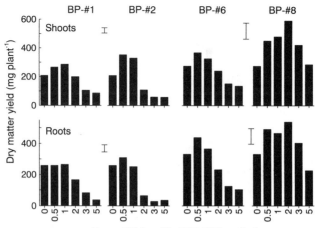

Fig. 7-3. Shoot and root DMY of maize grown with the high-sulfite FGD-BP's BP-#1, BP-#2, BP-#6, and BP-#8 on acid soil. Vertical bars are LSD (0.05) values—BP-#1 and BP-#2 had same LSD and BP-#6 and BP-#8 had same LSD.

DMY followed a sequence of BP-#8 > BP-#6 > BP-#2 = BP-#1. Root DMY was affected more than shoot DMY.

The CCBP's shown in Fig. 7-4 are unique in that BP-#4, BP-#10, and BP-#12 are components making up a final product BP-#5. Plants grown with BP-#4 (high-sulfite FGD-BP) had shoot and root DMY below that of plants grown in unamended acid soil at levels >1%. Plants grown with BP-#10 (a CaO material) at 1% had about twofold higher shoot and root DMY than plants grown in unamended acid soil, and DMY decreased as level increased. Plants grown with BP-#12 (Class-F FA) showed consistent increases in shoot

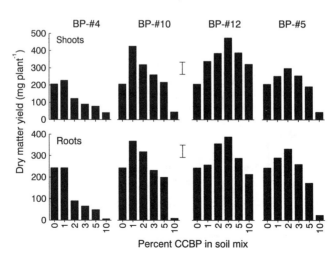

Fig. 7-4. Shoot and root DMY of maize grown with the selected CCBP's BP-#4, BP-#10, BP-#12 and BP-#5 (from the same location) on acid soil. Vertical bars are LSD (0.05) values.

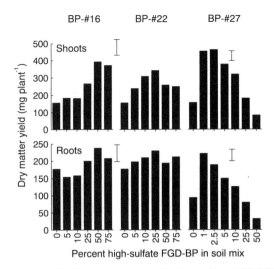

Fig. 7–5. Shoot and root DMY of maize grown with high-sulfate FGD-BP's BP-#16, BP-#22 and BP-#27 on acid soil. Vertical bars are LSD (0.05) values.

and root DMY up to 3% BP before decreasing at the higher levels to values similar to that of plants grown in unamended soil. Plants grown with BP-#5 had slight increases (about 1.5-fold) in shoot and root DMY up to 2% compared to those of plants grown in unamended soil, and decreased as level increased. Only limited growth occurred at the highest level (10%) of BP-#5, which was similar to plants grown with BP-#4 and BP-#10 at these levels.

Shoots showed no significant increase in DMY until BP-#16 (high-sulfate FGD-BP) was added at 25% (Fig. 7–5). An additional increase in shoot DMY occurred when 50% BP-#16 was added, and DMY was not further affected by 75%. Maximum shoot DMY at 50% BP-#16 was about 2.5-fold higher than for plants grown in unamended soil. Root DMY of plants grown with BP-#16 showed slight but insignificant decreases at 5 and 10%, and increased at higher levels (25, 50, and 75%). Maximum root DMY for plants grown with BP-#16 was only slightly above that of plants grown in unamended soil. Shoot DMY of plants grown with BP-#22 (high-sulfate FGD-BP) increased consistently as level increased to 25% before declining slightly at the highest levels (50 & 75%) (Fig. 7–5). Maximum shoot DMY for plants grown with BP-#22 was similar to maximum shoot DMY for plants grown with BP-#16. Root DMY of plants grown with BP-#22 were similar over all levels used, although slightly higher than that of plants grown in unamended soil. Shoot DMY increased 2.9-fold and root DMY increased 2.4-fold when BP-#27 was added to acid soil at only 1%, remained about as high up to 2.5% BP in soil, and declined consistently as level increased from 5 to 50% (Fig. 7–5). At the highest level of BP-#27, shoot DMY was 53% and root DMY was 34% of plants grown in unamended soil. Shoots were affected more than roots when plants were grown with BP-#16 and BP-#22, and shoots and roots were affected similarly when plants were grown with BP-#27.

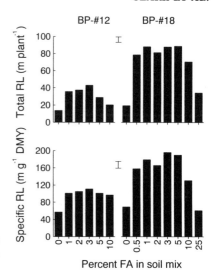

Fig. 7-6. Total and specific RL of maize grown
with the FA's BP-#12 and BP-#18 on acid soil.
Vertical bars are LSD (0.05) values.

Percent FA in soil mix

Leaf Areas and Root Lengths

Leaf areas of plants grown with the various CCBP's followed shoot
DMY sufficiently close that these data have not been reported. Leaf area
per gram shoot DMY was relatively constant over all experiments and the
mean value for plants grown in unamended soil was 35.8 m^2 kg^{-1} shoot
DMY.

Total RL (a developmental trait) did not particularly follow root DMY
patterns, and specific RL (a morphological trait) was considerably different
from total RL. Total RL for plants grown with FA increased about three-
to fourfold above that of plants grown in unamended soil, and decreased
at the highest levels (Fig. 7-6). Even at the highest levels of FA used, total
RL did not decrease below that noted for plants grown in unamended soil.
Plants grown with BP-#18 had higher total RL than plants grown with BP-#12
at comparable levels of BP added. Specific RL for plants grown with various levels of BP-#12 were about twofold higher than for plants grown in
unamended soil, and remained relatively constant over all levels used (Fig.
7-6). Specific RL for plants grown with 0.5 to 10% BP-#18 were about
threefold higher than for plants grown in unamended soil, and was relatively constant over these levels. Specific RL consistently decreased for plants
grown with 10 and 25% BP-#18, but these RL values were not below those
of plants grown in unamended soil. BP-#18 was more effective in enhancing
specific RL than BP-#12.

Total RL was three- to fivefold higher for plants grown with BP-#15
and BP-#21 at levels up to 1% than for plants grown in unamended soil,
and decreased as level of these BP's increased (Fig. 7-7). Total RL of plants
grown with BP-#15 was higher than that for BP-#21 added at similar levels
above 2%. Total RL of plants grown with BP-#26 increased about fourfold
above plants grown in unamended soil when 0.5% was added to soil, but

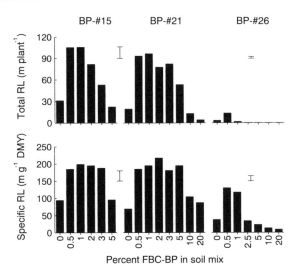

Fig. 7-7. Total and specific RL of maize grown with the FBC-BP-s BP-#15, BP-#21, and BP-#26 on acid soil. Vertical bars are LSD (0.05) values.

decreased extensively as level of this BP increased above the 0.5% level (Fig. 7-7). In fact, total RL was only about one-half that of plants grown in unamended soil at 1% BP-#26 and only 3 to 6% that of the unamended soil at the higher levels (0.5–20%). Total RL of plants grown with BP-#26 was markedly lower than that for BP-#15 and BP-#21. Specific RL was about twofold higher for plants grown with both BP-#15 and BP-#21 than for plants grown in unamended soil, and remained relatively constant over the fairly wide range of levels used before declining at the highest level (5% for BP-#15 and 10 and 20% for BP-#21) (Fig. 7-7). Specific RL for plants grown with 0.5 and 1% BP-#26 was about threefold higher than for plants grown in unamended soil, but decreased damatically to values below that of unamended soil when added at 2.5 to 20% (Fg. 7-7). BP-#15 and BP-#21 were similar in effectiveness for enhancing specific RL, but BP-#26 was not as effective as BP-#15 and BP-#21 for enhancing specific RL.

Total RL increased to a maximum at 1% levels of the high-sulfite FGD-BP's (BP-#1, BP-#2, and BP-#6) and decreased as level increased (Fig. 7-8). In contrast, total RL increased as level of BP-#8 increased to 2% before declining. The BP's to cause highest total RL followed a sequence of BP-#8 > BP-#6 ≥ BP-#2 > BP-#1. Specific RL increased slightly as level of these high-sulfite FBC-BP's increased, and remained higher than for plants grown in unamended soil even at the highest level of 5% (Fig. 7-8). Effectiveness of these FGD-BP's for enhancing specific RL followed the same sequence as for enhancing total RL.

Maximum total RL was obtained for plants grown with CCBP's BP-#4 and BP-#10 at 1% before declining as level of BP in the soil mix increased (Fig. 7-9). BP-#4 and BP-#10 showed dramatic increases in total RL when plants were grown at >1% of these BP's. Maximum total RL was 1.6-fold

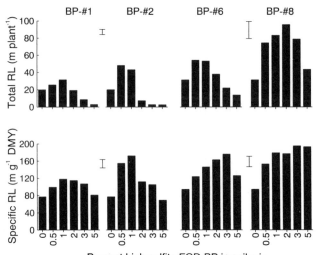

Fig. 7–8. Total and specific RL of maize grown with high-sulfite FGD-BP's BP-#1, BP-#2, BP-#6, and BP-#8 on acid soil. Vertical bars are LSD (0.05) values—BP-#1 and BP-#2 had same LSD and BP-#6 and BP-#8 had same LSD.

(BP-#4) and fourfold (BP-#10) higher than total RL for plants grown in unamended soil. Total RL for plants grown with 1 to 3% BP-#12 were about threefold higher than for plants grown in unamended soil, and declined as level increased to 5 and 10% (Fig. 7–9). Total RL of plants grown with BP-#5 increased to a maximum of about threefold higher than that of plants grown in unamended soil at 2% before declining at higher levels (Fig. 7–9).

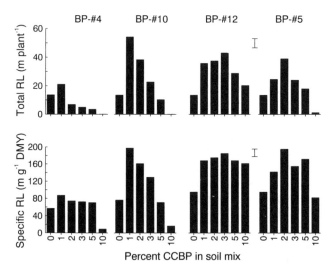

Fig. 7–9. Total and specific RL of maize grown with the selected CCBP's BP-#4, BP-#10, BP-#12, and BP-#5 (from the same location) on acid soil. Vertical bars are LSD (0.05) values.

The decrease in total RL at the highest level (10%) of BP-#5 was almost as dramatic as the decreases due to the highest level (10%) of BP-#4 and BP-#10. Specific RL increased slightly when plants were grown with the lowest level of BP-#4 (1%) and remained fairly constant even at levels as high as 5% before declining eightfold at the 10% level (Fig. 7–9). Specific RL increased and was highest at 1% BP-#10 (~2.5-fold higher than for plants grown in unamended soil) before declining as level increased. The specific RL values at the highest levels of BP-#4 and BP-#10 were only about 20% of those for plants grown in unamended soil. Specific RL for plants grown with BP-#12 was about twofold higher than for plants grown in unamended soil, and remained relatively constant over all levels. Specific RL for plants grown with BP-#5 increased consistently at 1 and 2% levels, remained fairly constant as level increased to 5%, and decreased to that of plants grown in unamended soil at the highest level (10%).

Total RL decreased slightly at the 5% level of the high-sulfate BP-#16 before increasing steadily as level increased to a maximum at 50% (about threefold higher than plants grown in unamended soil), and decreased slightly below the 50% maximum at the highest level (75%) (Fig. 7–10). Total RL of plants grown with high-sulfate BP-#22 increased to a maximum of 3.3-fold higher than that for plants grown in unamended soil at 25% before decreasing slightly as level increased to 50 and 75% (Fig. 7–10). Total RL of plants grown with 1 and 2.5% of the high-sulfate BP-#27 was over sevenfold higher than for plants grown in unamended soil, and decreased consistently as level of BP increased from 5 to 50% (Fig. 7–10). Total RL of plants grown with 50% BP-#27 was only about one-half that of plants grown with unamended soil. Specific RL increased only after 10% of BP-#16 was added and increased thereafter as level increased to a maximum (twofold increase above that of plants grown in unamended soil) at 50 and 75% (Fig. 7–10). Specific RL

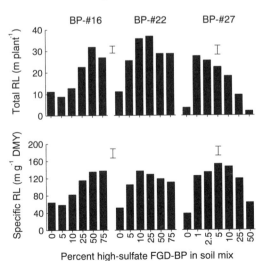

Fig. 7–10. Total and specific RL of maize grown with the high-sulfate FBC-BP's BP-#16, BP-#22, and BP-#27 on acid soil. Vertical bars are LSD (0.05) values.

consistently increased and reached a maximum (2.3-fold increase) at 10% BP-#22 and decreased only slightly as level increased to 75%. Specific RL of plants grown with 1 to 2.5% BP-#27 was similar (three to fourfold higher than for plants grown with unamended soil), but decreased dramatically at the highest level (50%). Except for BP-#27 at the highest level (50%) and BP-#16 at the lowest levels (5 and 10%), the three high-sulfate FGD-BP's were equally effective in enhancing specific RL for maize. BP-#27 and BP-#22 were more effective than BP-#16 for enhancing specific RL at the low levels added. Even at the highest level of BP-#27, specific RL was still above that of plants grown in unamended soil.

Soil pH and Electrical Conductivity

Soil pH and EC values were similar before and after plants were grown in pots. pH-CaCl$_2$ values were lower (pH 3.9) than pH-water (pH 4.3) values only in unamended soil after plants had grown in the pots. Because of similarities between pH and EC values before and after plants were grown, only pH-CaCL$_2$ and EC soil values after plants had been grown in pots have been reported. pH-CaCl$_2$ for unamended soil ranged from 3.82 to 4.03 in each experiment (Tables 7–8 to 7–11). Soil pH increased to 4.4 when BP-#12

Table 7–8. pH-CaCl$_2$ and EC of acid soil mixed with different levels of FA's and FBC-BP's.

Trait	By-product in soil mix	FA's		FBC-BP's		
		BP-#12	BP-#18	BP-#15	BP-#21	BP-#26
	%			pH		
pH-CaCl$_2$	0	3.91	4.03	4.00	4.03	3.82
	0.5		4.64	4.88	4.72	4.60
	1	4.02	5.06	5.70	5.26	5.12
	2	4.04	5.43	7.09	6.08	
	2.5					6.78
	3	4.13	5.88	7.61	6.54	
	5	4.44	6.33	8.19	6.74	7.80
	10	4.82	7.52		7.19	8.68
	20				8.25	10.40
	25		7.60			
	LSD (0.05)	0.05	0.05	0.05	0.05	0.27
				dS m^{-1}		
EC	0	0.17	0.12	0.09	0.12	0.10
	0.5		0.15	0.45	0.62	0.61
	1	0.35	0.19	0.98	1.34	1.10
	2	0.31	0.35	1.75	1.87	
	2.5					1.20
	3	0.34	0.56	1.65	2.00	
	5	0.56	0.97	1.83	2.09	1.34
	10	0.94	1.54		2.12	1.65
	20				2.50	1.96
	25		6.47			
	LSD (0.05)	0.09	0.28	0.20	0.20	0.21

Table 7-9. pH-CaCl$_2$ and EC of acid soil mixed with different levels of high-sulfite FGD-BP's.

Trait	FGD-BP in soil mix	FGD-BP's			
		BP-#1	BP-#2	BP-#6	BP-#8
	%			pH	
pH-CaCl$_2$	0	3.94	3.94	4.00	4.00
	0.5	4.13	4.24	4.10	4.23
	1	4.19	4.44	4.19	4.29
	2	4.38	5.13	4.29	4.58
	3	4.53	5.52	4.42	4.78
	5	4.94	6.46	4.64	5.38
	LSD (0.05)	0.09	0.09	0.05	0.05
				dS m^{-1}	
EC	0	0.11	0.11	0.09	0.09
	0.5	0.53	0.56	0.88	0.63
	1	0.99	1.46	1.37	1.13
	2	1.26	1.31	1.47	1.53
	3	1.31	1.47	1.74	1.78
	5	1.46	1.98	1.74	2.08
	LSD (0.05)	0.20	0.20	0.20	0.20

(Class-F FA) was added at the 5% level, and to 6.3 and 7.6 when BP-#18 (Class-C FA) was added at the 5 and 25% levels, respectively (Table 7–9). Soil pH increased to 8.2 when 5% BP-#15 and 20% BP-#21 were added and to 10.4 when 20% BP-#26 was added to soil (Table 7–8). Soil pH increased from addition of up to 5% high-sulfite FGD-BP's as follows: 4.9 for BP-#1, 6.5 for BP-#2, 4.6 for BP-#6, and 5.4 for BP-#8 (Table 7–9). Soil pH increases from addition of up to 10% of selected CCBP's were: 5.6 for BP-#4, 9.8 for BP-#10, 4.8 for BP-#12, and 5.9 for BP-#5 (Table 7–10). Soil pH

Table 7-10. pH-CaCl$_2$ and EC of acid soil mixed with different levels of selected CCBP's.

Trait	CCBP in soil mix	CCBP's			
		BP-#4	BP-#10	BP-#12	BP-#5
	%			pH	
pH-CaCl$_2$	0	3.91	3.91	3.91	3.91
	1	4.06	6.34	4.02	3.97
	2	4.34	7.71	4.04	4.17
	3	4.59	8.14	4.13	4.38
	5	5.06	8.41	4.44	4.97
	10	5.62	9.82	4.82	5.93
	LSD (0.05)	0.28	0.28	0.28	0.28
				dS m^{-1}	
EC	0	0.17	0.17	0.17	0.17
	1	1.08	0.38	0.35	1.06
	2	1.57	0.47	0.31	1.15
	3	1.25	0.60	0.56	1.48
	5	1.63	0.60	0.31	1.54
	10	3.00	0.75	0.94	2.31
	LSD (0.05)	0.09	0.09	0.09	0.09

Table 7-11. pH-CaCl$_2$ and EC of acid soil mixed with different levels of high-sulfate FGD-BP's.

Trait	FGD-BP in soil mix	FGD-BP's		
		BP-#16	BP-#22	BP-#27
	%		pH	
pH-CaCl$_2$	0	4.00	4.00	3.82
	1			4.22
	2.5			4.60
	5	4.07	4.29	5.38
	10	4.18	4.36	6.40
	25	4.26	4.78	7.50
	50	4.65	5.77	8.30
	75	5.52	6.54	
	LSD (0.05)	0.06	0.06	0.27
			dS m^{-1}	
EC	0	0.11	0.11	0.10
	1			1.22
	2.5			1.62
	5	1.48	1.22	2.05
	10	1.67	1.09	2.38
	25	1.62	1.14	2.72
	50	1.71	1.22	3.20
	75	1.68	1.20	
	LSD (0.05)	0.18	0.18	0.21

increased to 5.5 when 75% BP-#16 was added, to 6.5 when 75% BP-#22 was added, and to 8.3 when 50% BP-#27 was added (all high-sulfate FGD-BP's) (Table 7-11).

Soil EC of unamended soil ranged from 0.09 to 0.17 in the experiments (Tables 7-8 to 7-11). Soil EC increased to 0.6 at 5% BP-#12 and to 1.0 at 5% and to 6.5 at 25% BP-#18 (Table 7-8). Soil EC increased to 1.8 when BP-#15 was added at 5%, to 2.5 when BP-#21 was added at 20%, and to 2.0 when BP-#26 was added at 20% (Table 7-8). Increases in soil EC when high-sulfite FGD-BP's were added up to 5% in the soil mix were 1.5 for BP-#1, 2.0 for BP-#2, 1.7 for BP-#6, and 2.1 for BP-#8 (Table 7-9). Soil EC increased to 3.0 for BP-#4, 0.8 for BP-#10, 0.9 for BP-#12, and 2.3 for BP-#5 when these CCBP's were added to soil mixes at 10% (Table 7-10). Soil EC increased to 1.7 for BP-#16 and 1.2 for BP-#22 added at 75% and to 3.2 for BP-#27 added at 50% (Table 7-11). Soil EC values for BP-#16 and BP-#22 added from 5 to 75% were relatively constant (a large increase in soil EC occurred between 0 and 5% of added BP), but increased consistently as level of BP-#27 increased.

DISCUSSION

The maize hybrid used in these studies grew relatively well on the unamended acid Porters soil. This soil was chosen because of its known Al toxicity and Ca deficiency on several plant species. Sudangrass [*Sorghum bicolor*

(L.) Moench] and white clover (*Trifolium repens* L.) grown on this acid Porters soil did not grow well without added lime (V.C. Baligar & K.D. Ritchey, 1993, personal observations). No information was available about the tolerance of (PA329 × PA353P) maize to Al toxicity or acid soil conditions (W.M. Johnson, 1993, personal communication), but results of these experiments showed that this hybrid was fairly tolerant to this acid soil. This might be suspected since both inbreds making up the hybrid were developed in Pennsylvania where acid soils exist.

Levels of 10 to 25% FA added to soil mixes might be considered high, especially if they are for field application. Fly ashes are known to contain high levels of soluble salts (particularly B) and decreases in plant DMY have been reported at high levels of these materials (Carlson & Adriano, 1993). BP-#12 (Class-F FA) contained among the highest B concentrations noted for the CCBPs (Table 7–3).

The levels of the FBC-BP's used in these greenhouse studies appeared to be higher than those normally used in field and some other studies (Korcak, 1993; Stout et al., 1979; Terman et al., 1978), although high levels of FBC-BP used in one study (1.12 Mg ha^{-1} added twice or 560 Mg ha^{-1} added once or twice to field experiments) had detrimental effects on maize and soybean [*Glycine max* (L.) Merr.] growth (Mays et al., 1991). BP-#26 decreased maize dry matter and grain yields over twofold the 1st yr of field application (13.6 kg ha^{-1}) on an acid soil (K.D. Ritchey, 1993, unpublished data).

The high-sulfite FGD-BP's most likely contained sufficient sulfite to have detrimental effects on maize DMY. BP-#2 produced such pungent S odors that a special isolated area had to be used when this FGD-BP was dried. Volatile S compounds are common in many FGD-BP's (Adams & Farwell, 1981). Maize grown with chemical grade $CaSO_3$ at 0.25% in soil mix reduced shoot DMY 68% and root DMY 64% compared to plants grown in unamended acid Porters soil (Clark et al., 1994). Higher levels of chemical grade $CaSO_3$ caused dramatic DMY decreases, and plants grown with 4% $CaSO_3$ added to acid Porters soil produced toxic SO_2 when soil pH was low (4.2), but not when soil pH was near 7.0 (Ritchey et al., 1994). The results from our studies support the concept that the detrimental compound in the high-sulfite FGD-BP's tested was most likely sulfite.

Shoot and root DMYs were severely depressed when the highest level of BP-#4 (high-sulfite FGD-BP) was added, but DMY was not depressed as markedly when BP-#5 (final BP from a combination of BP-#4, BP-#10, and BP-#12) was added (Fig. 7–4). The DMY of plants grown with BP-#5 at its highest level was almost nil, and similar growth responses were noted for BP-#4 and BP-#10 at their highest levels. BP-#10 (a CaO material) and BP-#12 (Class-F FA) appeared to have some ameliorating effects on BP-#4 when these BP's were combined to produce BP-#5. Although the ratio of each BP added to make BP-#5 is unknown, it appeared that the DMY responses to BP-#5 were more closely related to those of BP-#4 and BP-#10 than to BP-#12.

The slight decrease in root DMY and no increase in shoot DMY at low levels of BP-#16 (5 & 10%) were real and verified when maize was grown with chemical grade $CaSO_4$ (Clark et al., 1994). Maize grown with chemical grade $CaSO_4$ at levels of 0.5 to 1.0% in soil mixes had shoot DMY about half and root DMY about one-third those of plants grown on unamended acid soil. Shoot and root DMY of plants grown with chemical grade $CaSO_4$ increased thereafter as level increased from 2 to 4% to become comparable to that of plants grown in unamended soil. The DMY above that of plants grown in unamended acid soil was not obtained until BP-#16 was added above 5% for shoots and 10% for roots (Fig. 7–5). Above 10% levels of BP-#16, shoot and root DMY continued to be fairly high and was similar to DMY for BP-#22. Reason for the reduced DMY at low levels of BP-#16 was likely because of displacement of Al by $CaSO_4$ from soil exchange sites into soil solution which could enhance Al toxicity. Wheat (*Triticum aestivum* L.) root enlongation decreased in an acid soil with addition of $CaCl_2$ and $CaSO_4$ which increased soil solution Al; $CaCl_2$ had greater effects than $CaSO_4$ (Wright et al., 1989). When BP-#16 or BP-#22 were incorporated or surface applied to an acid Porters soil in columns and leached with water, relatively high amounts of Al were noted in leachates (Wendell & Ritchey, 1993). Reasons for no reductions in DMY with BP-#22 at the 5 to 10% levels of application were likely because this BP contained sufficient alkaline elements to inactivate Al and by its relatively high $CaCO_3$ equivalency (Clark et al., 1994). Additional studies showed that maize grown with BP-#22 added at 0.5% in soil mixes had lower DMY than plants grown in unamended soil (R.B. Clark & S.K. Zeto, 1993, unpublished data).

The large enhancement of DMY at 1 and 2.5% BP-#27 was most likely because of added Mg to this BP. When Mg was added to acid Porters soil at low levels, DMY increased extensively above plants grown in unamended soil until Ca/Mg ratios became abnormal (R.B. Clark & S.K. Zeto, 1993, unpublished data). Magnesium can be a limiting nutrient for plants grown on many acid soils (see Adams, 1994). Once Mg became high and above normal levels in BP-#27, DMY decreased consistently as level increased (Fig. 7–10).

The responses of maize to the high-sulfate FGD-BP's used in our studies indicated that DMY enhancement was not caused by changes in water availability since water was not limiting in these studies (plants wre not allowed to undergo water stress). The enhanced growth effects of the high sulfate FGD-BP's were most likely because of enhanced beneficial nutrients or reduced Al toxicity.

Morphological traits are sometimes preferred to developmental traits to assess root growth characteristics (Keltjens, 1987; Tan et al., 1992). Total RL reflects developmental while specific RL reflects morphological traits. High specific RL indicates small diameter, fine, highly branched, and long roots while low specific RL indicates large diameter, stubby, poorly branched, and short roots. Except for BP-#26 above 1%, BP-#4 and BP-#10 at their highest levels (10%), and BP-#27 at 50%, the CCBP's used in these studies did not cause specific RL to be below that noted for plants grown in unamended soil. That is, the CCBP's tended to enhance or maintain favorable morphological root characteristics (fine, highly branched, and small diameter roots) even though total RL was reduced. Maize grown with Al will usually have stubby roots (Clark, 1977; Rhue & Gogan, 1977), and plants grown in this unamended acid soil known for its Al toxicity would likely have stubbier roots than plants grown in acid soil amended with CCBP's. Roots of plants grown in the unamended acid Porters soil showed some stubbing, but not as bad as for more Al sensitive plants like orchardgrass (*Dactylis glomerata* L.) and white clover (R.B. Clark & S.K. Zeto, 1993, personal observations).

Soil pH generally increased as level of CCBP increased, and the FBC-BP's (BP-#15, BP-#21, and BP-#26), the CaO material (BP-#10), the Class-C FA (BP-#18), and BP-#27 (FGD-BP) caused the greatest pH increases (above pH 7). High soil pH values also were noted when FBC-BP's (pH from 6.5->7) and LIMB FGD-BP (pH > 8) were added to an acid soil (pH 4.6) at levels of 2.8% in soil mixes (Sutton & Stehouwer, 1992). Increases in soil pH by most of the CCBP's used in our studies would probably not cause nutrient imbalances unless they contained heavy metals, toxic B, imbalanced Ca/Mg ratio, or insufficient mineral nutrients to sustain plant growth. Visual P-, Mg-, Ca-, and Zn-deficiency and Al toxicity symptoms were noted on leaves of plants grown with many of the BP's. Boron toxicity symptoms were noted on plants grown with high levels of FA, and FA's are known to contain high B (Carlson & Adriano, 1993). Some BP's had low P, Mg, Ca, and Zn and high B (Table 7–3). Magnesium deficiencies were particularly common when plants were grown with $CaCO_3$ (lime) (R.B. Clark & S.K. Zeto, 1993, visual observations).

Salinity problems do not usually occur in plants until soil EC values are greater than about 1.5 (salt sensitive), 3.5 (moderately salt sensitive), or 6.5 (moderately salt tolerant) dS m^{-1} (Maas, 1990). Most of the CCBP's at the highest levels used in our studies could potentially decrease yields of salt sensitive plants, but these high levels would not likely be used in the field. The BP's which raised soil EC values the most were the Class-C FA (BP-#18) to 6.5 when added at 25%, high-sulfate FGD BP-#27 to 3.2 when added at 50%, and high-sulfite FGD BP-#4 at 3.0 when added at 10%. Soil EC of BP-#18 was only about 1.0 when added at 5% in soil mixes. The high-sulfate FGD-BP's caused soil EC to rise to 1.7 (BP-#16) and to 1.2 (BP-#22) at 75%, and to 3.2 at 50% BP-#27. The high soil EC for BP-#27 was most likely caused by the high Mg [6% $Mg(OH)_2$] added to this BP. The soil EC values from the other two high-sulfate FGD-BP's would not likely be deleterious to growth of most salt sensitive plants.

These results provide information about the effects of CCBP's added to acid soil, especially FGD-BP's, on maize growth and on soil pH and EC under fairly optimal growth conditions. These experiments were conducted under relatively ideal conditions using disturbed soil compared to undisturbed soil and uncontrollable weather conditions in the field. Pots in which the plants were grown had water added manually to prevent leaching. Leaching would occur in the field, and some of these CCBP's likely contain compounds that reduce plant growth unless removed. These data about CCBP effects on plant growth and on soil pH and EC show that CCBP's are different from each other, and each CCBP will likely have different effects on plant growth and soil chemistry. High-sulfate FGD-BP's appear to have considerable promise for use even at very high levels (BP-#16 and BP-#22) on acid soil without causing deleterious effects on maize growth and on soil pH and EC, but the level of BP-#27 would need to be monitored because of its high Mg.

ACKNOWLEDGMENTS

We thank Dr. D.P. Bligh, Ms. B.A. White, Ms. J.D. Snuffer, and Mr. J. College for their assistance in chemical analyses of the CCBP's. We also thank Ms. S.S. Boyer for preparation of the figures.

REFERENCES

Adams, D.F., and S.O. Farwell. 1981. Sulfur gas emissions from stored flue gas desulfurization sludges. J. Air Pollut. Control Assoc. 31:557–564.

Adams, F. (ed.). 1984. Soil acidity and liming. 2nd ed. Agron. Monogr. 12. ASA, CSSA, and SSSA, Madison, WI.

Alcordo, I.S., and J.E. Rechcigl. 1993. Phosphogypsum in agriculture: A review. Adv. Agron. 49:55–118.

American Standard Test Methods. 1985. Standard specification for agricultural liming materials, Designation: C602-69 (reapproved 1985). p. 297–302. In Annual book of ASTM standards. ASTM, Philadelphia, PA.

Carlson, C.L., and D.C. Adriano. 1993. Environmental impacts of coal combustion residues. J. Environ. Qual. 22:227–247.

Clark, R.B. 1977. Effect of aluminum on growth and mineral elements of Al-tolerant and Al-intolerant corn. Plant Soil 47:653–662.

Clark, R.B., S.K. Zeto, K.D. Ritchey, R.R. Wendell, and V.C. Baligar. 1994. Effects of coal flue gas desulfurization by-products and calcium-sulfate, -sulfite, and -carbonate on maize grown in acid soil. In R.A. Date et al. (ed.) Plant-soil interactions at low pH: Principles and management. Proc. 3rd Int. Symp. Kluwer Acad. Publ., Dordrecht, the Netherlands. (In press.)

Electric Power Research Institute. 1988a. Method L4: Sulfate-total sulfur analysis by the gravimetric method. p. L4-1–L4-21. In FGD chemistry and analytical methods handbook. Vol. 2. EPRI, Palo Alto, CA.

Electric Power Research Institute. 1988b. Method M2: Sulfite analysis in scrubber liquors and slurry solids by iodine-thiosulfate titration. p. M2-1–M2-29. In FGD chemistry and analytical methods handbook. Vol. 2. EPRI, Palo Alto, CA.

Federal Register. 1990. Hazardous waste management system; identification and listing of hazardous waste; toxicity characteristic revision; final rule, Part V. Environmental Protection Agency. Fed. Reg. 55(126):26 986–26 998.

Keltjens, W.G. 1987. Nitrogen source and aluminum toxicity of two sorghum genotypes differing in aluminum susceptibility. J. Plant Nutr. 10:841–856.

Korcak, R.F. 1993. Utilization of fluidized bed combustion byproducts in horticulture. p. 21-1–21-9. In Proc. 10th Int. Ash Use Symp., Vol. 1, Orlando, FL. 18–21 January. Am. Coal Ash Assoc., Washington, DC.

Korcak, R.F., and W.D. Kemper. 1993. Long-term effects of gypsiferous coal combustion ash applied at disposal levels on soil chemical properties. Plant Soil 154:29–32.

Maas, E.V. 1990. Crop salt tolerance. p. 262–304. In K.K. Tanji (ed.) Agricultural salinity assessmment and management. Manuals & reports on engineering practice no. 71. ASCE, New York.

Mays, D.A., P.M. Giordano, and A.D. Behel, Jr. 1991. Impact of fluidized bed combustion waste on metal content of crops and soil. Water, Air, Soil Pollut. 57-58:307-317.

Mulhern, D.W., R.J. Robel, J.C. Furness, and D.L. Hensley. 1989. Vegetation of waste disposal areas at a coal-fired power plant in Kansas. J. Environ. Qual. 18:285-292.

Rhue, R.D., and C.O. Grogan. 1977. Screening corn for aluminum tolerance. p. 297-310. *In* M.J. Wright (ed.) Plant adaptation to mineral stress in problem soils. Cornell Univ. Agric. Exp. Stn.

Ritchey, K.D., T.B. Kinraide, R.R. Wendell, R.B. Clark, and V.C. Baligar. 1994. Strategies for overcoming temporary phytotoxic effects of calcium sulfite applied to agricultural soils. p. 457-462. *In* S.-H. Chiang (ed.) Proc. Int. Annu. Pittsburgh Coal Conf., 11th, Pittsburgh, PA. 12-16 September. Univ. Pittsburgh, PA.

Ritchey, K.D., J.E. Silva, and U.F. Costa. 1982. Calcium deficiency in clayey B horizons of Savanna Oxisols. Soil Sci. 133:378-382.

Shahandeh, H., and M.E. Sumner. 1993. Establishment of vegetation on by-product gypsum materials. J. Environ. Qual. 22:57-61.

Shainberg, I., M.E. Sumner, W.P. Miller, M.P.W. Farina, M.A. Pavan, and M.V. Fey. 1989. Use of gypsum on soils: A review. p. 1-111. *In* B.A. Stewart (ed.) Advances in soil science. Vol. 9. Springer-Verlag, New York.

Stehouwer, R.C., P. Sutton, and W.A. Dick. 1993. Fescue growth on acid mine spoil amended with FGD and sewage sludge. p. 50. *In* Agronomy abstracts, ASA, Madison, WI.

Stout, W.L., J.L. Hern, R.F. Korcak, and C.W. Carlson. 1988. Manual for applying fluidized bed combustion residue to agricultural lands. USDA-ARS No. 74. U.S. Gov. Print. Office, Washington, DC.

Stout, W.L., R.C. Sidle, J.L. Hern, and O.J. Bennett. 1979. Effects of fluidized bed combustion waste on the Ca, Mg, S, and Zn levels in red clover, tall fescue, oat, and buckwheat. Agron. J. 71:662-668.

Sutton, P., and R.C. Stehouwer. 1992. Use of flue gas desulfurization by-products as ag-lime substitutes. p. 293. *In* Agronomy abstracts. ASA, Madison, WI.

Tan, K., W.G. Keltjens, and G.R. Findenegg. 1992. Acid soil damage in sorghum genotypes: Role of magnesium deficiency and root impairment. Plant Soil 139:149-155.

Terman, G.L., V.J. Kilmer, C.M. Hunt, and W. Buchanan. 1978. Fluidized bed boiler waste as a source of nutrients and lime. J. Environ. Qual. 7:147-150.

U.S. Environmental Protection Agency. 1988. Wastes from the combustion of coal by electric utility power plants. EPA/530-SW-88-002. USEPA, Washington, DC.

U.S. Environmental Protection Agency. 1992. Method 3050A-Acid digestion of sediments, sludges, and soils (Revision 1 of 1986 version). p. 2050A-1-3050A-6. *In* Test methods for evaluating solid wastes, SW-846, 3rd ed. USEPA, Office of Solid Waste and Emergency Response, Washington, DC.

Wendell, R.R., and K.D. Ritchey. 1993. Use of high-gypsum flue gas desulfurization by-products in agriculture. p. 40-45. *In* S.-H. Chiang (ed.) 10th Annu. Int. Pittsburgh Coal Conf., Pittsburgh, PA. Univ. Pittsburgh, Pittsburgh, PA.

Wilson, G.W.T., B.A.D. Hetrick, and A.P. Schwab. 1991. Reclamation effects on mycorrhizae and productive capacity of flue gas desulfurization sludge. J. Environ. Qual. 20:777-783.

Wright, R.J., V.C. Baligar, K.D. Ritchey, and S.F. Wright. 1989. Influence of soil solution aluminum on root elongation of wheat seedlings. Plant Soil 113:294-298.

8 Improved Water and Nutrient Uptake from Subsurface Layers of Gypsum-Amended Soils

K. Dale Ritchey, C. M. Feldhake, and R. B. Clark

USDA-ARS
*Appalachian Soil and Water Conservation Research
Laboratory*
Beckley, West Virginia

D. M. G. de Sousa

EMBRAPA-CPAC
Planaltina, D.F. Brazil

In this chapter, we discuss the availability of by-product gypsum from the electric power generating industry and ways in which gypsum can mitigate subsurface soil acidity and enhance deeper plant rooting. Data from preliminary experiments on Appalachian Region soils and from long-term experiments on tropical soils of Brazil are used to illustrate potential benefits and problems from gypsum use. Recent efforts at diagnosing which soils will benefit from gypsum application and approaches to recommending application rates are reviewed.

AVAILABILITY OF COAL COMBUSTION BY-PRODUCT GYPSUM

Clean air legislation has resulted in coal-combustion power plant production of large volumes of high-Ca by-products. According to the American Coal Ash Association (ACAA, 1993), 16 400 000 t of flue gas desulfurization (FGD) residue were produced in 1991. Implementation of the Clean Air Act Amendments of 1990 will increase the amount of by-products. Phase I requires all electrical generating units larger than 100 MW which were emitting 2.5 lb SO_2 per million British Thermal Units or more in 1985 to halve their emissions by 1 Jan. 1995 (L. Pieper, personal communication, 1994). Phase II restrictions will apply to all electrical generating units greater than 25 MW by the year 2000.

Utilities have three choices for controlling the amount of SO_2 they emit: switching to a lower S coal, buying emission allowances, or retrofitting a scrubbing system. The most common scrubbers now in use are wet limestone or lime-based systems which treat flue gas after fly ash has been removed via electrostatic precipitators or baghouse systems. Wet limestone scrubbers with forced oxidation produce wallboard-quality gypsum. Several alternative scrubbing systems producing other types of by-products are under development, but there is a long lead time between their development and implementation. It is certain that considerable quantities of gypsum by-product will be available in the next decade from scrubbing systems currently operating or under construction.

At present, many high-Ca, high-S by-products are being stockpiled in lagoons or mixed with fly ash for storage in landfills. Utilization as a soil amendment would be a beneficial alternative to disposal. Clark et al. (1993) examined a number of coal combustion by-products, including Class-C and Class-F fly ashes, bottom ash, high calcium sulfite FGD residues, and fluidized bed combustion residues as possible soil amendments. They used maize grown in pots to investigate the effects of these by-products and found that wallboard-quality gypsums were among the most beneficial.

CROP PRODUCTION LIMITATIONS IMPOSED BY SUBSURFACE ACIDITY

Soil acidity is a serious limitation to crop productivity in many parts of the world including the eastern USA. The effects of low soil pH on plant growth arise from several factors. Two important components are Al toxicity and Ca deficiency. Both problems are relatively easy to correct in the plow layer of cultivated soils by incorporation of limestone if the land is not excessively steep. In established orchards or pastures, or on steep land, incorporation is more difficult. And, even on level land, incorporation of limestone into the plow layer will not readily correct subsurface acidity (Sumner, 1994). The principal results of high Al saturation and low Ca levels are decreased depth and density of plant root growth. Overcoming these limitations should result in deeper crop roots.

There are several important benefits of deep crop root systems:

1. During dry periods, roots growing in subsurface layers can exploit water from this part of the profile while shallower-rooted crops are limited to water uptake from the surface layer which is soon depleted of available moisture. Thus, deeper rooted plants suffer less from drought-induced water stress and associated yield reductions (Gonzalez-Erico et al., 1979).

2. Where deeper rooting has allowed the plow layer to remain moist, uptake of nutrients from this layer can continue. This is particularly important for P, the uptake of which depends mainly on diffusion and is therefore heavily controlled by soil water content.

3. Uptake of nutrients leached into subsurface layers will be greater with deep-rooted plants. This is especially relevant for N. During wet spring

weather, excessive rainfall can leach nitrate into the profile where it will per-colate into the groundwater unless deep roots can intercept it.

EFFECTS OF LIMESTONE AND GYPSUM IN OVERCOMING ACIDITY LIMITATIONS TO PLANT ROOT GROWTH

Limestone

Limestone ($CaCO_3$) incorporation is the most accepted way to over-come soil acidity. The reactions of the CO_3^{2-} with H^+ to form HCO_3^- and H_2CO_3 consume protons and raise the soil pH, resulting in precipitation of Al into relatively unavailable compounds. However, limestone is relatively immobile in soils and it tends to remain where it is applied.

Gypsum

Gypsum ($CaSO_4 \cdot 2H_2O$) differs from limestone in that it is consider-ably more soluble. Thus Ca and S can leach down through the soil profile after dissolving in rain water. The solubility of gypsum is 2.4 g L^{-1} (Weast, 1978, p. B-107). If maximum dissolution were obtained, 24 000 kg ha^{-1} gypsum could dissolve in 1020 mm rainfall, the average annual precipitation in the Appalachian Region. Although gypsum does not have the ability to neutralize Al in the same way as limestone, it does produce a number of ef-fects which can partially or fully mitigate the severity of several detrimental aspects of soil acidity (as discussed in recent reviews by Shainberg et al., 1989; McCray & Sumner, 1990; Alcordo & Rechigl, 1993).

Some ways in which gypsum decreases soil acidity are:

1. Gypsum incorporated into the plow layer dissolves allowing Ca to leach to subsoil horizons where native Ca levels are too low to support root growth. Calcium is not translocated downward within the plant, thus it must be present in the area around the root tip in adequate concentration for growth to occur (Bohnen, 1980). Exchangeable Ca levels as low as 0.02 $cmol_c$ kg^{-1} (meq 100 g^{-1}) have been reported in acid subsurface soils, such as in some forests of the southeastern USA and Oxisols of central Brazil and Colombia (Ritchey et al., 1989). This concentration is insufficient for root growth. In such soils, the downward leaching of a small amount of Ca can eliminate Ca deficiency (Ritchey et al., 1982). Soil solution levels of Ca are more ef-fectively increased with gypsum than with limestone, even within the plow layer (Shainberg et al., 1989). Peanut growers traditionally apply gypsum to provide the high levels of available Ca needed for development of under-ground peanut (*Arachis hypogea*) pods.

2. The higher levels of soluble Ca resulting from gypsum addition in-crease the ratio of Ca to Al activity in soil solution. Even at constant Al ac-tivity in soil solution, a higher concentration of Ca or other cations reduces Al toxicity to plants (Kinraide et al., 1992). One possible explanation for this is that the basic cations reduce cell-surface negativity and, consequent-

ly, the concentration of Al^{3+} at cell surfaces (Kinraide et al., 1992). Kinraide et al. (1994) indicated that Ca also may specifically mitigate H^+ toxicity.

3. Displacement of Al into soil solution by Ca supplied by gypsum may allow its physical removal from the profile by leaching, probably as aluminum sulfate. Oates and Caldwell (1985) showed over 40 mg mL^{-1} Al in leachates from a Gigger silt loam (Typic Fragiudalf) receiving 18 g kg^{-1} gypsum, and Wendell and Ritchey (1993) found similar levels for a Porters soil (Umbric Dystrochrept) treated with 10 g kg^{-1} gypsum. In general, this level of mobilization of Al has not been documented in the more oxidic soils such as those in the southeastern USA and central Brazil (Shainberg et al., 1989).

4. In the more oxidic soils, sorption of sulfate on the surfaces of hydrous oxides of Fe (Parfitt & Smart, 1978) or Al (Rajan, 1978) releases hydroxyl ions which raise the soil pH slightly and decrease Al solubility and availability. This "self-liming" effect of gypsum was described by Reeve and Sumner (1972) in Oxisols of South Africa.

5. Salt sorption (simultaneous sorption of both Ca and sulfate on soil surfaces) (Alva et al., 1991; Bolan et al., 1993) and exchange of Ca for Al raise the level of exchangeable Ca in subsoil. The increase in exchangeable Ca and the decrease in KCl-extractable Al result in lower effective Al saturation (ratio of KCl-extractable Al to the sum of exchangeable cations), which is probably the most generally useful parameter for characterizing the inhibition of plant root growth over a wide range of acid soils.

6. At high levels of sulfate and low pH levels, insoluble aluminum-sulfate compounds such as alunite, basalumite, and jurbanite may precipitate and thus reduce levels of toxic Al^{3+} in solution.

7. The presence of sulfate resulting from gypsum addition decreases the amount of free Al^{3+} present in soil solution because a portion of the soluble Al combines with sulfate to form ion pairs such as $AlSO_4^+$ (Pavan et al., 1982). The $AlSO_4^+$ is much less toxic to plants than Al^{3+} (Kinraide & Parker, 1987).

8. Gypsum increases the ionic strength of the soil solution and thus depresses the activity of Al^{3+}.

STUDIES IN APPALACHIAN REGION SOILS

Several gypsum effects can be demonstrated by a study using the Bt horizon of an acid Lily loam (Typic Hapludults, fine loamy, siliceous, mesic) from West Virginia (C.M. Feldhake & K.D. Ritchey, unpublished data, 1994). To simulate the effects of surface application of gypsum, columns filled to 15 cm with this soil were leached with 204 mm and 1020 mm of a saturated gypsum solution equivalent to 2.4 and 12 mo rainfall in the Appalachian region. The solution was prepared using wallboard-quality gypsum produced by an in-situ oxidation limestone-based FGD scrubber system. The quantity of gypsum added was the amount equivalent to 5 000 kg ha^{-1}

("5T") and 25 000 kg ha^{-1} ("25T"). An untreated control and a soil mixed with 4.7 cmol$_c$ calcium hydroxide kg^{-1} also were included.

Analysis of the soils, after termination of leaching, showed that the addition of gypsum in solution raised the electrical conductivity (EC) measured in 1:1 soil/water from 0.03 to 0.42 dS m^{-1} (Table 8-1). The gypsum treatments increased exchangeable Ca content in the soil by as much as 3.35 cmol$_c$ kg^{-1}. Part of the increase in "exchangeable" Ca could be attributed to Ca present in the soil solution, which contributes roughly 0.8 cmol$_c$ Ca kg^{-1} (multiplication of concentration of Ca in a saturated gypsum solution by the amount of water present in the soil at field capacity).

The KCl-extractable Al was decreased by 54% under the 25T gypsum leaching treatment (Table 8-1). Part of this decrease could be due to leaching out of Al displaced from the soil exchange complex. It could also be partially due to Al precipitation by hydroxyl ions released by sulfate from soil clays, or to precipitation of aluminum-sulfate minerals. According to equations presented by Sumner (1994), either of these last two mechanisms could explain the 0.19 to 0.48 unit increase in pH observed when soil reaction was measured in 0.01 M CaCl$_2$ to swamp out effects of Al displacement into solution by Ca. When measured in water, the pH of the 5T treatment soil was 0.17 units lower than the control. This pH drop could be attributed to the displacement of Al by Ca being greater than the displacement of hydroxyl by sulfate. At the 25T leaching rate, however, the pH measured in water was 0.07 units higher than in the control.

The net result of the gypsum addition was a marked decrease in Al saturation (Table 8-1) and an improvement in the ability of the soil to support root growth (Table 8-2). The improvement in the soil was shown using a bioassay method (Ritchey et al., 1989) in which the rate of root growth of sudangrass [*Sorghum bicolor* (L.) Moench] seedlings was measured over a 4-d period in small quantities of the treated soil. Growth improved 51 and 422% compared to the control for the 5T and 25T gypsum leaching treatments.

After leaching treatment, the soils were placed in 26-cm-deep pots and a 5-cm layer of limed and fertilized topsoil placed on the surface. The pots were sown to orchardgrass (*Dactylis glomerata* L.) and allowed to grow for

Table 8-1. Effects of treatment on pH, electrical conductivity, exchangeable Ca and Al, and Al saturation of the effective CEC in Lily Bt soil. Treatments consisted of control, soil leached with gypsum solution equivalent to 5 000 kg ha^{-1} (5T) or 25 000 kg ha^{-1} (25T) and soil mixed with 4.7 cmol$_c$ calcium hydroxide kg^{-1} (C.M. Feldhake & K.D. Ritchey, 1994, unpublished data).

Treatment	pH		EC 1:1 water	Ca	Al	Effective Al saturation
	0.01M CaCl$_2$	1:1 water				
			dS m^{-1}	cmol$_c$ kg^{-1}		%
0	3.84	4.29	0.03	0.1	4.6	95
5T	4.03	4.12	0.42	2.1	3.4	60
25T	4.32	4.36	0.42	3.4	2.1	33
Lime	5.13	5.36	0.12	4.0	0.5	11

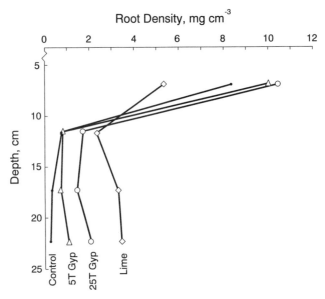

Fig. 8-1. Root density of orchardgrass grown in Lily Bt soil leached with gypsum solution equivalent to 5 000 kg ha^{-1} (5T) or 25 000 kg ha^{-1} (25T) and in soil mixed with 4.7 cmol$_c$ calcium hydroxide kg^{-1} (C.M. Feldhake & K.D. Ritchey, 1994, unpublished data).

15 wk to develop a uniform canopy. After subjecting the plants to drought by discontinuing watering, the root and water distributions were measured. The lime treatment (Fig. 8-1) gave the best distribution of roots, followed by the 25T gypsum leaching treatment. Mean root density was 62% higher in the 25T treatment than in the control. Water uptake efficiency was related to root distribution, with the 25T gypsum and lime treatments leaving the least amount of residual water in the lower levels of the pots (Fig. 8-2).

Foliar P was 34% higher in the orchardgrass leaves under the 25T gypsum treatment and foliar Mn was increased by 174% compared to the control (Table 8-2). This may reflect increased availability of these elements as well as higher rooting density. Increased P levels in maize (*Zea mays* L.) grown on gypsum-treated soils also have been observed (R. Clark, unpublished data, 1993). Phosphate and sulfate anions are sorbed by similar sites on soil mineral surfaces, although phosphate is held more tightly than sulfate. High levels of sulfate in soil solution in the gypsum-treated soil may increase competition for sorption sites, raising the level of phosphate in solution and allowing it to be taken up by plants more readily.

The improved rooting density and water uptake data from this experiment demonstrated that gypsum can help overcome limitations imposed by subsurface acidity of soils of the Appalachians. The mechanism of leaching of displaced Al may play a greater part in acidity mitigation in the Appalachian soils than in more oxidic soils. The USDA-ARS is now testing the effects of gypsum application on subsurface rooting in field experiments being conducted in the mountainous region of the eastern USA.

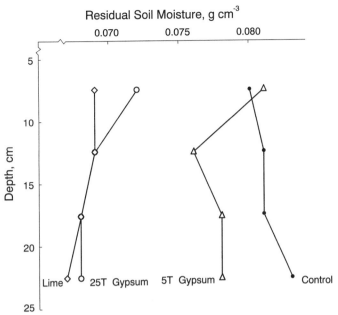

Fig. 8–2. Residual soil moisture in pots of Lily Bt soil leached with gypsum solution equivalent to 5 000 kg ha^{-1} (5T) or 25 000 kg ha^{-1} (25T) and in soil mixed with 4.7 cmol$_c$ calcium hydroxide kg^{-1} (C.M. Feldhake & K.D. Ritchey, 1994, unpublished data).

STUDIES IN HIGHLY WEATHERED SOILS

Gypsum effects on relatively highly weathered soils of the southeastern USA have been extensively studied by M.N. Sumner and his colleagues. Their results [summarized by Shainberg et al. (1989) and Sumner (1994)] showed that gypsum application generally improved rooting and yields.

The earliest and most extensive field research on calcium sulfate (CS) utilization following the publication by Reeve and Sumner (1972) of the

Table 8–2. Effects of treatment on 4-day bioassay sudangrass root growth and orchardgrass foliar concentrations of Mn, P, and K in Lily Bt soil. Treatments consisted of control, soil leached with gypsum solution equivalent to 5 000 kg ha^{-1} (5T) or 25 000 kg ha^{-1} (25T) and soil mixed with 4.7 cmol$_c$ calcium hydroxide kg^{-1} (C.M. Feldhake & K.D. Ritchey, 1994, unpublished data).

Treatment	Sudangrass root growth	Orchardgrass foliar concentration		
		Mn	P	K
	mm	mg kg^{-1}	——g kg^{-1}——	
0	11a†	208c	1.13c	11.0b
5T	17a	444b	1.52a	12.2b
25T	46b	570a	1.51a	15.6a
Lime	64c	213c	1.28b	14.6a

† Numbers in the same column followed by identical letters are not significantly different by the Tukey HSD test at 5%.

laboratory research showing its potential was conducted in Brazil (Shainberg et al., 1989). In the late 1970s, investigators at the EMBRAPA-CPAC Cerrados Agricultural Research Center observed that considerable improvements in drought tolerance and maize yield were exhibited in experiments where ordinary superphosphate (OSP) was compared to the same P application in the form of triple superphosphate. On investigation they found that the CS in the OSP had reduced subsurface soil acidity and improved deep rooting (Ritchey et al., 1980). A number of field trials on CS effectiveness have since been conducted by EMBRAPA and other Brazilian organizations, and, in general, economically beneficial plant responses were observed during drought years in the dystrophic Oxisols region in the center of the country (Vitti et al., 1992; Dematte, 1992). On the other hand, in the higher cation exchange capacity (CEC) soils of the more temperate soils in the south of Brazil, where droughts are less frequent, and soil water retention capacity higher, few responses to gypsum were found (Ernani et al., 1992).

The Oxisols of the Brazilian Cerrado are, in fact, particularly amenable to improvement by CS addition. The subsoil exchangeable Al content ranges from 0 to 3 $cmol_c$ kg^{-1}, exchangeable Ca levels are as low as 0.02 $cmol_c$ kg^{-1}, and the soils have high levels of iron and aluminum oxide minerals which produce pH increases when sulfate anions exchange with hydroxyls on their surfaces. Annual precipitation (1200 mm) falls during a 6-mo period, encouraging rapid downward movement of dissolved solutes. Oxidic mineralogy of the soil clays facilitates sulfate retention and resultant higher CEC. Calcium levels in virgin Cerrado profiles are often extremely low, thus a significant portion of the increase in plant root growth associated with CS addition may be due to meeting the nutritional requirement for Ca (Ritchey et al., 1982). The low CEC of the Oxisols of the Cerrados also means that the increases in Ca and decreases in Al levels caused by the addition of CS will have a proportionally larger effect than they would in a high CEC soil with greater absolute amounts of Al.

As a result of the widespread experimentation on gypsum use in Brazil, two volumes of research results containing 31 articles on gypsum use have been published, and gypsum application is now part of the official recommendation for low-CEC soils of the Cerrados (Sousa et al., 1993). Because little of the long-term research conducted in Brazil has been intensively reviewed in the USA, we will present results of several Brazilian experiments in some detail.

According to Sousa et al. (1992a), the rate of CS movement downward through Oxisol profiles and its persistence are affected by the amount of CS added, soil pH, amount and mineralogy of the clay fraction, pore size distribution, and the net rate of downward water movement. An example of base distribution resulting from CS addition is shown in Fig. 8–3, where Ca + Mg was measured by EDTA titration 5 yr after addition of 70, 140, and 873 kg ha^{-1} P as OSP, which contained CS equivalent to 436, 872, and 5438 kg ha^{-1} of $CaSO_4 \cdot 1/2 H_2O$ (Ritchey et al., 1980). The base status of the soil was greatly improved down to 105 cm. Sousa and Ritchey (1986) stated that leaching of sulfate was rapid at first but then slowed down. Figure 8–4

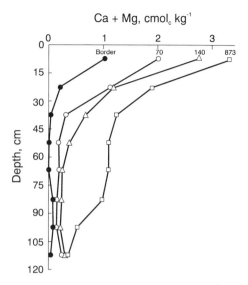

Fig. 8-3. Distribution of Ca + Mg in a clayey Typic Haplustox 5 yr after addition of various rates of P in the form of OSP, which supplied the equivalent of 436, 872, and 5438 kg ha^{-1} $CaSO_4 \cdot 1/2\ H_2O$ (Ritchey et al., 1980).

indicates relatively rapid downward movement of sulfate with the addition of 1154 and 2276 mm of water but little additional movement thereafter. The retention of sulfate is low at high pH levels (Couto et al., 1979), and his may help explain why little sulfate remained in the limed plow layer.

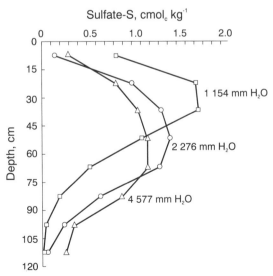

Fig. 8-4. Distribution of sulfate-S in a clayey Typic Haplustox receiving 6000 kg ha^{-1} $CaSO_4 \cdot 1/2\ H_2O$ as affected by increasing amounts of rain and irrigation water (Sousa & Ritchey, 1986).

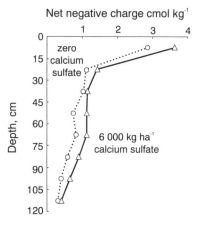

Fig. 8–5. Net negative charge as a function of depth in a clayey Typic Haplustox 11 mo after the addition of 6000 kg ha^{-1} CaSO$_4$ • 1/2 H$_2$O and 2276 mm of water (Sousa & Ritchey, 1986).

Souse and Ritchey (1986) found that the sorption of sulfate resulted in increased net negative charge in the subsurface layers of an Oxisol profile (Fig. 8–5), permitting higher levels of retained Ca. Calcium sulfate addition to Cerrado Oxisols generally resulted in pH (1:2 soil/H$_2$O) increases of 0.1 to 0.8 units (Ritchey et al., 1980), indicating that the clay mineralogy was such that hydroxyl displacement by sulfates exceeded Al displacement by Ca.

The cation distribution in a Typic Haplustox 8 yr after the addition of 6000 kg ha^{-1} CS (Fig. 8–6) illustrates the persistence of the effects of CS added in the form of Plaster of Paris, CaSO$_4$ • 1/2 H$_2$O (Sousa et al., 1992a). The exchangeable Ca levels in the 50- to 100-cm layers were more than eight times greater under the CS treatment. Aluminum was consistently lower in the CS plots except in the 20- to 40-cm layer where it was higher,

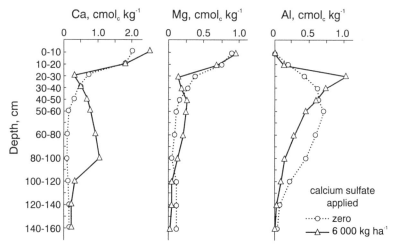

Fig. 8–6. Distribution of Ca, Mg and Al in a clayey Typic Haplustox 100 mo after application of 6000 kg ha^{-1} CaSO$_4$ • 1/2 H$_2$O (Sousa et al., 1992a).

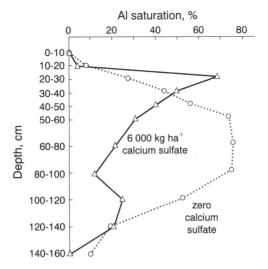

Fig. 8-7. Aluminum saturation in a clayey Typic Haplustox 100 mo after application of 6000 kg ha^{-1} CaSO$_4$ · 1/2 H$_2$O (Sousa et al., 1992a).

perhaps caused by intense cation uptake by crop roots or accumulation of protons produced by N fixation occurring just below the limed plow layer. Except for this layer, the overall change in the profile resulting from lower extractable Al and higher Ca was a great reduction in Al saturation (Fig. 8-7), with mean Al saturation in the 40- to 120-cm layers dropping from 66 to 24% in the treated soil. Mean CEC in this zone increased by 39% (Fig. 8-8). Root distribution of leucaena (*Leucaena leucocephala* cv. Cunningham) at the time of sampling reflected the improved growing conditions with three

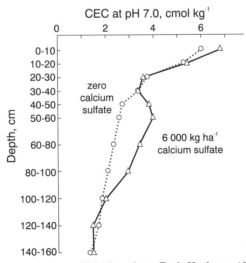

Fig. 8-8. Cation exchange capacity at pH 7.0 in a clayey Typic Haplustox 100 mo after application of 6000 kg ha^{-1} CaSO$_4$ · 1/2 H$_2$O (adapted from Sousa et al., 1992a).

Table 8-3. July 1990 air-dry weight of roots in 0.01 m^3 soil from various depths below leucaena (cv. Cunningham) planted in December 1986 as affected by CaSO$_4$ · 1/2 H$_2$O applied March 1982 (unpublished data of D.M.G. Sousa, 1993).

Depth	Calcium sulfate applied, kg ha^{-1}	
	0	6 000
cm	——————— mg cm^{-3} ———————	
0–10	18.3	32.9
10–20	6.7	13.1
20–30	1.9	3.2
30–40	2.1	4.4
40–50	0.4	4.4
50–60	0.4	3.2
60–80	0.6	7.9
80–100	0.9	5.4
100–120	0.5	3.5
120–140	0.5	2.8
140–160	0.6	2.6

to nine times more roots present in the 40- to 160-cm layers (Table 8-3). Leucaena yields measured over a period of 4 yr showed that the high CS rate in some cases had more than doubled productivity (Table 8-4).

The effects of CS application on other crop species can be illustrated by results of a long-term experiment presented by Sousa and colleagues (1986, 1992a). Treatment levels of 0, 2000, 4000, and 6000 kg ha^{-1} CS in the form of Plaster of Paris (21.7% S and 25.8% Ca), were incorporated into a clayey Dark Red Latosol (Typic Haplustox, fine, kaolinitic, isohyperthermic) in March 1982 at Planaltina, D.F. Brazil (15 °S, 47 °W). A blanket application of S was applied at 30 kg ha^{-1} to meet the nutritional needs of the crop. Thus, responses observed were not due to S, but to improved access to water and nutrients made available by deeper crop rooting resulting from overcoming Al toxicity and Ca deficiency in the subsurface horizons. Maize root distribution was greatly improved by CS (Fig. 8-9). The percentage of roots found below 45-cm depth increased at least 600% with the addition of 2000 kg ha^{-1} or more CS. There also was greater extraction of water at lower depths. For the 6000 kg ha^{-1} rate, water content at the 60- to 100-cm depth was 0.061 kg kg^{-1} less than that in the zero-CS treatment (Fig. 8-10), in-

Table 8-4. Yield of leucaena (cv. Cunningham) planted in December 1986 as affected by CaSO$_4$ · 1/2 H$_2$O applied in March 1982 (Sousa et al., 1992a).

Calcium sulfate applied	Yield				
	1987	1988	1989	1990	Total
	——————— 1000 kg ha^{-1} ———————				
0	1.5	8.6	3.1	5.4	18.6
2	2.8	9.7	4.8	9.5	26.8
4	3.8	10.9	4.5	10.6	29.8
6	3.6	10.8	6.8	12.4	33.6
LSD (5%)	1.8	1.6	1.5	2.1	

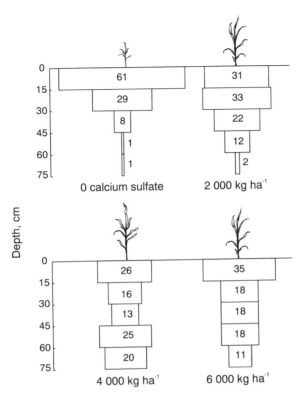

Fig. 8–9. Relative maize root distribution with depth (%) in the dry season of 1983 as affected by rates of $CaSO_4 \cdot 1/2\ H_2O$ applied March 1982 (Sousa et al., 1992a).

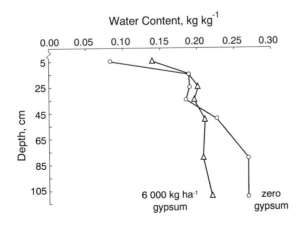

Fig. 8–10. Water content in a clayey Typic Haplustox sown to maize after 25 d without irrigation. $CaSO_4 \cdot 1/2\ H_2O$ was applied 17 mo previously (adapted from Sousa et al., 1992a).

Table 8-5. Maize grain yields as affected by levels of $CaSO_4 \cdot 1/2\ H_2O$ applied March 1982 (Sousa et al., 1992a).

| Calcium sulfate applied | Dry season Days without irrigation | | | | Rainy season | |
| | 1982 | | 1983 | | | |
	0	21	0	25	1984–1985	1987–1988
				$1000\ kg\ ha^{-1}$		
0	3.9	1.9	4.7	3.2	2.8	5.1
2	--	3.1	--	3.6	2.8	5.6
4	--	3.3	--	4.6	2.9	6.1
6	4.8	3.5	6.8	5.5	3.5	6.0
LSD (5%)	0.8	0.6	0.4	1.0	NS	0.6

dicating that the deeper rooted crop had been able to utilize the water unavailable to the plants in the shallower rooted treatment.

In the treatments where plants were subjected to drought by suspending irrigation for 21 d, CS considerably improved yields, with the 6000 kg ha^{-1} treatment plots yielding 72% more than the zero treatment in 1983 (Table 8-5). However, even with 4000 kg ha^{-1} CS, yield in the stressed treatment was not as great as in the zero-CS treatment which did not suffer drought. This indicates that CS amelioration of subsoil root growth reduces, but does not necessarily eliminate decreases in yield associated with droughts. The resulting improvement in yield stability from year to year is beneficial to farmers.

The 1983 yields in the plots which were not subjected to drought also were improved by the 6000 kg ha^{-1} CS treatment, with an increase of 45% over the control (Table 8–5). This may be partly explained by increased N uptake from the deeper rooting associated with the CS treatment. Sousa and Ritchey (1986) observed that the total N contained in grain and stover from the 6000 kg ha^{-1} treatment was 135 kg ha^{-1}, 44 kg ha^{-1} greater than that from the zero-CS treatment. This increase was due both to higher N concentration in the plant (Fig. 8-11) and higher biomass production. The amount of increased N taken up by the crop was nearly equivalent to the decrease of 48 kg ha^{-1} in extractable N measured in the soil profile (Fig. 8-12), indicating that the denser root system in the CS treatment was able to extract N from the profile that otherwise would not have been absorbed by the crop. This has important implications for U.S. agriculture where gypsum improvement of rooting depth may augment the uptake of nitrate leached below the plow layer, thus reducing possible contamination of groundwater.

The uptake of other nutrients also was improved by the CS treatment. Foliar concentrations of P, Ca, Cu, and Mn in the drought-stressed maize were increased in the CS treatments (Fig. 8-11). Phosphorous does not leach in these soils, and the increased P uptake was probably due to higher moisture contents in the plow layer of the CS treatment. The data in Fig. 8-10 illustrate that the plow layer tended to be dryer in the absence of CS compared to the high-CS treatment (0.08 vs. 0.14 kg kg^{-1}). Because the crop was able

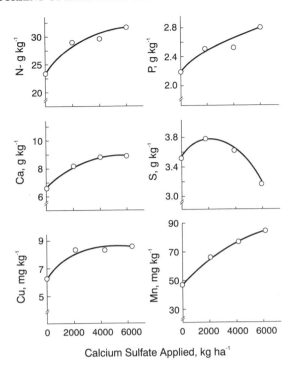

Fig. 8–11. Effect of $CaSO_4 \cdot 1/2 \ H_2O$ rate on foliar mineral levels of maize subjected to 25 d without irrigation in the dry season of 1983 (Sousa et al., 1992a).

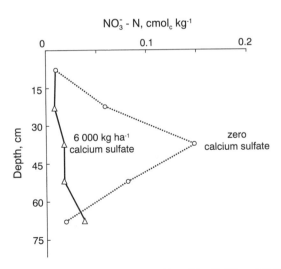

Fig. 8–12. Distribution of nitrate–N with depth as affected by $CaSO_4 \cdot 1/2 \ H_2O$ applied to a clayey Typic Haplustox. The profile was sampled 74 d after emergence of irrigated maize in the 1983 dry season (Sousa & Ritchey, 1986).

Table 8-6. Wheat grain yield during 1985 dry season as affected by $CaSO_4 \cdot 1/2 \ H_2O$ applied in March 1982 (Sousa et al., 1992a).

Calcium sulfate applied	Days without irrigation	
	0	15
	1000 kg ha^{-1}	
0	3.0	2.2
2	--	3.4
4	--	3.3
6	4.5	3.5
LSD (5%)	0.4	0.4

to extract water from deep in the profile, the soil in the CS plow layer did not dry out as much as that in the plow layer of the shallow-rooted zero-CS treatment. This higher level of moisture would have allowed continued diffusion of P to root surfaces longer than in the dryer soil.

Wheat yields in the 1985 dry season responded to the improved subsoil conditions resulting from CS addition (Table 8-6) with increased productivity of over 50%. Total nutrient uptake in wheat (*Triticum aestivum* L.) grain and straw also were improved (Table 8-7). The increase in N of 30 kg ha^{-1} in the 6000 kg ha^{-1} CS treatment over the zero treatment was similar to the 40 kg ha^{-1} decrease measured in soil N content down to 75-cm depth in the 6000 kg ha^{-1} treatment vs. the zero treatment where 67 kg ha^{-1} nitrate-N was present (Fig. 8-13).

Soybean (*Glycine max* L. Merr.) was grown in the experiment during the 1982-1983 and 1983-1984 rainy seasons and the 1986 dry season. There were no responses to CS during the two rainy seasons. Even in the irrigated crop, the difference between production obtained in the fully irrigated treatment and the treatment left unirrigated for 21 d was small (Table 8-8). The lack of marked response to CS may indicate that the soybean cultivar was somewhat tolerant to drought. Thus, gypsum use should not be expected to greatly increase yields for crop species and cultivars that are little affected by moderate water stress. Even in stress-sensitive cultivars, droughts occurring before initiation of the reproductive phase are less damaging than after flowering.

Table 8-7. Total nutrient content in wheat grain and straw grown during 1985 dry season as affected by $CaSO_4 \cdot 1/2 \ H_2O$ applied in March 1982. Irrigation was suspended 15 d during the flowering period (Sousa et al., 1992a).

Calcium sulfate applied	Mineral uptake					
	N	P	K	Ca	Mg	S
1000 kg ha^{-1}			kg ha^{-1}			
0	80	15	53	12	11	7
2	128	22	81	16	16	12
4	122	23	80	17	17	11
6	110	22	78	16	16	12

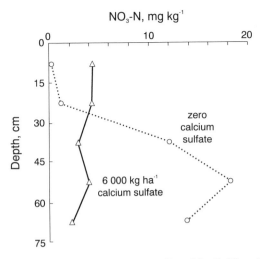

Fig. 8-13. Distribution of nitrate-N with depth as affected by $CaSO_4 \cdot 1/2\ H_2O$ applied to a clayey Typic Haplustox March 1982. The profile was sampled 71 d after emergence of irrigated wheat in the 1985 dry season (Sousa et al., 1992a).

Rice (*Oryza sativa* L.) was grown in the 1986–1987 wet season and did not respond positively to CS (Table 8-9). The upland rice varieties used in the Cerrado are normally tolerant to acid, low fertility conditions, and the subsoil in the experimental area after seven crops apparently was sufficiently ameliorated to supply the modest demands of this crop. These results underline the fact that gypsum should not be expected to have much beneficial effect in situations where root development at depth is already satisfactory.

Sorghum [*Sorghum bicolor* (L.) Moench], a drought-tolerant but Al-sensitive crop, was planted during the dry season of 1988. Yields were increased 24% by the residual effects of the high-CS treatment in the treatment where water stress was imposed, but the effect was not statistically significant, perhaps partially because of variability caused by bird damage (Table 8-10). Nitrogen concentration in the grain was significantly increased

Table 8-8. Soybean grain yield as affected by $CaSO_4 \cdot 1/2\ H_2O$ applied in March 1982 (Sousa et al., 1992a).

Calcium sulfate applied	Rainy season		1986 dry season Days without irrigation		
	1982–1983	1983–1984	0	21	42
			1000 kg ha^{-1}		
0	3.1	2.8	2.2	2.1	1.5
2	3.0	2.9	--	2.2	1.6
4	3.0	3.0	--	2.3	1.5
6	3.0	2.8	2.6	2.4	2.2
LSD (5%)	NS†	NS	NS	0.2	0.4

† NS = not significant.

Table 8-9. The 1986-1987 rainy season rice grain yield as affected by rates of $CaSO_4 \cdot 1/2\ H_2O$ incorporated in March 1982 (Sousa et al., 1992a).

Calcium sulfate applied	Routh rice grain yield
1000 kg ha^{-1}	
0	2.7
2	2.6
4	2.3
6	2.4
LSD (5%)	0.3

by the CS treatment (Table 8-10), and the total N uptake in the grain was up to 55% higher in the CS treatments (Sousa et al., 1992a).

MANAGEMENT OF PROBLEMS ASSOCIATED WITH APPLICATION OF GYPSUM

The majority of the effects of gypsum addition are beneficial, but problems may arise from its use in some situations.

1. The high concentration of Ca resulting from addition of gypsum may displace Al ions from the exchange complex into the soil solution. This effect would tend to cause pH to decrease, but it may at least partially be counterbalanced by a concomitant release of hydroxyls from adsorption of sulfate ions. Addition of gypsum at 20 to 30 g kg^{-1} to Porters soil caused small decreases (28%) in dry matter production of maize grown in unleached pots (Clark et al., 1993). This could be attributed to elevated levels of Al in soil solution. In the field where natural leaching would occur, the removal of Al from the profile would presumably eliminate such effects in the subsurface zone and plant rooting would increase. This was verified by improved bioassay plant root growth observed by Wendell and Ritchey (1993) in Porters soils after leaching.

2. After application of gypsum in agricultural situations, the Mg levels in the plow layer tend to decrease (Ritchey et al., 1980; Shainberg et al., 1989; Sumner, 1994). This occurs because the high concentration of Ca in soil so-

Table 8-10. The 1988 sorghum grain yield as affected by $CaSO_4 \cdot 1/2\ H_2O$ applied March 1982 (Sousa et al., 1992a).

	Days without irrigation		
Calcium sulfate applied	0	18	Grain N concentration
1000 kg ha^{-1}			g kg^{-1}
0	2.8	2.3	13.5
2	--	2.6	14.9
4	--	2.9	16.5
6	3.2	2.8	16.1
LSD (5%)	NS†	NS	2.3

† NS = not significant.

lution arising from gypsum incorporation tends to displace exchangeable Mg into the soil solution, where it can form uncharged ion pairs with sulfate ions. These ion pairs are readily leached down through the profile. If the Mg reserves are low, the loss of some of the exchangeable Mg and the increased competition for uptake caused by high levels of Ca may induce Mg deficiency in crops. In forages, this can cause a Ca–Mg imbalance in grazing livestock (Mayland & Grunes, 1979). Thus, it is recommended that dolomitic limestone or some other Mg source be applied at the same time as gypsum. An additional benefit of lime incorporation is that it causes gypsum to leach more rapidly (Ritchey et al., 1980), since sulfate adsorption is decreased as soil pH increases (Couto et al., 1979). Decreases in exchangeable K in the plow layer are less common than decreases in Mg. One reason for this is probably the greater role played by plants. Crop roots absorb K from the soil in relatively large quantities and accumulate excess amounts in aboveground dry matter. During senescence, K is washed out of the dead plant material and deposited on the soil surface (da Silva & Ritchey, 1982). This biocycling helps reduce K leaching losses (Sousa et al., 1992a; Quaggio, 1992).

3. Increased Mn levels in aboveground plant tissue such as found in the orchardgrass pot study of Feldhake and Ritchey (Table 8-2), and the field maize study reported by Sousa et al. (1992a) (Fig. 8–11), also have been observed by others (Fried & Peech, 1946; Shainberg et al., 1989). Part of this effect is apparently due to increased Mn availability. In addition, when gypsum allows plant roots to grow deeper into acid subsoils they may encounter soil with higher soluble Mn levels than that to which they had been exposed before. However, according to Shainberg et al. (1989), the increased foliar Mn levels observed have not resulted in toxicity or yield decreases.

CRITERIA FOR IDENTIFICATION OF GYPSUM-RESPONSIVE SOILS AND RECOMMENDATION OF APPLICATION RATES

The agronomic conditions where beneficial effects of gypsum can be expected include use on those soils where: (i) subsurface soil is sufficiently high in available Al or low in available Ca to limit root penetration by plant cultivars grown in the region, (ii) periodic droughts occurring during the growing season cause water stress to crops or excess rainfall leaches N below the plow layer, (iii) subsoil exchangeable Al levels are low enough that increases of 1 to 3 $cmol_c$ Ca kg^{-1} resulting from gypsum additions will significantly decrease acidity effects, and (iv) adequate levels of Mg can be economically maintained.

In order to make efficient use of gypsum, it is necessary to be able to identify which soils will respond to its application and to determine the amount of gypsum that should be applied. Dematte (1992) stated that the Brazilian soils which have the greatest potential for response to gypsum are acid soils with low CEC. This includes the Oxisols, oxidic Ultisols, and low-CEC acid Entisols and Inceptisols (collectively these comprise 60% of Brazil's area). Within such soils, long-term use of lime and OSP may have already mitigated detrimental effects of acidity, and diagnostic techniques are needed.

Sousa et al. (1992b), working entirely with low-CEC soils from the Cerrado region of highly weathered soils in central Brazil, proposed several tests for predicting which of these soils would respond to gypsum application and for recommending application rates. In cases where information on the exchangeable Ca and Al contents of the subsoils is available, expected responses to gypsum treatment are shown in Fig. 8–14 (Sousa et al., 1992b). This figure is based on equations developed from bioassay root growth rates measured in 120 subsoil horizons from the Cerrado region. When subsoil exchangeable Ca levels are below 0.1 cmol$_c$ kg^{-1}, there is a high probability of response to gypsum, regardless of Al levels, because of root growth response to Ca as a nutrient. At higher levels of Ca where the nutritional needs of the plant roots are exceeded, the probability of response is controlled principally by the Al saturation of the effective CEC; the responses are usually high for Al saturation levels of 65% or greater and low where Al saturation is less than 35%.

An alternative to the use of soil Ca and Al levels is direct application of the bioassay method (Sousa et al., 1992b). This test has the advantage that the prediction is based on the response of the particular cultivar that a farmer is using. Given the considerable differences that exist between crop cultivars and species in regard to their tolerance to low levels of Ca and high levels of Al, the biological test may be more accurate than generalized predictions based on the information in Fig. 8–14. To decide whether a given subsoil has a good probability of response, 4-d root lengths of seedlings of the actual cultivar being used on the farm are compared in subsoil with and without the addition of 1% by volume of gypsum. According to the authors, where growth in the gypsum treated soil is improved by 15% or less, limited benefit is expected from field use of the material. Where the improvement is 30% or more, gypsum application may be beneficial.

After identifying a subsoil as being a good candidate for improvement by gypsum treatment, it is necessary to know the amount of gypsum that should be added. Sousa et al. (1992b) reviewed several guidelines already published in Brazil that included recommendations based on clay content, extractable Al, or extractable Al and exchangeable Ca + Mg. For low CEC soils of the Cerrado, Sousa et al. (1992b) proposed several new methods for estimating appropriate levels of gypsum application. These methods were evaluated by comparing the recommendations with the actual amount of gypsum that would need to be sorbed onto a soil to maintain a level of 10 mg mL^{-1} S in solution when 2 g of subsoil were shaken 18 h with 20 mL of 0.75 mM CaSO$_4$ • 2H$_2$O solution. For the subsurface soil layer between 20- and 60-cm depth the actual amount of S needed to maintain 10 mg mL^{-1} in solution varied from 86 to 518 kg S ha^{-1}.

The most satisfactory prediction for the gypsum requirement (GR) for a 20- to 60-cm depth was obtained using the equation

$$GR \text{ (kg S ha}^{-1}) = -114 + 82.773 \, A_s - 2.739 \, A_s^2$$

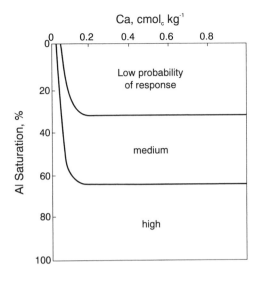

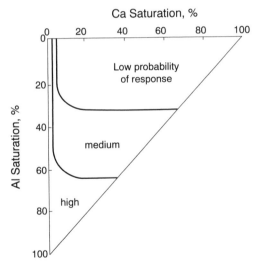

Fig. 8–14. Probability of response to Ca sulfate addition in relation to soil exchangeable Ca (*a*) or Ca saturation (*b*) and effective Al saturation levels, based on bioassay root response of 112 subsurface samples from low CEC soils of the Cerrados (Sousa et al., 1992b).

where A_s = (S sorbed)/(S in solution) after 2 g of soil were shaken for 18 h with 20 mL of 0.75 mM CaSO$_4$ • 2H$_2$O solution (r^2 = 0.97). In the absence of laboratory capability for S analyses, the best equation was

$$GR = -89.18 + 8.96 \text{ CLAY-PCT} + 23.11\, A_{Ca}$$
$$- 0.459\, (\text{CLAY-PCT} \times A_{Ca}),$$

where CLAY-PCT = percentage of clay present in soil passing a 2-mm sieve and A_{Ca} = the ratio of sorbed Ca to solution Ca when the soil is equilibrated with 20 mL of 0.75 mM CaSO$_4$ • 2H$_2$O solution (r^2 = 0.82).

A much simpler recommendation equation was obtained based on the percentage of clay in the soil

$$GR \text{ (kg S ha}^{-1}) = 17 + 6.508 \text{ CLAY-PCT } (r^2 = 0.79).$$

Sumner (1994) recently published a simple test to predict whether a subsoil will respond to gypsum. It is based on the capacity of the soil to sorb sulfate and raise the pH through displacement of hydroxyl groups. If the pH of the supernatant of a soil shaken intermittently overnight with 0.005 M Ca sulfate (10 g soil to 25 mL solution) is higher than the pH when shaken with 0.005 M Ca chloride, and the EC of the 0.005 M Ca sulfate solution is lower after shaking than before (indicating sorption of Ca and sulfate), then the subsoil is considered a good candidate for gypsum treatment.

The work of both Sumner et al. (1994) and Sousa et al. (1992b) was based on highly weathered soils with low CEC levels and a significant amount of oxidic minerals. Ernani et al. (1992) discussed the applicability of gypsum treatment for higher CEC clayey soils in the more temperate southern part of Brazil, where the 1400 mm annual precipitation is well distributed and droughts are infrequent. Calcium levels in these soils are well above the range where responses associated with overcoming Ca deficiencies could be expected. Crop responses to field application of gypsum were surveyed in a number of experiments on these high-CEC soils. In general, little or no beneficial response to gypsum was observed. The lack of benefit is reasonable given the already high exchangeable Ca levels and generally well-distributed rainfall. This underlines the importance of developing diagnostic criteria for identifying subsoils where gypsum application is beneficial.

Another situation where gypsum may not improve yields is where subsoil Al levels are high (van Raij, 1992) and the amount of Ca already present is sufficient to meet nutritional needs of roots. The amount of gypsum that can dissolve in the soil solution is on the order of 0.8 cmol$_c$ kg^{-1} and increases in exchangeable Ca seldom exceed 3–4 cmol$_c$ kg^{-1}. In soils with over 10 cmol$_c$ kg^{-1} Al, it is unlikely that gypsum treatment will have an appreciable beneficial effect on the chemical environment. Yamada (1988), as discussed in Dematte (1992), found that in two Red-Yellow Podzolic soils with Al levels of 5 cmol$_c$ kg^{-1} Al, and Al saturations on the order of 80%, gypsum addition did not improve bioassay root growth. R.R. Wendell (personal communication, 1992) similarly found no improvement in bioassay root growth in unleached Appalachian soils treated with gypsum, except those with low exchangeable Ca levels. More research is needed on higher CEC soils to ascertain plant responses to gypsum.

CONCLUSIONS

Many of the power plants which burn medium- and high-S coal are located in the eastern USA, in or near the Appalachian Region and the coastal plains of the South, where many of the soils are acid. As a result of scrubbing SO_2 from flue gas from these facilities, wallboard-quality gypsum is becoming increasingly available. This source of inexpensive gypsum opens the possibility that crop productivity on soils amenable to gypsum amelioration can be increased at relatively low cost. By-product gypsum has been shown to be effective in improving deep rooting of crops grown on low-CEC acid soils in the southeastern USA and oxidic soils of Brazil. Because of its relatively high solubility it can leach below the zone of application and supply Ca and reduce Al toxicity in acidic subsoils, increasing rooting depth and improving access to water and nutrients. Initial results of studies on soils of the Appalachian region, which are less weathered and have a different mineralogy, indicate that although gypsum addition is beneficial, the mechanism of amelioration may be different. Gypsum use apparently results in higher rates of Al displacement into soil solution than in southeastern soils. Plant growth after leaching is improved, but more information is needed about short-term effects on plant root growth and possible environmental effects.

Methods for identifying which soils will benefit from gypsum addition and development of criteria for application rates are needed. The USDA-ARS Appalachian Soil and Water Conservation Research Laboratory is currently studying the reactions of these soils to gypsum addition in relation to soil properties and evaluating pasture and row crop responses to field application of gypsum in the Appalachian region.

REFERENCES

Alcordo, I.S., and J.E. Rechigl. 1993. Phosphogypsum in agriculture: A review. Adv. Agron. 49:55–118.

Alva, A.K., M.E. Sumner, and W.P. Miller. 1991. Salt absorption in gypsum amended acid soils. p. 93–97. In R.J. Wright et al. (ed.) Proc. Int. Symp. Plant-Soil Interactions at Low pH, Beckley, WV. 24–29 June 1990. Dordrecht, the Netherlands.

American Coal Ash Association. 1993. Coal combustion by-product—production and consumption. ACAA, Washington, DC.

Bohnen, H. 1980. Calcium, pH and aluminum in solution and corn growth. Ph.D. diss. Cornell Univ., Ithaca, NY (Diss. Abstr. 810 2905).

Bolan, N.S., J.K. Syers, and M.E. Sumner. 1993. Calcium-induced sulfate adsorption by soils. Soil Sci. Soc. Am. J. 57:691–696.

Clark, R.B., K.D. Ritchey, and V.C. Baligar. 1993. Dry matter yields of maize grown with coal combustion by-products. p. 15-1 to 15-11. In Proc. 10th Int. Ash Use Symp. Vol. 1. Washington, DC. 18–21 January. Electric Power Res. Inst., Palo Alto, CA.

Couto, W., D.J. Lathwell, and D.R. Bouldin. 1979. Sulfate sorption by two Oxisols and an Alfisol of the tropics. Soil Sci. 127:108–116.

da Silva, J.E., and K.D. Ritchey. 1982. Lateral redistribution of potassium in Oxisols caused by rain washing from maize plants to the soil (In Portuguese.) Rev. Bras. Cien. Solo 6:183–188.

Dematte, J.L.I. 1992. Agronomic land capability of soils and the use of gypsum. (In Portuguese.) p. 307–324. In 2nd Seminar on the Use of Gypsum in Agriculture, Uberaba, Brazil. IBRAFOS, Brazil.

Ernani, P.R., P.C. Cassol, and G. Peruzo. 1992. Agronomic efficiency of agricultural gypsum in the south of Brazil. p. 307–324. In 2nd Seminar on the Use of Gypsum in Agriculture, Uberaba, Brazil. 24–26 March. IBRAFOS, Brazil.

Fried, M., and M. Peech. 1946 The comparative effects of lime and gypsum upon plants grown on acid soils. J. Am. Soc. Agron. 38:614–625.

Gonzalez-Erico, E., E.J. Kamprath, G.C. Naderman, and W.V. Soares. 1979. Effect of depth of lime incorporation on the growth of corn on an Oxisol of central Brazil. Soil Sci. Soc. Am. J. 43:1155-1158.

Kinraide, T.B., and R.D. Parker. 1987. Non-phytotoxicity of the aluminum sulfate ion, $AlSO_4^+$. Plant Physiol. 83:546-551.

Kinraide, T.B., P.R. Ryan, and L.V. Kochian. 1992. Interactive effects of Al^{3+}, H^+, and other cations on root elongation considered in terms of cell-surface electrical potential. Plant Physiol. 99:1461-1468.

Kinraide, T.B., P.R. Ryan, and L.V. Kochian. 1994. Al^{3+}-Ca^{2+} interactions in aluminum rhizotoxicity. II. Evaluating the Ca^{2+}-displacement hypothesis. Planta 192:104-109.

Mayland, H.F., and D.L. Grunes. 1979. Soil-climate-plant relationships in the etiology of grass tetany. p. 123-175. In V.V. Rendig and D.L. Grunes (ed.) Grass tetany. ASA Spec. Publ. 35. ASA, CSSA, and SSSA, Madison, WI.

McCray, J.M., and M.E. Sumner. 1990. Assessing and modifying Ca and Al levels in acid subsoils. p. 45-75. In B.A. Stewart (ed.) Advances in soil science. Vol. 14. Springer-Verlag, New York.

Oates, K.M., and A.G. Caldwell. 1985. Use of by-product gypsum to alleviate soil acidity. Soil Sci. Soc. Am. J. 49:915-918

Parfitt, R.L., and R.S. Smart. 1978. The mechanism of sulfate adsorption on Fe oxides. Soil Sci. Soc. Am. J. 42:48-50.

Pavan, M.A., F.T. Bingham, and P.R. Pratt. 1982. Toxicity of aluminum to coffee in Ultisols and Oxisols amended with $CaCO_3$, $MgCO_3$ and $CaSO_4$-$2H_2O$. Soil Sci. Soc. Am. J. 46:1201-1207.

Quaggio, J.A. 1992. Maize and soybean crop responses to application of limestone and gypsum and ion movement in soils of the State of Sao Paulo. (In Portuguese.) p. 241-262. In 2nd Seminar on the Use of Gypsum in Agriculture. Uberaba, Brazil. 24-26 March. IBRAFOS, Brazil.

Rajan, S.S.S. 1978. Sulfate adsorbed on hydrous alumina, ligands displaced, and changes in surface charge. Soil Sci. Soc. Am. J. 42:39-44.

Reeve, N.G., and M.E. Sumner. 1972. Amelioration of subsoil acidity in Natal Oxisols by leaching of surface applied amendments. Agrochemophysica 4:1-6.

Ritchey, K.D., J.E. Silva, and U.F. Costa. 1982. Calcium deficiency in clayey B horizons of savanna Oxisols. Soil Sci. 133:378-382.

Ritchey, K.D., D.M.G. Sousa, E. Lobato, and O. Correa. 1980. Calcium leaching to increase rooting depth in a Brazilian Savannah Oxisol. Agron. J. 72:40-44.

Ritchey, K.D., D.M.G. Sousa, and G. Costa Rodrigues. 1989. Inexpensive biological tests for soil calcium deficiency and aluminum toxicity. Plant Soil 120:273-282.

Shainberg, I., M.E. Sumner, W.P. Miller, M.P.W. Farina, M.A. Pavan, and M.V. Fey. 1989. Use of gypsum on soils: A review. p. 1-111. In B.A. Stewart (ed.) Advances in soil science. Vol. 13. Springer-Verlag, New York.

Sousa, D.M.G., E. Lobato, and L.N. Miranda. 1993. Soil liming and fertilization for soybean cultivation. (In Portuguese.) p. 137-158. In N.E. Arantes and P.I.M. de Souza (ed.) Symp. on the culture of soybean in Cerrados, Uberaba, 1992. Piracicaba, POTAFOS.

Sousa, D.M.G., E. Lobato, K.D. Ritchey, and T.A. Rein. 1992a. Response of annual crops and leucaena to gypsum in the Cerrado. (In Portuguese.). p. 277-306. In 2nd Seminar on the Use of Gypsum in Agriculture, Vol. 2. Uberaba, Brazil. 24-26 March. IBRAFOS, Brazil.

Sousa, D.M.G., E. Lobato, K.D. Ritchey, and T.A. Rein. 1992b. Suggestions for diagnosis and recommendation of gypsum application for soils of the Cerrados. (In Portuguese.) p. 139-158. In 2nd Seminar on the Use of Gypsum in Agriculture, Uberaba, Brazil. 24-26 March. IBRAFOS, Brazil.

Sousa, D.M.G., and K.D. Ritchey. 1986. Use of gypsum in Cerrado soils. (In Portuguese.) p. 119-144. In Seminar on the Use of Phosphogypsum in Agriculture, Brasilia, Brazil. 11-12 June 1985. Dep. de Difusao de Tecnol., EMBRAPA, Brasilia, Brazil.

Sposito, G. 1989. The chemistry of soils. Oxford Univ. Press, New York.

Sumner, M.E. 1994. Amelioration of subsoil acidity with minimum disturbance. p. 147-185. In N.J. Jayawardane and B.A. Steward (ed.) Subsoil management techniques. Lewis Publ., Inc., Boca Raton, FL.

van Raij, B. 1992. Reactions of gypsum in acid soils. (In Portuguese.) p. 105-119. In 2nd Seminar on the Use of Gypsum in Agriculture, Uberaba, Brazil. 24-26 March. IBRAFOS, Brazil.

Vitti, G.C., J.A. Mazza, H.S. Pereira, and J.L.I. Dematte. 1992. Experimental results from the use of gypsum in sugar-cane-agriculture. p. 191–224. *In* 2nd Seminar on the Use of Gypsum in Agriculture, Uberaba, Brazil. 24–26 March. IBRAFOS, Brazil.

Weast, R.C. (ed.) 1978. CRC handbook of chemistry and physics. CRC Press, Inc., West Palm Beach, FL.

Wendell, R.R., and K.D. Ritchey. 1993. Use of high-gypsum flue gas desulfurization by-products in agriculture. p. 40–45. *In* S.-H Chiang (ed.) Proc. 10th Ann. Int. Pittsburgh Coal Conf. 20–24 Sept. Univ. Pittsburgh, PA.

Yamada, T. 1988. Maximum sulfate adsorption capacity as an additional parameter for the recommendation of gypsum. M.S. thesis. E.S.A. Piracicaba, Sao Paulo, Brazil.

9 Environmental Considerations in Land Application of By-Product Gypsum

W. P. Miller

University of Georgia
Athens, Georgia

Disposal of industrial by-products is becoming an increasingly serious constraint on many industries; public environmental concern, awakened by research findings on the hazards of many such wastes, has been codified into regulation at local, state, and national levels. Stockpiling, burial, and ocean dumping of wastes is no longer tolerated for many such by-products, and for many industries the fate of their waste products is a major concern dictating their continued operation or possible expansion of operations.

Gypsum ($CaSO_4 \cdot 2H_2O$) is a relatively ubiquitous industrial waste product derived from a number of processes involving the use of neutralization of acids. Many industries produce it in a relatively pure form ($>90\%$ purity), and as such it is a product that has potential uses in other industrial processes, such as wall board manufacture, or in agriculture as a soil amendment. The rationale for utilization of such materials includes both the avoidance of disposal costs in environmentally approved landfills, and the potential to market the by-product at some minimal charge to the user. From the user's point of view, such material is often low enough in cost to replace traditional mined gypsum, as long as transport costs from the point of origin are not excessive.

Since it is derived from neutralization of waste acids that may contain trace impurities, by-product gypsum material typically contains elements of environmental concern that may limit the utilization of the material as a soil amendment. The type and concentrations of these elements vary widely among the various sources of by-product gypsum, and obviously must be individually evaluated in order to assess their potential for environmental impact. Both producers and users of by-product gypsums must be assured that the utilization of this material will not result in soil or water contamination, either in the short or long term. The objective of this chapter is thus to assess current information on the potential for by-product gypsum use on land from an environmental point of view, particularly as to the likelihood for soil and water contamination.

SOURCES AND PRODUCTION OF BY-PRODUCT GYPSUM

By-product gypsum is produced by industries such as mineral processors, plating, and electronics by neutralization of spent sulfuric acid solutions with Ca alkalis such as $CaCO_3$ or $Ca(OH)_2$. The general reaction for this neutralization is

$$CaCO_3 + H_2SO_4 + H_2O \rightarrow CaSO_4 \cdot 2H_2O + CO_2 \qquad [1]$$

in which calcite (limestone) is used as the alkali. Such acid solutions often contain soluble metals, such as Fe, Al, and trace metals such as Zn, Cd, and Cr, depending on the specific industrial process: mineral processing acids are often high in Fe and Al from mineral dissolution, while plating bath wastes may contain variable levels of plating metals. These metals hydrolyze with alkali addition to form solid-phase hydroxide or oxide compounds, which are retained with the by-product gypsum. No estimates of total production of such materials are available, but they may be locally significant.

A significant source of by-product gypsum is the benefication of rock phosphate in the manufacture of phosphoric acid and P fertilizer, derived from the treatment of the rock with sulfuric acid. The general reaction for acidulation of apatite is

$$Ca_{10}(PO_4)_6F_2 + 10\ H_2SO_4 + 20\ H_2O \rightarrow 10\ CaSO_4 \cdot 2H_2O$$

$$+ 6\ H_3PO_4 + 2\ HF \qquad [2]$$

which produces phosphoric acid, a basic industrial feedstock and precursor to P fertilizer synthesis, as well as waste gypsum and hydrofluoric acid. Approximately 500 million tonnes of this by-product ("phosphogypsum") are stockpiled in central Florida and in the Mississippi Valley, where P fertilizer processing is or has been concentrated. Annual production of phosphogypsum in Florida alone is roughly 30 million tonnes per year (Anonymous, 1993). It has been increasingly marketed for agricultural use in the southeastern USA, but such use has recently been restricted by the USEPA due to the presence of trace contaminants.

An increasingly available source of by-product gypsum is associated with the desulfurization of flue gas from coal-fired electric generating utilities. Air quality standards demanding reduction in emissions of SO_2 from such plants are forcing utilities to install scrubber facilities which typically use $CaCO_3$ to neutralize acidity produced from the oxidation of SO_2. Various technologies are emerging which produce gypsums of various qualities. Fluidized bed methods inject limestone directly into the furnace, and produce a mixed waste product composed of gypsum, excess $CaCO_3$, and ash. Wet scrubber methods pass the flue gas through slurries of fine limestone, and depending on the degree of oxidation of the SO_2, may produce material ranging from predominately $CaSO_3$ (calcium sulfite) to nearly pure $CaSO_4$, the later obviously produced in scrubbers more efficient at SO_2 oxidation ("forced oxidation" technologies). In such scrubbers the process might be shown as a two-step reaction

$$SO_2(g) + H_2O \rightarrow H_2SO_3(aq), \quad H_2SO_3(aq) + 1/2\ O_2$$
$$\rightarrow H_2SO_4(aq) \rightarrow 2H^+ + SO_4^{2-} \qquad [3]$$

$$CaCO_3(s) + 2H^+ + SO_4^{2-} + H_2O \rightarrow CaSO_4 \cdot 2H_2O(s)$$
$$+ CO_2(g) \qquad [4]$$

where the $SO_2(g)$ first dissolves in the aqueous phase and is oxidized to H_2SO_4, which is then neutralized by the $CaCO_3$.

Current production of such flue-gas desulfurization gypsum (or FDG) is small, but in the eastern USA, where regulation will demand significant reduction of SO_2 emissions in the next decade, large amounts of FDG may be produced. Approximately 0.10 t of gypsum is produced per tonne of coal burned, assuming 2% S coal and complete capture and oxidation of S in the coal. With nearly 2 billion tonnes of coal burned per year in the USA, annual FDG production could increase to 50 million tons per year over the next few decades, depending on how much of the generating capacity is desulfurized using this technology (Carlson & Adriano, 1993).

BENEFITS OF LAND APPLICATION OF GYPSUM

With its high content of Ca and S, gypsum has been used as a soil amendment for improvement of fertility relations on a wide range of soils. It was used as early as the eighteenth century in Virginia, proving beneficial on soils that were likely eroded, acidic, and low in Ca (Rinehart et al., 1953). Lime ($CaCO_3$) has replaced gypsum in this use, being superior in acid neutralization and a good source of available Ca. Gypsum is still used as a S source, typically bulk blended with other fertilizers, on S-deficient soils (usually very sandy, low organic matter soils). It is also top-dressed on peanuts (*Arachus hypogea*) in the southeastern USA as a supplemental Ca source during pod set, at rates of 0.1 to 0.5 t ha^{-1}. Total use for this market is approximately 2 to 3 million tonnes per year; in the past this was largely met by crushed mined gypsum, although phosphogypsum is increasingly used as a more economical alternative.

Gypsum also has been extensively used to improve soil physical properties in arid region soils having high exchangeable sodium percentages (ESP). The relatively high solubility of gypsum [30 mmol$_c$ L^{-1}] supplies electrolyte to the soil solution, thus promoting flocculation of these highly dispersive soils, and also replaces Na with Ca on the exchange complex (Oster, 1982; Shainberg et al., 1989). Rates of application are based on levels of exchangeable Na present in the soil. Phosphogypsum has been used for such remediation of sodic soils in Israel and elsewhere, and results in substantial increases in rain and irrigation water infiltration, even on soils with ESP as low as five (Kazman et al., 1983). Gypsum applications on soils of very low ESP (<2) in more humid regions also have been demonstrated to decrease runoff and the tendency of soils to erode (Miller, 1987; Shainberg et al., 1989). This topic will be discussed later in this chapter.

Recent research in the humid regions has demonstrated substantial benefits of gypsum application to acidic soils (Shainberg et al., 1989; Ritchey et al., 1995, see Chapter 8). In these soils, the acidic, low-Ca subsoils limit root growth and often plant yield, typically due to water stress (Sumner et la., 1986). Gypsum additions allow Ca to move into the subsoil, thereby increasing rooting and water extraction from subsoil horizons and promoting plant growth. Research in the southeastern USA, Brazil, South Africa, and Australia has shown yield responses on acid soils to gypsum (either mined or phosphogypsum) on a range of annual and perennial crops, with one-time application rates of 5–10 t ha^{-1} (Shainberg et al., 1989; Ritchey et al., 1995, see Chapter 8).

PROPERTIES OF BY-PRODUCT GYPSUMS

For the purposes of this discussion, by-product gypsum will be defined as materials containing a high proportion (>80% by weight) $CaSO_4$ • $2H_2O$; some waste streams contain significant admixtures of other materials that constitute an essentially "mixed" waste. Some desulfurization materials, for example, are composed of 20 to 50% unreacted lime and/or ash in addition to gypsum. The uses and environmental impacts of these materials will likely be different than the more pure gypsum materials, and thus will not be discussed further here. As the major sources of relatively pure by-product gypsum are phosphate processing and flue-gas desulfurization (via forced oxidation technology), these types of materials will be emphasized in the following discussions.

Selected properties of by-product phosphogypsum (PG) and FDG are presented in Table 9-1. As noted previously, they are greater than 80% $CaSO_4$ • $2H_2O$; quartz sand tends to be the major bulk contaminant in PG (resulting in higher SiO_2 contents), while fly ash or unreacted lime may be

Table 9-1. Bulk properties of by-product gypsum materials.†

Property	PG	FDG
$CaSO_4$ $2H_2O$ (%)	85–95	80–95
SiO_2 (%)	3–10	0–10
$CaCO_3$ (%)	0	0–10
Crystalline size (μm)	40–250	40–250
Surface area (m^2 kg^{-1})	8 000–15 000	8 000–15 000
pH (saturated paste)	4.8–5.6	7.0–8.2
Electric conductivity (saturated paste) (dS m^{-1})	2.1–2.3	2.0–18.0

† From May and Sweeney (1983); Miller, unpublished data.

present in variable amounts in FDG. Both types of materials are composed of fine (silt and fine sand-sized) gypsum crystallites with surface areas in the ranges of 6 000 to 15 000 m^2 kg^{-1}; studies of PG solubility characteristics have shown it to be as rapidly soluble as mined gypsum or "reagent grade" $CaSO_4 \cdot 2H_2O$ (Shainberg et al., 1989). Phosphogypsum typically has an acidic pH due to residual H_2SO_4 present in the pore fluids, but little buffering capacity. The presence of soluble salts in the process water often leads to higher ionic strengths (measured as electrical conductivity) in FDG materials than in PG. Some FDG produced using dolomitic limes may have very high EC due to the formation of soluble $MgSO_4$ during desulfurization.

Calcium and S are obviously the major elements present in by-product gypsums; representative analyses are shown in Table 9-2. Other major elements are typically present at $< 1\%$. Phosphorus and F contents of PG may be 0.5 to 1%, which may be important agriculturally at high rates of addition.

Trace element concentrations of gypsum materials tend to be low compared with other wastes such as fly ash or sewage sludge, or even background levels in uncontaminated soils (Table 9-3). Phosphogypsum in particular has very low levels of most trace elements, since heavy metals tend to be retained in the phosphoric acid product rather than the by-product gypsum. In FDG, trace contaminant levels tend to be higher, due to the capture of volatile elements in the stack gas within the desulfurization reactor, and to the presence of these elements in the process water used in the reactor, which often originates from decant water from ash ponds with high levels of some contaminants. Boron (B), As, Se, and Mo are often cited as potentially hazardous elements in fly ash (Adriano et al., 1980), but of these only B is significantly elevated in FDG. Mercury (Hg) emissions from power plants are also a concern (Kaakinen et al., 1975); one study showed that desulfurization removed 60% of the gaseous Hg from a stack stream, resulting in approximately 2 mg Hg kg^{-1} in the by-product gypsum (Behrens & Hargrove, 1980).

Both PG and FDG contain radionuclides originating from the ^{238}U decay series; there has been some environmental debate concerning the sig-

Table 9-2. Major element contents of by-product gypsums and fly ash.

Element	PG†	FDG‡	Fly ash§
		% by weight	
Ca	20.1	0.95	1–18
Mg	0.04	0.05	0.2–8
Fe	0.1	0.22	1–29
K	0.05	0.01	0.1–3
Al	0.1	0.15	2–20
Na	0.15	0.15	0.2–2
P	0.6	0.02	0.05–1
F	0.6	ND¶	ND

† PG values from May and Sweeney (1983); Miller, unpublished data.
‡ FDG data from Miller, unpublished data.
§ Fly ash values from Straughan et al. (1980); Page et al. (1979); Adriano et al. (1980).
¶ ND = not determined.

Table 9-3. Trace element contents of by-product gypsums, fly ash, and soils.

Element	PG†	FDG‡	Fly ash§	Soil
			mg kg^{-1}	
As	4	4	2–300	0.1–30
B	ND¶	80	15–1000	2–100
Ba	5	10	100–1000	50–1000
Cd	4	4	0.5–75	0.01–10
Co	ND	7	5–75	5–300
Cr	20	25	10–200	10–500
Cu	9	16	20–200	21–100
Hg	ND	ND	0.02–1	
Mn	ND	70	60–3000	100–4000
Mo	ND	5	6–50	2–8
Ni	10	10	10–200	10–100
Pb	15	40	10–1000	2–200
Se	ND	ND	1–100	0.01–2
Zn	ND	25	20–500	10–300

† PG values from May and Sweeney (1983); Miller, unpublished data.
‡ FDG data from Miller, unpublished data.
§ Fly ash values from Straughan et al. (1980); Page et al. (1979); Adriano et al. (1980).
¶ ND = not determined.

nificance of the levels in PG particularly. This topic will be discussed later in this chapter.

Toxic organic contaminants such as polyaromatic hydrocarbons (PAH) have not been extensively studied in gypsum by-products. Phosphogypsum is unlikely to contain any such materials, given its chemistry of production; FDG may contain volatile organics derived from the flue gas stream, but Bannwarth et al. (1992) argue that the chemical environment within the furnace is unsuitable for the production of chlorinated compounds.

POTENTIAL ENVIRONMENTAL IMPACTS OF GYPSUM USE ON LAND

Based on the composition of by-product gypsum materials described above, potential negative impacts of land application of these materials have been reported in the literature, and in on-going studies by the author. For FDG, trace anionic contaminants such as As, B, Mo, and Se are generally considered to be of most concern, due to relatively higher concentrations and greater solubility in agricultural soils. These elements may have direct phytotoxic effects on plants, may accumulate in plant tissue and have adverse effects on animals or humans consuming them, and/or leach or runoff to water supplies used by animals or humans. The radionuclide content of PG is probably the central issue with respect to utilization of this material on land, specifically bioaccumulation of radionuclides and the potential for water contamination. The appreciable PO$_4$ content of PG, which is readily water soluble, may cause some concern if it is transported through runoff to surface waters where eutrophication may be induced.

Runoff and percolate from gypsum-treated soils contain appreciable salt loads in the form of $CaSO_4$. This higher soluble salt level may have beneficial effects environmentally by enhancing flocculation of soil and sediment colloids in regions where salt contents of natural waters are low, such as in the southeastern USA. This flocculating effect may enhance infiltration, thereby reducing runoff, and also decrease erosion and sediment delivery to surface water; these beneficial results of gypsum use on land will be discussed in some detail in the following section. The high soluble SO_4 levels associated with gypsum use may be an issue in some instances: saturated gypsum solutions contain up to 15 mM SO_4 L^{-1}, or approximately 1500 mg L^{-1}. Transport of this SO_4 to poorly drained soils or sediments will enhance H_2S production, thereby changing the chemistry of such soils (Bannwarth et al., 1992). The potential environmental impact of such changes has not been documented in research findings.

EFFECTS OF BY-PRODUCT GYPSUM ON RUNOFF AND EROSION

The Role of Electrolytes in Runoff and Erosion Processes

The importance of soil chemical parameters such as exchangeable Na, pH, and level of soluble electrolytes in determining the hydraulic properties of soils has been documented on a wide variety of soil materials (Quirk & Schofield, 1955; Frenkel et al., 1978; Shainberg et al., 1981). Application of soluble Ca salts such as gypsum improves saturated hydraulic conductivity of many soils by flocculating the colloid fraction, thereby preventing the clogging of water transmission pores by dispersed clay particles (Shainberg et al., 1989). In the case of Na-affected soils this flocculating effect is due to both replacement of Na by Ca on the exchange sites of the soil colloids, and to the increase in ionic strength of the soil solution; both mechanisms compress the colloid electrical double layer and decrease repulsive forces between clay particles, thereby allowing van der Waals attractive forces to initiate flocculation (van Olphen, 1977).

Many soils with low ESP (≤ 5) also are dispersive at low ionic strength (Rengasamy et al., 1984; Miller & Baharuddin, 1986), and may show reduced permeability due to colloid mobilization, particularly if mechanical energy is input (e.g., raindrop impact energy) to disrupt colloid interaction (Sumner, 1992). Several early studies by Rinehart et al. (1953) and Kemper and Noonan (1970) demonstrated improved infiltration with gypsum use on nonsodic soils. The phenomenon of soil crusting or sealing at the soil surface thus may have a considerable chemical component on many soils around the world that are either traditionally "sodic," or that have sufficiently low salt content to allow them to be dispersive (Shainberg, 1992; Sumner, 1992).

Soil erosion rates also are related to dispersion processes through both the effect on infiltration and on the size of erodible particles at the soil surface (Miller & Baharuddin, 1986, 1987). Dispersive soils generate more runoff

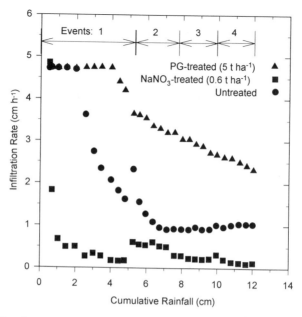

Fig. 9-1. Infiltration rates over simulated rainfall event (48 mm h^{-1}) for Greenville sandy clay loam (Rhodic Kandiudult) Ap horizon soil as affected by PG (5 t ha^{-1}) or NaNO$_3$ (0.6 t ha^{-1}) treatments. Four successive rainfall events were applied to each treatment (Miller & Scifres, 1988).

and finer eroded partricles, thus leading to greater erosion rates and a higher potential for sediment delivery to surface waters. Chemical treatment of soils to reduce dispersion therefore has the potential to reduce peak runoff flows from agricultural soils and to reduce sediment loads and fine particle contents in runoff.

Effect of Gypsum Applications on Infiltration

The effect of Ca salt additions on infiltration rate under rainfall is dramatic and immediate on many soils due to the inhibition or delay of surface seal formation (Agassi et al., 1981; Rengasamy et al., 1984; Miller, 1987). For low-ESP soils, critical flocculation concentrations (minimum salt concentrations required to flocculate a clay suspension) are typically only 1 to 5 mM CaSO$_4$ L^{-1} (Miller et al., 1990), which is readily exceeded in runoff waters from soils which have received surface application of gypsum.

Figure 9-1 shows the infiltration rate over time for a sandy clay loam Ultisol from the Coastal Plain of Georgia as affected by NaNO$_3$ and CaSO$_4$ treatments applied to the soil surface immediately preceding rainfall (Miller & Scifres, 1988). With no treatment the soil begins to generate runoff after 2.2 cm of applied rainfall (at 4.8 cm h^{-1}), and reaches a final infiltration rate (FIR) of 1 cm h^{-1} after 12 cm of rain. With 5 mt PG ha^{-1} applied to the soil surface, runoff is delayed until roughly 4 cm of applied rain, and the FIR is 2.5 cm h^{-1}. Sodium treatment results in near immediate runoff with FIR of <0.2 cm h^{-1}. Similar findings have been reported for a range of soils, as summarized in Table 9-4, which shows runoff as percentage of rainfall for untreated and gypsum-treated soils. The magnitude of response varies among soils, with more dispersive, crusting soils with higher runoff coefficients (% of rainfall as runoff) tending to show greater reductions with gypsum applications. The effect of gypsum on infiltration also tends to be less at high slope and at high rainfall intensity, as gypsum particles are increasingly lost from the soil surface at surface flows which occur under these conditions. As suggested by Rengasamy et al. (1984), soils that readily disperse in water are most likely to respond positively to gypsum applications. Sodic soils that contain swelling minerals often show the greatest response, but predominately kaolinitic soils with low ESP's also show reduced runoff (Stern et al., 1991).

It should be noted that most of the studies of by-product gypsum use in runoff control have used PG or mined gypsum; FDG has been less studied in this regard, although some data exists on fluidized bed material (Norton et al., 1995, see Chapter 6). Some FDG's have calcium carbonate equivalent values of 10 to 40 and are very reactive, thus tending to increase soil pH dramatically when surface applied. Such pH increases may enhance dispersion despite higher electrolyte levels (Miller et al., 1990), and therefore lead to greater runoff and erosion rates than untreated soils.

Experience in Australia, Israel, and South Africa suggests that rates of 3 to 10 t gypsum ha^{-1} applied on the soil surface at planting will continue to supply electrolytes and enhance infiltration for a growing season of 300 to 500 mm rainfall, or until tillage mixes the remaining gypsum into the soil. For crusting soils that may have average runoff coefficients of 60% of total rainfall for untreated soil, and which may show a 50% reduction in runoff with gypsum treatment, 500 mm of rainfall will generate 300 mm runoff, but only 150 mm with gypsum treatment. The additional infiltrated water will likely be used in transpiration, with some possibly moving through the root zone to groundwater. The overall environmental impact of such changes in hydrology are likely to be favorable, with lower storm water peak flows and lessened transport capacity for sediment to surface water (see below). This favorable environmental impact will apply only as long as trace contaminants present in the by-product gypsum are not transported to surface or groundwaters, as will be discussed later.

Effect of Gypsum on Erosion and Sediment Production

Reductions in runoff volumes tend to be accompanied by lesser rates of soil erosion and sediment delivery to streams; since many contaminants are transported via surface processes, either soluble or adsorbed to soil colloids, the overall effect is likely to be improved water quality. Studies of gypsum effects on soil loss have generally borne out the above principle, showing reduced soil loss with gypsum application. Some example data is shown in Table 9-4, including both laboratory (rainfall simulator) and field (small plot)

Table 9-4. Runoff and erosion measurements from literature studies of gypsum-treated soils under natural or simulated rainfall.

Reference†	Soil‡	Rainfall amount	Intensity	Slope	Gypsum rate	Runoff		Soil loss	
						Control	Gypsum	Control	Gypsum
		mm	mm h⁻¹	%	t ha⁻¹	— % of rain —		— t ha⁻¹ —	
1	Cecil sal (p)	125	50	9	5	65	33	0.72	0.40
1	Worsham 1 (p)	125	50	9	5	87	76	3.70	1.90
1	Greenville sacl (p)	125	50	9	5	58	24	0.94	0.27
2	Wedowee sal (p)	120	48	9	5	71	31	2.90	1.35
3	Grumusol c (f)	250	(N)§	12	5	50	13	3.46	0.31
3	Grumusol c (f)	250	(N)	32	5	53	18	14.37	1.71
4	Paleudalf 1 (f)	340	(N)	5	5	33	18	ND¶	ND
4	Rhodustalf sal (f)	316	(N)	5	5	64	36	ND	ND
4	Haplustalf sal (f)	5010	(N)	5	5	67	45	ND	ND
5	Hamra sal (p)	60	40	20	5	70	25	55.0	10.0
6	Orthox sicl (p)	80	42	5	5	55	32	2.3	1.5
6	Rhodudalf 1 (p)	80	42	5	5	63	41	9.0	3.6
6	Paleudalf sicl (p)	80	42	5	5	82	54	9.6	4.0

† 1 = Miller (1987); 2 = Miller and Scifres (1988); 3 = Agassi et al. (1990); 4 = Stern et al. (1991); 5 = Warrington et al. (1989); 6 = Ben-Hur et al. (1992).
‡ Textural class abbreviations: c = clay, si = silt, sa = sand, l = loam, p = pan study, f = field plot data.
§ (N) = natural rainfall.
¶ ND = not determined.

studies from the USA, Israel, and South Africa. While these results are quite variable, significant reductions in erosion have been shown from gypsum use on a wide variety of soils.

Often soil loss reductions due to gypsum use are proportionately greater than runoff reductions on the same soil: runoff may be halved on gypsum-treated soils compared to controls, while sediment production may be lessened by fivefold or greater. There are several possible explanations for this result. The flocculating effect of gypsum on soil colloids tends to enhance aggregation and thereby increase the "apparent" size of sediment particles/aggregates, which are then less transportable in the lessened runoff volume of the gypsum-treated soil. Miller (1987) observed a decrease in percentage of fine (clay-sized) particles in the sediment of PG-treated soils compared to untreated soils. In addition, lower runoff volumes from gypsum-treated soil reduce the potential for rill formation; soil loss is often proportionately less at higher slopes with gypsum treatment, since rilling greatly increases soil loss at high slopes on untreated soils (Ben-Hur et al., 1992). The increase in flocculation and microaggregation due to gypsum amendment should result in reduced sediment transport and delivery to surface waters; however, as runoff water from treated areas is diluted with low electrolyte water from untreated areas, dispersion may again be favored. No watershed scale studies have been performed to evaluate these effects on actual sediment delivery on a large scale.

TRACE METALS AND OTHER CONTAMINANTS:
FOOD CHAIN AND WATER QUALITY

For many by-product gypsum materials, the largest potential negative environmental impact of land application is the potential for trace contaminants to leach to groundwater, run-off to surface waters, or be taken up by plants in sufficient quantity to either cause plant or animal toxicities (Adriano et al., 1980). The total loading of such contaminants is determined by both the concentration in the gypsum and its application rate, both annual and cumulative amounts. While many contaminants are present in relatively low concentrations, large applications repeated over time on the same area obviously will increase the total loading of such contaminants.

Solubilities of added contaminants, and hence mobilities in soils, depend on the initial form of the contaminants in the byproduct gypsum and on their chemical reactions within the soil system. Significant rates of gypsum amendment will modify the chemical environment of the soil by adding large amounts of soluble Ca and SO_4, and potentially modifying soil pH (particularly FDGs containing residual alkali). Assessment of environmental impacts of trace contaminants in gypsum-amended soils must account for such effects on the range of soils that may be used for land application.

It should be noted that FDG materials contain varying proportions of fly ash, which tends to contain significantly higher levels of contaminants than purer gypsum materials. Fluidized bed gypsums often contain residual ash, and some wet scrubber technologies may use the gypsum reactor vessel to collect ash rather than electrostatic precipitation. This latter approach could result in a wet scrubber material composed of 50% ash/50% gypsum. The environmental properties of these materials will be heavily influenced by this ash component, and the extensive literature on fly ash (e.g., Adriano et al., 1980; Carlson & Adriano, 1993; Ferrailo et al., 1990; Page et al., 1979; Sharma et al., 1989) should be consulted in evaluating land application of such materials.

Trace Metal Cations in By-Products Gypsums

Metallic cations, including Cd, Pb, Cr, Ni, Cu, Zn, and Hg, are environmentally important in land application of waste products due to their ubiquity in such wastes, and their potential toxicity to plants and animals if allowed to contaminate water or food supplies. Copper and Zn are micronutrients, and their additions may be beneficial on some soils, as long as their solubility does not increase to levels of plant toxicity or excessive uptake.

Total concentrations of most metallic cations of concern in by-product gypsums are relatively low (Table 9-3). Compared to other land-applied materials such as sewage sludge or fly ash, FDG and PG metal concentrations are lower by an order of magnitude or more; even high loadings (10–50 t ha^{-1}) at these low concentrations have little impact on the total metal load in most agricultural soils (Straughan et al., 1980). Plant uptake of these elements should not therefore be impacted by gypsum applications, unless the particular gypsum being used is very high in a given metal, or the gypsum used causes a significant pH decrease when mixed with the soil (which may be possible with PG amendment at high rates on a very sandy soil). Solubilities of metals in gypsum materials appear to be highly pH dependent, as is true in soil systems; in FDG, metals are probably precipitated as the oxide or hydroxide form, and become soluble through dissolution at low pH. Lead solubility in several FDG and fly ash samples is shown in Fig. 9-2, illustrating the dominant effect of pH. The initial alkaline pH of these materials maintains metals in a relatively insoluble state, and unless soil pH falls to levels below 4.5 to 5, metals remain relatively insoluble.

Direct measurements of trace metal uptake from by-product gypsum-amended soils are not common. In one study, Mays and Mortvedt (1986) found no increased uptake of Cd or yield effects on corn (*Zea mays* L.), soybean (*Glycine max* L.) or wheat (*Triticum aestivum*) when 112 t FDG ha^{-1} containing 0.2 mg Cd kg^{-1} was applied to a pH 6 silt loam soil. In general the considerable experience with sewage sludge (King, 1986; Page et al., 1987; Anonymous, 1989) shows that if total metals are below certain critical levels (18 mg Cd kg^{-1}, 300–500 kg^{-1} for Pb and Ni, and 1000 mg kg^{-1} for Cr, Cu, and Zn), food chain effects of land application at moderate rates are not likely. Such guidelines suggest by-product gypsums may be safely land-applied from a plant uptake perspective. Similar guidelines are not available for Hg at this time, and the potential for Hg to accumulate

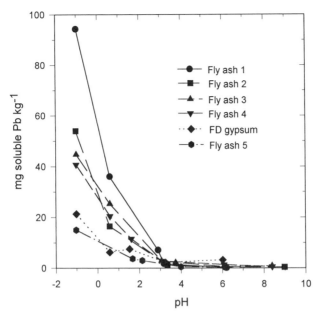

Fig. 9–2. Solubility of Pb from fly ash and FDG materials in water as a function of pH (determined in 1:5 solid/solution extracts).

in FDG (Kaakinen et al., 1975) should be investigated with respect to utilization of this byproduct.

Leaching of metals has been addressed in several studies involving by-product gypsums. May and Sweeney (1983) found little evidence of excessive metal leaching under PG stacks in central Florida, despite relatively low pH (4.5) in the leachate plume. Acidic PG mixed 1:1 with an alkaline soil showed increased levels of metals in the pore water due to decreased pH, but such levels were still below drinking water standards (Luther & Dudas, 1993). In a column experiment, Zhu and Alva (1993) observed a decrease in Cu and Zn in percolates from sandy soil amended with 10 t FDG ha^{-1} compared to controls; dissolved organic carbon (DOC) also was decreased by FDG treatment, suggesting that the higher ionic strength due to the gypsum flocculated the DOC which was enhancing the transport of the trace metals. This later effect may be an important observation, as DOC and dispersed colloids have been suggested to facilitate transport of contaminants through soils (McCarthy, 1990; Kaplan et al., 1993). If transport of contaminants at a particular site is due to their association with mobile colloids, gypsum treatment may aid in flocculating the clay and reducing contaminant transport.

Anionic Contaminants

Most authors agree that anionic trace contaminants are the major environmental threat in land application of coal combustion-derived, by-product

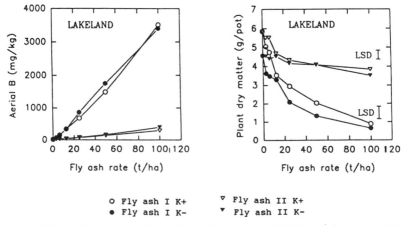

Fig. 9-3. Effect of fly ash rate on tissue B levels (A) and corn yield (B) on Lakeland sand (Typic Quartzipsamment) amended with high-B ash (750 mg B kg^{-1}) or low B ash (80 mg B kg^{-1}), K fertilizer was either added (+) or omitted (−) in separate treatments in the greenhouse pot study (Kukier et al., 1994)

gypsums (Adriano et al., 1980; Straughan et al., 1980). Boron, As, Mo, and Se are volatile elements that may either condense on fly ash particles in the exhaust stack, or be carried into the desulfurization vessel to be captured there (Kaakinen et al., 1975). While their concentrations often are not as high as in fly ash, they may in some cases be sufficiently concentrated to present a hazard when gypsum is applied at high rates. Typically they are much more water soluble than trace metals, and are adsorbed to soil surfaces to a lower degree, particularly in high-sulfate, higher pH environments. Amounts of these elements in PG are generally much lower than in coal combustion by-products (Table 9-3).

Boron has been widely studied in fly ash-amended soils due to its high concentrations and acute toxicity to most plants; it is, however, relatively nontoxic to animals, and is therefore of concern primarily in retarding plant growth. The element also is quite high in many FGD materials, both from wet scrubber and fluidized bed processes, ranging from 50 up to nearly 1000 mg B kg^{-1}. Greenhouse pot studies often show severe phytotoxic effects of B on fly ash and FDG amended soils, particularly if fresh (unleached) ash is used and no drainage is allowed from the pots. Rates of 10 to 20 t ha^{-1} typically seriously reduce yields under these conditions, with B concentrations reaching >1000 mg B kg^{-1} plant tissue (Walker & Dowdy, 1980; Kukier et al., 1994). Yield response of corn to fly ash applications on a Lakeside sand (Psamment) in a pot experiment is shown in Fig. 9-3B, demonstrating a strong yield reduction with a high-B ash (650 mg kg^{-1} total B, pH 11), but a much lesser effect of a lower B, acidic ash (80 mg B kg^{-1}, pH 5.5). Boron in plant tissue showed a strong correlation to yield reduction, reaching over 3000 mg kg^{-1} in corn tissue (Fig. 9-3A). Sandy soils are more prone to show toxicities than silty or clayey soils, as water soluble (and available) B is reduced by sorption; B solubility also is increased at lower

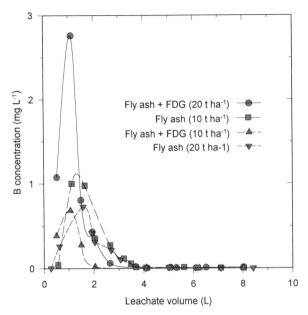

Fig. 9–4. Breakthrough curves for B in soil columns of Appling loamy sand (Typic Kand-haphudult) Ap horizon soil (repacked 50-cm long columns, 1 pore volume = 950 m, fly ash B = 650 mg kg⁻¹, pH = 11.0).

pH, presumably due to enhanced dissolution of the ash or gypsum matrix (Kukier et al., 1994). In field experiments, however, B toxicity is much less likely to occur due to the leaching of excess B out of the root zone. Ransome and Dowdy (1987) observed B toxicity to soybean on a silty soil at 40 t FDG ha⁻¹ the 1st yr after planting, but obtained a positive yield response the following 2 yr to the residual B in the soil. The author's observation (unpublished data) is that applications of up to 30 t ha⁻¹ of either high-B fly ash or FDG had an initial stunting effect on soybean seedlings on sandy and silty soils, but no final effect on yields, as plants recovered from the initial phytotoxic effect due to high B and soluble salt levels induced by fresh ash additions just prior to planting.

Boron readily leaches from FDG- and fly ash-amended topsoils to lower horizons in the soil profile due to its rapid solubility, and its low adsorption potential by soil surfaces. In a leaching study using 50-cm-long columns of Appling loamy sand (Typic Halpludult) with alkaline fly ash and/or FDG gypsum applied to the column surface, water-soluble B was quantitatively moved out of the soil within four pore volumes of deionized water leaching (Fig. 9–4). Initially very high concentrations (1–3 mg B L⁻¹), which are extremely phytotoxic in the root zone, are reduced to nearly zero levels within this leaching period. This minimal leaching requirement suggests that B toxicity can be avoided in field applications by allowing several rainfall events to move B out of the root zone. Groundwater quality is not an issue due to the low toxicity of B to animals.

Arsenic in fly ash and in FDG is relatively less soluble than is B; in three FDG materials tested by the author, only 1 to 11% of the total As was water extractable. This element is quite toxic to animals at levels of 1 to 5 mg kg^{-1}. Native plants growing near fly ash ponds have been found to contain >100 mg As kg^{-1} (Cherry & Guthrie, 1979), while uptake experiments from agricultural soil (pH 5.5) amended at 10% ash gave only 0.2 mg As kg^{-1} in bean (*Phaseolus* spp.) leaf tissue (Furr et al., 1976). Uptake by corn on a Lakeland sand soil from southern Georga amended with high rates of fly ash reached 1.5 mg As kg^{-1} in plant tissue for one high-As ash, but showed considerably less uptake from a lower As ash (Fig. 9–5A); associated extractable As from these soils (using an anion resin extraction method) showed a nearly linear increase in extractable soil As with increasing ash rate (Fig. 9–5B). Arsenic uptake from FDG-amended soils has not been measured, but should be considerably less than for fly ash at similar rates, due to lower As contents in FDG.

Leaching of As to groundwater in situations such as under ash ponds or ash landfill areas has been suggested as a hazard (Roy et al., 1981); however, studies have not documented such contamination (Cherry & Guthrie, 1977; Spencer & Drake, 1987). In the case of land application, oxidized As (V) predominates in the form of AsO_4^{3-}, which is strongly adsorbed by soil colloids, particularly by hydrous oxides at low pH. Arsenic adsorption is shown in Fig. 9–6 for a Cecil sandy loam Ap (Typic Kandhapludult; pH 5.5) and clay Bt horizon (pH 4.8); the isotherms for AsO_4^{3-} are similar to PO_4, which is a very strongly adsorbed anion, particularly on the Bt horizon sample. The AsO_4^{3-}, which occurs under reducing conditions in soils and sediments, is considerably less strongly sorbed. The high concentrations of SO_4 present in gypsum-amended soils are not likely to compete for sorption sites with As, given the weak adsorption of this species (Fig. 9–6). Leaching column experiments performed by the authors using columns of Appling loamy sand Ap soil (Typic Kandhapludult; pH 5.5) with surface-applied fly ash or mixed fly ash-FDG (1:1) did not detect As in effluents over 10 pore volumes of leaching (detection limit = 10 μg As L^{-1} on 0.45-μm filtered solutions). Vertical distribution of As in the soil following leaching showed no increase in extractable As below the surface layer of the amended soil (Fig. 9–7). This strong adsorption of As suggests little potential for groundwater pollution, unless strongly preferential flow occurs or the anion moves in association with mobile colloids. Surface water contamination, which has not been experimentally assessed, is probably more likely in the form of As transported to streams adsorbed to colloids eroded from ash- or FDG-amended soils.

Molybdenum and Se are less toxic to animals, lower in concentration in FDG, and relatively strongly adsorbed by soil surfaces (Mo shown in Fig. 9–6). Roy et al. (1981) and Bannwarth et al. (1992) have suggested these elements may more often benefit soil fertility than pose an enviornmental hazard. Acidic, leached soils are often low in these elements, and Se is chronically low in human and animal diets of many regions around the world (Adriano et al., 1980). Very high Mo availability may induce Cu deficiency in grazing

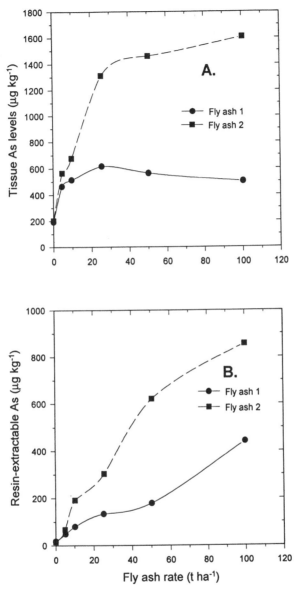

Fig. 9-5. Effect of fly ash rate on As content of corn leaves (*A*) and resin-extractable soil As (*B*) on Lakeland sand (Typic Quartzipsamment) Ap soil (greenhouse pot study; Kukier et al., 1994)

animals, but documented cases are not common (Adriano et al., 1980). Selenium and Mo uptake was enhanced by fly ash applications (10% rate) for bean, increasing Mo levels from 1 to 3 mg kg^{-1} and Se from 0.01 to 0.5 mg kg^{-1} (Furr et al., 1976). Molybdenum increased from 4 to 12 mg kg^{-1} in alfalfa (*Medicago sativa*) grown on a calcareous soil amended with 8% fly ash, while

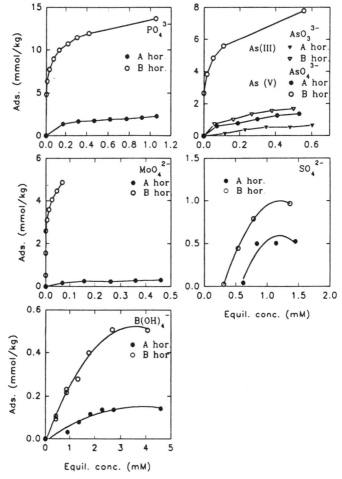

Fig. 9–6. Adsorption isotherms (equilibrium concentrations vs. adsorbed ions) for selected oxy-anions on Cecil (Typic Kandhapludult) Ap and Bt soil samples determined in batch equilibration studies.

Se increased from 0.2 to 4.6 mg kg^{-1} (Straughan et al., 1980). Selenium also was found at elevated levels (0.1 mg kg^{-1}) in a leachate plume under an abandoned fly ash landfill site (Spencer & Drake, 1987). These levels are not considered hazardous, however, and unless the material to be land applied contains extremely high levels of these elements, they should not be of major environmental concern (Carlson & Adriano, 1993).

RADIONUCLIDES IN BY-PRODUCT GYPSUMS

Nearly all geologic materials contain some measurable level of radioactivity emanating from decay of naturally occurring radionuclides found as

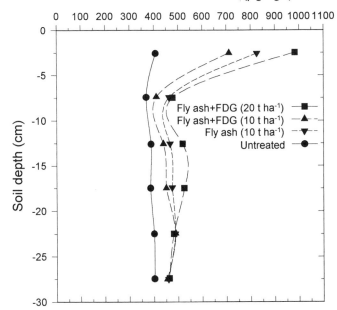

Fig. 9–7. Distribution of dilute double acid-extractable As with depth in fly ash and FDG-amended Appling soil columns after leaching with deionized water.

trace contaminants in minerals. In many rocks the major source of such radio-activity is the ^{238}U decay series, which produces a number of radioactive daughters. Fly ash shows an enrichment in radioactivity of two to five times compared to the activity in the burned coal, which is highly variable (Coles et al., 1978; Sharma et la., 1989). Measured activities of ^{236}Ra in fly ash are in the range of 150 to 200 Bq kg^{-1}, and 100 to 400 Bq kg^{-1} for ^{210}Pb; ra-dionuclides are more enriched in fine ash fractions, some of which may not be captured in electrostatic precipitators (Coles et al., 1978). Some isotopes are also relatively volatile at furnace temperatures (e.g., ^{210}Po), and are like-ly to be transported to desulfurization vessels and captured there (Sharma et al.,1989). No data is available on radiation levels in FDG materials, although Bannwarth et al. (1992) suggests both fly ash and FDG contain no more radionuclides than typical clay soils or mined gypsum.

Radionuclides in Phosphogypsum

Phosphogypsum has received considerable attention with respect to the environmental impacts of its radionuclide content. The USEPA has inves-tigated potential health effects of PG in response to ocean dumping and stack-ing of this material, and has restricted the transport and use of the material at various times, largely based on inhalation hazards associated with gas-eous radionuclide emissions. Phosphogypsum has been used for some years as a Ca amendment for peanut in the southeastern USA; however, this use is currently restricted to material containing less than 400 Bq ^{236}Ra kg^{-1}, which excludes much of the PG stockpiled and being produced in central Florida, which contains 100 to 2000 Bq ^{236}Ra kg^{-1} (Fed. Reg., 1992).

Plant uptake and water contamination with Ra and other radionuclides have been cited as possible pathways for human effects, in addition to inhalation risks. In addition to ^{236}Ra, ^{214,210}Pb, ^{214}Bi, and ^{210}Po are important decay products in the ^{238}U series that may have environmental consequences if PG is land applied at high rates, particularly if reapplied over a number of years. Computation of potential plant uptake using hypothetical radionuclide loadings from PG amendment and uptake coefficients for agronomic plants suggests there is not a hazard from such use (Roessler, 1986). Limited field results are available on this risk, but they generally bear out this hypothesis.

Plant Uptake of Radionuclides

Mays and Mortvedt (1986) studied the effect of varying rates (up to 112 t ha^{-1}) of PG on radionuclide content of agronomic crops in field experiments on a silty soil. The highest PG rate increased the total ^{236}Ra content from 35 Bq kg^{-1} to 75 Bq kg^{-1} in the topsoil. Uptake of Ra in the plant tissue was, however, not significantly affected by PG treatment at any addition level for wheat, soybean, or corn.

In an experiment on Cecil sandy loam (Typic Hapludult) and Tifton loamy sand (Plinthic Kandiudult) with 10 t PG ha^{-1} applied (^{236}Ra = 660 Bq kg^{-1}), uptake of radionuclides was measured in alfalfa tissue and soybean leaves and seeds (Miller & Sumner, 1992). For alfalfa, there were no significant differences in any of the measured radionuclides (^{40}K, ^{214}Pb, ^{214}Bi, ^{226}Ra, ^{210}Po) between PG treated and untreated soils (Table 9-5). The predominant radionuclide was, in fact, ^{40}K, originating largely from fallout from nuclear testing. The values for ^{236}Ra are within the range of 50 to 300 Bq kg^{-1} in grains cited as background levels in Russia (Drichko & Lisachenko, 1984). Similar results were obtained for soybean leaf tissue (sampled at flowering) and in seed collected at harvest (Table 9-6). Analyses of soils from the treated plots showed only marginal increases in total radio-

Table 9-5. Alfalfa uptake of radionuclide from PG-amended soils.†

	Cecil sandy loam		Tifton loamy sand	
Element	Control	PG-treated	Control	PG-treated
	——————————— Bq kg^{-1} dry tissue ———————————			
^{40}K	440 (41)	500 (55)	554 (74)	437 (104)
^{214}Pb	0.56 (0.81)	0.87 (0.75)	3.07 (0.19)	2.74 (1.07)
^{214}Bi	1.15 (0.30)	1.54 (0.50)	3.26 (0.19)	2.92 (1.70)
^{226}Ra	32.9 (4.2)	25.4 (13.1)	73.5 (4.4)	73.6 (24.4)
^{210}Po	7.0 (1.1)	7.5 (0.8)	4.7 (0.7)	4.7 (0.7)

† From Miller and Sumner (1992), mean values reported, standard deviations in parentheses.

nuclides in treated plots, due to the relatively low amounts added compared to background levels in the soils (25 Bq ^{226}Ra kg^{-1} in Tifton, 70 Bq kg^{-1} in Cecil Ap). The lack of plant uptake on these sandy soils suggests limited environmental impact, although the total loading rate of PG was not as high as in the Mays and Mortvedt (1986) study.

Leaching and Profile Distribution of Radionuclides

Leaching of radionuclides from PG-treated areas also has been investigated as an environmental hazard. Sampling of soils and water under PG stacks in central Florida did not show ^{236}Ra levels greater than background levels, suggesting that Ra had not moved out of the stacks to an appreciable degree (May & Sweeney, 1983). Overall background levels in Florida have been shown to be relatively high, however, due to the higher content of radionuclides in the sedimentary rocks of the area (Michel & Jordana, 1987). With respect to PG-treated agricultural soils, Miller and Sumner (1992) found no movement of radionuclides applied to the soil surface in 10 t PG ha^{-1} below the plow layer in a sandy loam soil in Georgia; similarly Mays and Mortvedt (1986) detected only nominal increase in ^{236}Ra below the plow layer after application of 112 t PG ha^{-1}.

In a leaching column experiment on a sandy loam Cecil soil and loamy sand Tifton soil, movement of radionuclides from 30-cm long soil columns was measured over approximately eight pore volumes of leachate (Miller & Sumner, 1992). Levels of ^{236}Ra were not different between control and PG-treated (10 t ha^{-1} equivalent) columns for the Cecil soil, but were higher for Tifton soil (Fig. 9-8). The highest levels, approaching 0.04 Bq L^{-1}, are still within the range of naturally occurring ^{236}Ra in groundwaters in the southeastern region (Michel & Jordana, 1987), but are evidence of the mobility of this radionuclide in sandy soils. No other activity was present in leachates of either soil in significant amounts.

Perspective on Radionuclides in Phosphogypsum

The above experimental evidence points to a preliminary conclusion that radionuclides are not a significant environmental hazard in land application

Table 9-6. Radionuclide uptake by soybean (leaf and seed tissue) on Cecil soil amended with 10 t/ha PG.

Element	Leaf tissue		Seed tissue	
	Control	PG-treated	Control	PG-treated
		Bq/kg dry tissue		
^{40}K	339 (16)	342 (11)	597 (18)	614 (15)
^{214}Pb	3.81 (0.1)	4.18 (0.7)	3.59 (2.26)	5.25 (1.99)
^{214}Bi	3.70 (0.15)	3.74 (0.52)	2.70 (3.11)	5.29 (1.92)
^{226}Ra	87.7 (19.6)	91.4 (13.7)	37.4 (2.9)	55.5 (34.4)
^{210}Po	3.85 (0.52)	5.14 (0.30)	ND‡ (--)	1.52 (1.37)

† From Miller and Sumner (1992), mean values reported, standard deviations in parentheses.
‡ ND = not determined.

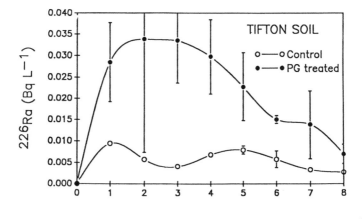

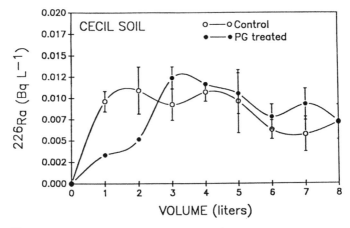

Fig. 9–8. ^{226}Radon leaching from PG amended (10 t ha^{-1} equivalent) columns of Cecil sandy loam and Tifton loamy sand (Miller & Sumner, 1992).

of by-product gypsums; plant uptake and groundwater quality concerns do not appear at this time to be borne out by the evidence collected on PG-amended soils. Unless high rates (>10–50 t ha^{-1}) of such materials are repeatedly applied to soils that are quite sandy and with little sorptive capacity, the relatively low loading rates of radioactivity and adsorption processes occurring in soils appear to mitigate the potential of plant or water contamination. Radionuclide contents of most fly ashes (and presumably FDG) are considerably lower than PG, and similary should pose little threat to the environment. Inhalation exposure pathways are not considered here, however, and obviously must be evaluated separately in formulating recommendations and regulations concerning the handling and use of these materials.

CONCLUSIONS

The impetus for considering land application of by-product materials is to both enhance productivity of waste producer and user, and to minimize environmental hazards and costs associated with disposal of such materials. Only if land application can be accomplished in an economically viable and environmentally safe way will it be adopted by users, and facilitated by the producers. The economics of such practices will depend on marketplace and political forces, which are beginning to favor recycling over disposal alternatives. Much still hinges on the question of whether land application is at least as environmentally safe as approved disposal practices such as lined and monitored landfills.

Our current information is far from complete with respect to by-product gypsum materials. There is tremendous variation in composition, particularly with respect to trace contaminants, that will probably require analyses in order to ascertain the quality of such material for land application uses. Trace metal cation concentrations are low in most of these materials, and can be maintained in relatively insoluble states by proper soil management. Mercury is one contaminant that needs further study, in terms of concentrations of FDG and fate in amended soils. Anionic contaminants such as B, Mo, As, and Se pose the major threat environmentally, and a good deal less is known of their chemistry in waste materials and soils. Most by-product gypsums are not seriously contaminated with these elements, but FDG mixed with fly ash may be an instance where rates and management of application will need to be considered based on contaminant levels.

Gypsum materials may be environmentally beneficial in many cases; the potential to reduce surface crusting and thereby decrease runoff and erosion is clearly an application where use on croplands, urban soils, and disturbed soil areas may significantly improve water quality and maintain soil productivity. The capacity to enhance root growth through Ca addition in acid soils (Ritchey et al., 1995, see Chapter 8) suggests its use on eroded soils and in forests where acid rain and base depletion has degraded soil productivity (Bannwarth et al., 1992). The general flocculating effect of gypsum-containing solutions on soil colloids also may minimize colloid-assisted contaminant transport in the subsurface. Results on environmental impacts to date are sufficiently optimistic to suggest that research on land application of by-product gypsums should move forward, in conjunction with further studies on the overall environmental impacts of such uses.

REFERENCES

Anonymous. 1993. Annual report of research. Florida Inst. Phosphate Res., Bartow, FL.

Anonymous. 1989. Peer review of USEPA proposed standards for the disposal of sewage sludge. USDA-SCS, Washington, DC.

Adriano, D.C., A.L. Page, A.A. Elseewi, A.C. Chang, and I. Straughan. 1980. Utilization and disposal of fly ash and other coal residues in terrestrial ecosystems: A review. J. Environ. Qual. 9:333–344.

Agassi, M., I. Shainberg, and J. Morin. 1990. Slope, aspect, and phosphogypsum effects on runoff and erosion. Soil Sci. Soc. Am. J. 54:1102–1106.

Agassi, M., I. Shainberg, and J. Morin. 1981. Effect of electrolyte concentration and soil sodicity on infiltration rate and crust formation. Soil Sci. Soc. Am. J. 45:848–851.

Bannwarth, H., S. Schmelz, and I. Wllke. 1992. Ecological utilization possibilities of brown coal ash and flue gas gypsum (FGD gypsum). VGB Kraftwerkstechnik. 72:596–608.

Behrens, G.P., and O.W. Hargrove. 1980. Evaluation of Chiyoda Thoroughbred 121 flue gas desulfurization process and gypsum stacking. Final Rep. Proj. 536-4. Electric Power Res. Inst., Palo Alto, CA.

Ben-Hur, M., R. Stern, A.J. van der Merwe, and I. Shainberg. 1992. Slope and gypsum effects on infiltration and erodibility of dispersive and nondispersive soils. Soil Sci. Soc. Am. J. 56:1571–1576.

Carlson, C.L., and D.C. Adriano. 1993. Environmental impacts of coal combustion residues. J. Environ. Qual. 22:227–247.

Cherry, D.S., and R.K. Guthrie. 1977. Toxic metals in surface waters from coal ash. Water Resour. Bull. 13:1227–1236.

Cherry, D.S., and R.K. Guthrie. 1979. The uptake of chemical elements from coal ash and settling basin effluent by primary producers. II. Relation between concentrations in ash deposits and tissues of grasses growing on the ash. Sci. Total Environ. 13:27–31.

Coles, D.G., R.C. Ragaini, and J.M. Ondov. 1978. Behavior of natural radionuclides in western coal-fired power plants. Environ. Sci. Technol. 12:442–446.

Drichko, V.F., and E.P. Lisachenko. 1984. Background concentrations of ^{226}Ra, ^{227}Th, and ^{40}K in cultivated soils and agricultural plants. Sov. J. Ecol. 15:81–85.

Ferraiolo, G., M. Zilli, and A. Converti. 1990. Fly ash disposal and utilization. J. Chem. Tech. Biotechnol. 47:281–305.

Frenkel, H., J.O. Goertzen, and J.D. Rhoades. 1978. Effects of clay type and content, exchangeable sodium percentage, and electrolyte concentration on clay dispersion and soil hydraulic conductivity. Soil Sci. Soc. Am. J. 42:33–39.

Furr, A.K., W.C. Kelly, C.A. Bache, W.H. Gutemann, and D.J. Lisk. 1976. Multielement uptake by vegetables and miller grown in pots on fly ash amended soil. J. Agric. Food Chem. 24:885–888.

Kaakinen, J.K., R.M. Jorden, M.H. Lawasani, and R.E. West. 1975. Trace element behavior in coal-fired power plant. Environ. Sci. Technol. 9:862–869.

Kaplan, D.I., P.M. Bertsch, D.C. Adriano, and M.E. Sumner. 1993. Soil-borne mobile colloids as influenced by water flow and organic carbon. Environ. Sci. Technol. 27:1193–1200.

Kazman, Z., I. Shainberg, and M. Gal. 1983. Effect of low levels of exchangeable Na (and phosphogypsum) on the infiltration rates of various soils. Soil Sci. 135:184–192.

Kemper, W.D., and L. Noonan. 1970. Runoff as affected by salt treatment and soil texture. Soil Sci. Soc. Am. Proc. 34:126–130.

King, L.D. 1986. Introduction to the agricultural use of municipal and industrial sludges. p. 1–10. In L.D. King (ed.) Agricultural use of municipal and industrial sludges in the southern United States. Southern Coop. Bull. no. 314, North Carolina State Univ., Raleigh, NC.

Kukier, U., M.E. Sumner, and W.P. Miller. 1994. Boron release from flyash and its uptake by corn. J. Environ. Qual. 23:596–603.

Luther, S.M., and M.J. Dudas. 1993. Pore water chemistry of phosphogypsum-treated soil. J. Environ. Qual. 22:103–108.

May, A., and J.W. Sweeney. 1983. Evaluation of radium and toxic element leaching characteristics of Florida phosphogypsum stockpiles. U.S. Dep. of Interior/Bureau of Mines. Rep. 8776. U.S. Dep. Interior, Washington, DC.

Mays, D.A., and J.J. Mortvedt. 1986. Crop response to soil applications of phosphogypsum. J. Environ. Qual. 15:78–81.

McCarthy, J.F. 1990. The mobility of colloidal particles in the subsurface. Hazard Mater. Control 3:38–43.

Michel, J., and M.J. Jordana. 1987. Nationwide distribution of Ra-228, Ra-226, Rn-222, and U in groundwater. p. 227–240. In Radon in groundwater, Sommerset NJ. April. Proc. Natl. Water Well Assoc. Lewis Publ., Worthington, OH.

Miller, W.P. 1987. Infiltration and soil loss of three gypsum-amended Ultisols under simulated rainfall. Soil Sci. Soc. Am. J. 51:1314–1320.

Miller, W.P., and M.K. Baharuddin. 1986. Relationship of soil dispersibility to infiltration and erosion of southeastern soils. Soil Sci. 142:235–240.

Miller, W.P., and M.K. Baharuddin. 1987. Particle size of interrill-eroded sediments from highly weathered soils. Soil Sci. Soc. Am. J. 51:1610-1615.

Miller, W.P., H. Frenkel, and K.D. Newman. 1990. Flocculation concentration and sodium/calcium exchange of kaolinitic soil clays. Soil Sci. Soc. Am. J. 54:346-351.

Miller, W.P., and J. Scifres. 1988. Effect of sodium nitrate and gypsum on infiltration and erosion of a highly weathered soil. Soil Sci. 145:304-309.

Miller, W.P., and M.E. Sumner. 1992. Impacts from radionuclides on soil treated with phosphogypsum. USEPA Final Proj. Rep., USEPA, Washington, DC.

Norton, L.N. 1995. Mineralogy of high calcium/sulfur-containing coal combustion by-products and their effect on soil surface sealing. p. 87-106. In D.L. Karten et al. (ed.) Agricultural utilization of urban and industrial by-products. ASA Spec. Publ. 58. ASA, CSSA, and SSSA, Madison, WI.

Oster, J.D. 1982. Gypsum usage in irrigated agriculture: A review. Fert. Res. 3:73-89.

Page, A.L., A.A. Elseewi, and I.R. Straughan. 1979. Physical and chemical properties of fly ash from coal-fired power plants with reference to environmental impacts. Residue Rev. 71:83-120.

Page, A.L., T.J. Logan, and J.A. Ryan. 1987. Land application of sludge: Food chain implications. Lewis Publ., Chelsea, MI.

Quirk, J.P., and R.K. Schofield. 1955. The effect of electrolyte concentration on soil permeability. J. Soil Sci. 6:163-178.

Ransome, L.S., and R.H. Dowdy. 1987. Soybean growth and boron distribution in a sandy soil amended with scrubber sludge. J. Environ. Qual. 16:171-175.

Rengasamy, P., R.S.B. Greene, G.W. Ford, and A.H. Mehanni. 1984. Identification of dispersive behavior in the management of red-brown earth. Aust. J. Soil Res. 22:413-422.

Rinehart, J.C., G.R. Blake, J.C.F. Tedrow, and F. Bear. 1953. Gypsum for improving drainage of wet soils. New Jersey Agric. Exp. Stn. Bull. 77.

Ritchey, K.D., D.M.G. de Sousa, C.M. Feldhake, and R.B. Clark. 1995. Improved water and nutrient uptake from subsurface layers of gypsum-amended soils. p. 157-181. In D.L. Karlen et al. (ed.) Agricultural utilization of urban and industrial by-products. ASA Spec. Publ. 58. ASA, CSSA, and SSSA, Madison, WI.

Roessler, C.E. 1986. Radiological assessment of the application of phosphogypsum to agricultrural land. p. 5-23. In J.M. Lloyd, Jr. (ed.) Florida Inst. Phosphate Research, Miami, FL. 10-12 December. Proc. 2nd Int. Symp. Phosphogypsum. Florida Inst. Phosphate Res., Barton, FL.

Roy, W.R., R.G. Thiery, R.M. Schuller, and J.J. Suloway. 1981. Coal fly ash: A review of the literature and proposed classification system with emphasis on environmental impacts. Environ. Geol. Notes 96:1-69.

Shainberg, I. 1992. Chemical and mineralogical components of crusting. p. 33-54. In M.E. Sumner and B.A. Stewart (ed.) Soil crusting: Chemical and physical processes. Lewis Publ., Boca Raton, FL.

Shainberg, I. J.D. Rhoades, and R.J. Prather. 1981. Effect of low electrolyte concentration on clay dispersion and hydraulic conductivity of a sodic soil. Soil Sci. Soc. Am. J. 45:273-277.

Shainberg, I., M.E. Sumner, W.P. Miller, M.P. Farina, M.A. Pavan, and M.V. Fey. 1989. Use of gypsum on soils: A review. Adv. Soil Sci. 9:1-111.

Sharma, S., M.H. Fuelkar and C.P. Jayalakshmi. 1989. Fly ash dynamics in soil-water systems. Crit. Rev. Environ. Control 19:251-275.

Spencer, L.L., and L.D. Drake. 1987. Hydrogeology of an alkaline fly ash landfill in eastern Iowa. Groundwater 25:519-526.

Stern, R., M.C. Laker, and A.J. van der Merwe. 1991. Field studies on the effect of soil conditioners and mulch on runoff from kaolinitic and illitic soils. Aust. J. Soil Res. 29:249-261.

Straughan, I., A.A. Elseewi, and A.L. Page. 1980. Mobilization of selected trace elements in residues from coal combustion with special reference to fly ash. p. 389-402. In D.D. Hemphill (ed.) Trace substances in environmental health. Vol. 12. Univ. Missouri, Columbia.

Sumner, M.E. 1992. The electrical double layer and clay dispersion. p. 1-32. In M.E. Sumner and B.A. Stewart (ed.) Soil crusting: Chemical and physical processes. Lewis Publ., Boca Raton, FL.

Sumner, M.E., M.V. Fey, and P.W. Farina. 1986. Amelioration of acid subsoils with phosphogypsum. p. 211-230. In J.M. Lloyd, Jr. (ed.) Florida Inst. Phosphate Res., Miami, FL. 10-12 December. Proc. 2nd Int. Symp. Phosphogypsum. Florida Inst. Phosphate Res., Bartow, FL.

U.S. Environmental Protection Agency. 1992. National emissions standards for hazardous air pollutants; national emissions standards for radon emissions from phosphogypsum stacks. Fed. Reg. 75:23,305.

van Olphen, H. 1977. An introduction to clay colloid chemistry. 2nd ed. John Wiley & Sons, New York.

Walker, W.J., and R.H. Dowdy. 1980. Elemental composition of barley and ryegrass grown on acid soils amended with scrubber sludge. J. Environ. Qual. 9:27-30.

Warrington, D., I. Shainberg, M. Agassi, and J. Morin. 1989. Slope and phsophogypsum's effects on runoff and erosion. Soil Sci. Soc. Am. J. 53:1201-1205.

Zhu, B., and A.K. Alva. 1993. Trace metal and cation transport in a sandy soils with various amendments. Soil Sci. Soc. Am. J. 57:723-727.

10 The Alkaline Stabilization with Accelerated Drying Process (N-Viro): An Advanced Technology to Convert Sewage Sludge into a Soil Product

Terry J. Logan

The Ohio State University
Columbus, Ohio

Jeffrey C. Burnham

BioCheck Laboratories
Toledo, Ohio

Since the advent of nationwide advanced wastewater treatment in the late 1960s in the USA, much attention has been placed on treatment and disposal or utilization of the residual solids of wastewater processing. Those solids, referred to as sewage sludge, and more recently biosolids, have traditionally been placed in municipal landfills, incinerated, ocean dumped, or applied to agricultural land. Lesser amounts have been used for reclamation of disturbed lands, and given away or sold for gardening or commercial horticulture. Prior to the 1980s, most sludge was biologically digested as a means of stabilizing the sludge organics and to partially kill pathogens. In the 1980s, more advanced technologies for sludge treatment emerged that produced a pathogen-free product and stabilized sludge organic matter. The two most widely used approaches are biological composting and alkaline stabilization. While composting relies on biological degradation, heat and drying to kill pathogens and stabilize sludge organic matter, alkaline stabilization utilizes a combination of high pH, heat and drying to achieve the same purpose. This paper summarizes the N-Viro alkaline sludge stabilization process, the chemical and physical properties of the final product, N-Viro Soil, and identifies beneficial uses for the product.

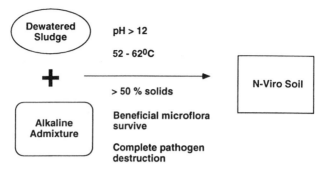

Fig. 10-1. Basic components of the advanced ASAD process for sewage sludge treatment (N-Viro).

THE N-VIRO ALKALINE STABILIZATION
WITH ACCELERATED DRYING PROCESS

The basis of the Alkaline Stabilization with Accelerated Drying (ASAD) Process (Fig. 10-1) is to destroy pathogens through a combination of the following stresses: alkaline pH, accelerated drying, high temperature, high ammonia, salts, and indigenous microflora. These stresses are produced in the sludge/alkaline admixture (AA) through the unique properties of the AA. Typical properties of AAs are their alkali content (Stresses 1, 3, and 4), fine particle size (Stress 2), and low moisture content (Stress 2). As with compost, mesophilic temperatures (52–62 °C) and a soil-like environment result in complete pathogen kill and contribute to the growth of indigenous microorganisms that suppress the regrowth of pathogens or putrifying organisms (Fig. 10–2).

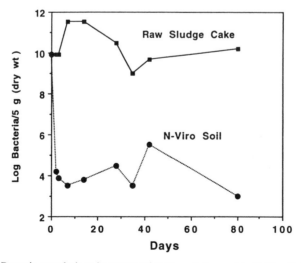

Fig. 10-2. Bacteria populations in untreated sewage sludge and in N-Viro Soil following the ASAD process. Residual bacteria are nonpathogenic heterotrophs (Burnham et al., 1992b).

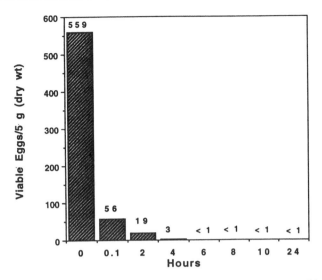

Fig. 10-3. Destruction of *Ascaris* ova in the ASAD process. *Ascaris* were added to the raw sludge (Burnham et al., 1992b).

The relatively high population of microorganisms in the final product (10^3-10^4/g) produces a slow composting in the material in which degradable organics from the sludge are stabilized. The ASAD Process meets current USEPA regulations (USEPA, 1993) for complete (Class A) pathogen destruction. For example, the ASAD Process reduced viable Helminth ova from >500 to <1 per 5 g dry sludge (Fig. 10-3).

A number of waste alkaline materials are used singly or together as the AA in the ASAD Process. These include cement kiln dust (CKD), lime kiln dust (LKD), various coal combustion ashes that include flue gas desulfurization byproduct (FGD), wood ash, or alkaline fly ash. If the AA contains enough free lime [CaO, Ca(OH)$_2$ or other strong alkali] to give a pH rise to >12 and an exothermic reaction, necessary to achieve desired temperatures (52-62 °C), no other additive is needed. The CaO is added to supplement the free lime content of the AA if it is not "hot" enough.

Raw primary/waste activated, waste activated, or digested sludge, with solids content of 18 to 40% is used. The use of raw primary/waste activated sludge is preferred to conserve N in the final product and to reduce NH$_3$ emissions from the alkali addition. Anaerobically digested sludges can be treated, but NH$_3$ loss is greatest. Sludge and AA are mixed with a pug mill or screw blender (Fig. 10-4). The ratio of AA to sludge solids varies primarily with the solids content of the sludge, with a higher ratio being used for sludges with lower solids content. In general, AA dose ranges from 25 to 50% of the dewatered sludge wet weight. There is a tradeoff here between the cost of sludge dewatering and the residual lime content of the final product. Beneficial use options for the N-Viro product should be considered together with sludge dewatering costs in designing a given system. With proper mixing speeds, the resultant product is a granular, easy-to-handle soil-like

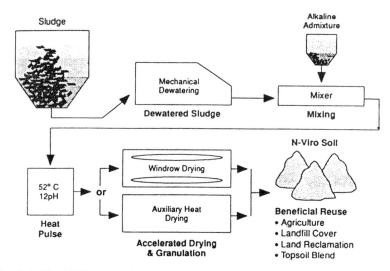

Fig. 10–4. The ASAD process for sludge treatment using mechanical or auxiliary heat drying.

material that is further processed by one of two methods; Alternative 1 or Alternative 2.

Alternative 1

The sludge/AA mixture is air dried while the pH remains above 12.0 for at least 7 d. The N-Viro Soil must be held for at least 30 d and until solids content is at least 65% by weight. Ambient air temperatures during the first 7 d of processing must be above 5 °C.

Alternative 2

The sludge/AA mixture is heated while the pH exceeds 12.0 using exothermic reactions from the alkali in the AA. Temperatures must be 52 °C throughout the mixture. The material must be stored in such a way (e.g., in a bin) so as to maintain uniform minimum temperatures for at least 12 h. Following this heat pulse, the N-Viro Soil is air dried (while pH remains above 12.0 for at least 3 d) by windrowing, or by rotary kiln dryer, until the solids content is 60 to 65% by weight. This alternative has no ambient air temperature requirements.

The advantages of the rotary kiln dryer over windrowing are shorter drying time (30 min vs. 3–7 d), less space (hundreds of m^3 instead of thousands), and greater odor control (it is easier to scrub the discharge from the dryer than from a windrow building). Combined capital, operating and maintenance costs are about the same. The dryer is more likely to be used at large facilities in urban areas where space and odor control are more critical.

An important feature of the ASAD Process in producing a marketable sludge-derived product is odor control. The AA used in the ASAD Process

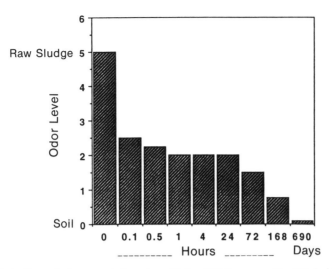

Fig. 10-5. Control of sewage sludge odor with the ASAD process. The 690-d old product was stored out of doors (Burnham et al., 1992a).

controls odor by: (i) adsorption to the AA, (ii) destruction of putrefying organisms, (iii) drying, and slow composting of the sludge organics. Raw sludge odor was reduced immediately on addition of AA (Fig. 10-5) as a result of reactions of sludge with the AA and with drying, while continued improvement in odor was achieved when the final product was stored outdoors for 690 d (Fig. 10-5).

PROPERTIES OF N-VIRO SOIL

A study of 28 N-Viro Soil materials from various operating facilities in the USA, the United Kingdom and Australia was initiated in 1992 to develop baseline characteristics of N-Viro Soil materials that would reflect differences in process variables such as sludge type and quality, type and dose rate of AA and age of the material. For example, N-Viro Soils from Toledo, Ohio, were made with either LKD or FGD byproduct, while the Oakland, California, N-Viro Soil was made with alkaline fly ash. The other 25 materials were made with CKD or mixtures of CKD and other alkaline reagents, such as CaO or various combination ashes such as bottom ash or fluidized bed ash. Several of the materials (Tomball, Tx, Sydney, Australia, and Durham, NC) had been aged with extended windrowing or by extended storage while the majority were freshly made. These materials were studied as received and variability in their properties reflects the range of operating conditions to be found at the more than 30 N-Viro facilities worldwide.

Soil physical characterization data are also presented for four Ohio mineral soils for comparison. These soil had been selected for a separate study and the characterizations were performed at a different time from that of the N-Viro Soils. However, identical methods and equipment were employed in both cases. The soils were selected to give a range of texture and structure and included: Miamian silt loam (fine, mixed, mesic Typic Hapludalf), Kokomo silty clay loam (fine, mixed mesic Typic Argiaquoll), the B_2 horizon of a Paulding clay (very-fine, illitic, nonacid, mesic Typic Haplaquept), and the B_2 horizon of a Hazleton sandy loam (loamy-skeletal, mixed, mesic Typic Dystrochrept).

Chemical Properties

The pH of N-Viro Soil measured in a water suspension is determined by the residual $Ca(OH)_2$ in the product because of the much higher water solubility of $Ca(OH)_2$ in the product because of the much higher water solubility of $Ca(OH)_2$ relative to that of $CaCO_3$. The pH of 28 N-Viro Soil materials ranged from 7.2 to 12, with 26 of the 28 in a narrow range of 11.5 to 12.2 (Table 10–1). The two materials with pH around seven have been carbonated by storage and aging.

The pH of N-Viro Soil when added to soil is very different than that of the material itself. Because of the higher reactivity of the $Ca(OH)_2$ component of N-Viro Soil than that of $CaCO_3$, the initial acid neutralization reaction in an acid soil is

$$Ca(OH)_2 + 2H^+ = Ca^{2+} + 2H_2O$$

This reaction is very rapid; in the laboratory it is observed in a matter of minutes, and in soil, observations as recent as a few days after N-Viro Soil application show the same reaction. Following this initial reaction, acid neutralization occurs by calcite dissolution

$$CaCO_3 + H^+ = Ca^{2+} + HCO_3^-$$

This is the reaction that occurs in soil when limestone is added. If N-Viro Soil is added to a calcareous soil for purposes other than liming, the reaction that occurs is with $Ca(OH)_2$ alone; since the soil already is in equilibrium with $CaCO_3$, there will be little or no reaction of the $CaCO_3$ in N-Viro Soil (Fig. 10–6). The final pH that will be achieved in a calcareous soil will be that of $CaCO_3$ unless the amount of $Ca(OH)_2$ applied is greater than can be neutralized by the soil. It is important to note that even soils with pH > 8 have some base-neutralizing capacity. Field experience with N-Viro Soil applied to calcareous soil at rates up to 40 Mg/ha show that soil pHs never exceeded eight. This suggests that the $Ca(OH)_2$ component of the acid neutralizing capacity of N-Viro Soil is not the dominant form.

Soluble salts are high in N-Viro Soil (Table 10–1), as is the case with sludge composts or heat-dried sludges, since these are all processes involving the evaporation of water and concentration of salts from the sludge. Some salt also is contributed from the AA in N-Viro Soil.

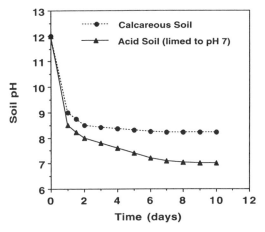

Fig. 10-6. Schematic representation of the N-Viro Soil pH reaction in acidic and calcareous soils. N-Viro Soil added to acidic soil to achieve a target pH of 7 (Logan, 1992a).

Total C content of N-Viro Soil is about 12%, giving an estimated organic matter content of 20% or so (Table 10-1). This is about half that of a digested sludge, the difference a result of the added mineral AA. The main factor determining the organic C content of N-Viro Soil is the dose rate of AA which, as discussed previously, increases as the solids content of the dewatered sludge decreases.

N-Viro Soil contains about 1% total N (Table 10-1), almost entirely as organic N because the high pH drives off free NH_3 from the sludge. Labora-

Table 10-1. Summary of chemical properties of the 28 N-Viro Soils (Logan & Harrison, 1995).

Parameter	Maximum	Minimum	Mean
pH	12.8	7.24	11.9
EC, dS/m	15	3	9
Total			
		%	
C	20.6	5.6	12.1
N	1.93	0.22	0.89
P	1.14	0.21	0.41
Na	0.59	0.06	0.18
K	4.31	0.09	1.26
Mg	9.45	0.20	1.00
Ca	39.8	10.5	25.0
CCE*	77	18	53
		mg/kg	
As	35.8	2.4	7.6
Cd	4.02	<0.01	0.83
Cu	294	43	134
Mo	3.45	0.02	1.38
Ni	563	6.6	55
Pb	452	<0.01	48
Se	3.67	0.58	1.69
Zn	426	39	186

tory incubation studies suggest that about 15 to 20% of the N is mineralized in 60 d (Table 10-2). Actual N mineralization was about 30% of sludge N, similar to that observed for digested sludges, but it was observed that about half of the liberated NH_3 was strongly adsorbed to the AA, perhaps by the formation of $(NH_4)_2CO_3$ (Logan et al.,1989). In a 20-wk incubation of two soils from Minnesota, N mineralization ranged from 12 to 80% on one soil and from 30 to 46% on the other (Smith et al., 1993).

Total P content of N-Viro Soil varies from 0.2 to 1.1% (Table 10-1), with a mean of 0.4%. This is much lower than that of digested sludge which is typically 1 to 3%. All of the P is contributed by the sludge and probably exists in the N-Viro Soil as a combination of organic P and Ca phosphates. X-ray diffraction indicated the presence of apatite in N-Viro Soil from Toledo, Ohio, manufactured with LKD as AA (Logan, 1993). The LKD diffractogram showed no evidence of apatite.

Potassium content of N-Viro Soil varied widely (Table 10-1), but on average contained more than 1% K, in contrast to digested sewage sludges which usually contain <0.1% K. The K in N-Viro Soil is contributed by CKD which can contain as much as 5 to 7% K. More than 90% of the K in N-Viro Soil is water soluble.

Magnesium content of N-Viro Soil is about the same as that of K (Table 10-1), the high value of 9.5% coming from a material made with a Mg LKD. Sodium content is low, usually <0.2% (Table 10-1).

The chemistry of N-Viro Soil is dominated by Ca, with total Ca content ranging from 10 to 40% (Table 10-1). A combination of x-ray diffraction and titration data suggest that the majority of Ca is in the form of $CaCO_3$, with lesser amounts as $Ca(OH)_2$ or other Ca salts like gypsum (Logan, 1993). About 20% of the Ca in N-Viro Soil is water soluble.

The N-Viro Soil has a high acid neutralizing capacity, ranging from 18 to 77% with a mean of 53% (Table 10-1). The CCE is primarily a function of the AA dose rate. Dilute acid titration of N-Viro Soil shows that about one-third of the acid neutralizing capacity is more alkaline than $CaCO_3$, probably in the form of $Ca(OH)_2$, or Na_2CO_3 (Logan, 1993). Some preliminary data, using the sucrose titration test for $Ca(OH)_2$ (ASTM, 1993), suggests that fresh N-Viro Soil contains about 4 to 8% $Ca(OH)_2$. As discussed

Table 10-2. Observed and predicted N mineralization from N-Viro Soil based on laboratory incubation (Logan et al., 1989).

Incubation time	Nitrogen mineralized		
		Predicted	
	Observed	Uncorrected	Corrected for NH_3 fixation
d		% of TKN†	
30	12	32	17
60	16	36	20

† TKN = Total Kjeldahl N.

previously, the strong alkalinity in N-Viro Soil is responsible for its high pH, but pH's fall rapidly in soil to that buffered by $CaCO_3$, between 7.5 and 8.2.

Trace element contents of N-Viro Soil are low compared to that of U.S. sludges (USEPA, 1990). Trace elements are primarily contributed by the sludge, but the oxyanions As, Mo, and Se tend to be higher in the AA, particularly the combustion ashes. The N-Viro Soil trademark requires that the product meet the pollutant concentration limits for 10 trace elements in the 503 sludge regulations (USEPA, 1993). This requires a continuous quality control program to ensure that neither the sludge nor the AA contributes trace elements that would cause the N-Viro Soil to exceed the regulated trace element limits. The ASAD Process also immobilizes the metallic trace elements by coprecipitating them as carbonate, hydroxide, phosphate, sulfate, and silicate minerals. It is likely that metals also are bound by sorption or complexation reactions to oxides and organic matter in the N-Viro Soil. X-ray diffraction identified calcite, gypsum, portlandite and apatite as crystalline phases in Toledo N-Viro Soil (Logan, 1993), and trace elements could have coprecipitated with any of these minerals. An important question regarding the long-term fate of the trace metals in N-Viro Soil is their solubility under acid conditions, i.e., after the alkalinity in the material has been neutralized. In the absence of long-term field studies (the ASAD technology only dates from the mid-1980s), a simulated acid leaching test was conducted on soils from a field experiment in northwest Ohio (Bennett, 1989). The N-Viro Soil from Toledo had been applied at a rate of 22.4 Mg/ha, and surface soil samples were taken from the fertilizer control plot and from the plot receiving N-Viro Soil. The N-Viro Soil contained 9, 87, 433, 118, and 120 mg/kg of Cd, Cu, Ni, Zn, Cr, and Pb, respectively (Logan et al., 1989). Both soils had a pH of 7.5 to 8.0. They were acid treated to lower pH to 5.0 and then leached with the USEPA toxicity leaching procedure for hazardous wastes. Leachable metals in the two soils were either not significantly different (Cd, Cu, Cr, Zn) or were lower in the N-Viro plots (Fig. 10-7), suggesting that the metal forms in the N-Viro Soil plot were more acid stable than in the soil receiving fertilizer. Of the coprecipitated solid phases likely to form in N-Viro Soil, only the metal phosphates (Ma et al., 1993) are likely to be stable under acid conditions.

Physical Properties

The N-Viro Soils have an average solids content of 62%, particle density of 1.96 Mg/m^3, bulk density of 0.59 Mg/m^3, and total porosity of 70% (Table 10-3). Mean moisture retention was 66, 58, 34 and 31% by volume at saturation, 5.9 kPa, 33 kPa, and 1.5 MPa, respectively. Available water content was 27% by volume (5.9 kPa–1.5 MPa). Uncompacted and compacted saturated hydraulic conductivities were high, 3×10^{-2} and 9×10^{-4} cm/s, respectively. An average of 20, 20, 12, 7, 6 and 35% of N-Viro Soil, respectively, were in >5-, 5- to 2-, 2- to 1-, 1- to 0.5-, 0.5- to 0.25-, and <0.25-mm water stable aggregates. Fifty-six percent of the dry materials were >2 mm in diameter, almost all of this (88%) in sizes <16 mm. Of

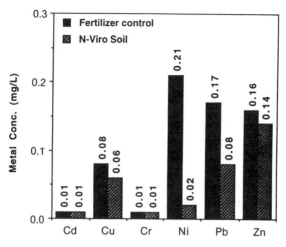

Fig. 10-7. EP toxicity leachate metal concentrations from field plot soils receiving commercial fertilizer or N-Viro Soil (22.4 Mg/ha). Soil pH was decreased to five with acid addition prior to the leaching test (Bennett, 1989).

the <2-mm fraction, 69% was >0.5 mm. The N-Viro Soils have a shrinkage of 58%, and Atterberg liquid limit, plastic limit and plasticity index of 76, 64 and 18% H_2O by weight, respectively. Parameter values for the N-Viro Soils were similar to those of the four comparison mineral soils, and suggest that these materials are similar in physical attributes to those of medium to fine textured, porous soils with granular, stable aggregates and nonplastic consistency. The moisture retention characteristics are those of a fine textured soil dominated by fine pores and medium available water holding capacity, and these materials are highly permeable, even when compacted.

USES OF N-VIRO SOIL

The N-Viro Soils are currently produced at more than 30 locations in the USA, United Kingdom, and Australia. Because of its unique attributes (Table 10-4), it can be used in a variety of ways. Existing markets, or markets undergoing developmental research include: agricultural limestone substitute/low analysis fertilizer, land reclamation, soil amendment/urban soils, soil blend ingredient, and landfill cover material.

Agricultural Limestone Substitute/Low Analysis Fertilizer

This is the most developed N-Viro Soil market, particularly in Ohio, New York, Florida, Minnesota, Iowa, Australia and the United Kingdom. Application rates are based on a calcium carbonate equivalency (CCE) analysis, with the average CCE about 50% (Table 10-1) on a dry weight basis, or about 30 to 40% on a product basis. As a limestone substitute, N-Viro

Table 10-3. Summary of physical properties of 28 N-Viro Soils (Logan & Harrison, 1994).

Parameter	Units	N-Viro Soil		Sandy loam	Silty clay loam	Silt loam	Clay
		Mean	SD				
Solids	% by wt	62	8	ND†	ND	ND	ND
Particle density	Mg/m³	1.96	0.18	2.68	2.62	2.65	2.60
Bulk density	Mg/m³	0.59	0.11	1.62	1.16	1.32	1.12
Porosity	% by volume	69.9	3.5	40	56	50	57
Moisture retention	% by volume						
Saturation		66	6	34	52	47	57
5.9 kPa		58	7	19	40	37	48
33 kPa		34	4	7	34	31	44
1.5 MPa		31	5	3	21	18	30
Avail. H_2O		27	7	16	21	19	18
Saturated hydraulic conductivity	cm/s	0.028	0.05	0.019	0.025	0.015	0.00
Aggregate stability mean weight diameter	mm	1.99	0.54	0.73	1.58	1.26	1.48
Liquid limit	% by wt.	76.0	20.8	ND	40.7	31.2	54.9
Plastic limit	% by wt.	63.7	16.0	ND	23.6	20.9	35.0
Particle size	% by wt.						
>2 mm		55.6		ND	ND	ND	ND
2–0.05 mm		44.4		ND	ND	ND	nD
<0.05 mm		0.0		ND	ND	ND	ND

† ND = not determined.

Table 10-4. Significant product attributes of N-Viro Soil.

Pathogen free	N-Viro Soil is USEPA-certified, Class A, complete pathogen destruction process.
Community acceptable appearance	The product has a more pleasing appearance than cake of liquid sludge
Acceptable odor	Odor is controlled in the product as a result of adsorption by the alkaline reagent and organic matter decomposition by indigenous microorganisms
Dry and granular properties	N-Viro Soil product has >60% solids and maintains a granular structure
Easily spread	The material can be spread with existing lime, fertilizer or manure spreaders
Stable under storage	The product can be stored in the open without causing odor, does not absorb much water and stacks well
Balanced nutrients	N-Viro Soil made from cement kiln dust provides a higher K content than sludge
Use options	N-Viro Soil can be used as a limestone substitute, low analysis fertilizer, soil amendment, synthetic topsoil ingredient, or topsoil substitute

Soil would typically be applied at a rate of 10 Mg/ha dry weight every 2 to 4 yr, depending on location.

At an application rate of 10 Mg/ha and a total N-P-K content of 1, 0.4 and 1.3% (Table 10-1), 100, 40 and 130 kg/ha N, P and K would be applied. Assuming that 20% of the N is available in the year of application (Table 10-2), that 50% of the P and 100% of the K are bioavailable, the available nutrients would be 20, 20 and 130 kg/ha of N, P and K, respectively. While this would not satisfy the total nutrient needs of most crops, the farmer can realize a substantial savings in input cost even if he is paying a competitive price for the limestone equivalent.

Land Reclamation

N-Viro Soil has many of the attributes necessary for use in reclamation of degraded lands, particularly acidic minespoil or mine tailings: acid neutralizing capacity, organic matter, nutrients, soil-like physical properties, and high biological activity. N-Viro Soil is currently used in Kentucky and Tennessee for mine reclamation. In a greenhouse study of an acidic (pH < 3) pyritic minespoil from southeastern Ohio, N-Viro Soil at rates of 50, 100 and 200 Mg/ha was as effective in perennial grass growth as equivalent amounts of lime and fertilizer (Fig. 10-8). At the 200 Mg/ha rate, root biomass was significantly greater with N-Viro Soil.

In a greenhouse study on metal-contaminated alluvial soil in Kansas, Pierzynski and Schwab (1993) found that 5 Mg/ha N-Viro Soil was as effective as 1.12 Mg/ha limestone and more effective than 5 Mg/ha poultry or cattle manure, or 100 kg P/ha as inorganic fertilizer, in lowering Cd, Zn, and Pb contents of soybean (*Glycine max*) (Fig. 10-9). The N-Viro Soil was particularly effective in binding Pb, while limestone more effectively immobilized Cd and Zn, findings similar to our field study of trace metal uptake by vegetables (Logan & Burnham, unpublished data).

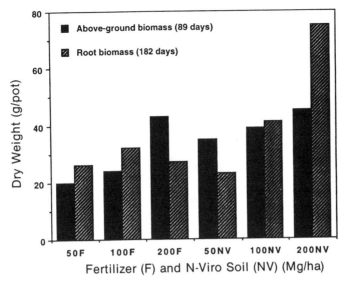

Fig. 10-8. Aboveground and root biomass of perennial grasses in acidic minespoil amended with N-Viro Soil at 50, 100 and 200 Mg/ha. Limestone and NPK fertilizer were added in controls at rates equivalent to the N-Viro Soil (Logan, 1992a).

Soil Amendment/Urban Soils

This is a relatively new market for N-Viro Soil. In a field experiment at Ohio State University Turf Farm, Columbus, N-Viro Soil at 50 and 100 Mg/ha increased growth of bluegrass (*Poa pratensis*), measured at 75 d after seeding, from 45 kg/ha in the control to 53 and 129 kg/ha, respectively

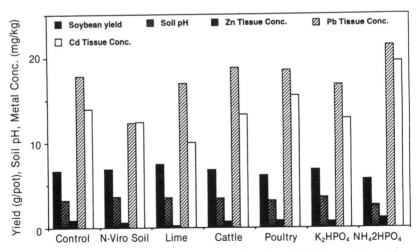

Fig. 10-9. Soil pH, yield and trace element uptake of soybeans (*Glycine max*) in metal contaminated soil amended with limestone, N-Viro Soil, manures, or P fertilizer (Pierzynski & Schwab, 1993).

(Logan & Bargar, 1992). A standard turf fertilizer was applied uniformly to all plots. In demonstration projects in Toledo and Minneapolis-St. Paul, N-Viro Soil has been used to build city parks. At these rates of application, N-Viro Soil improves turf growth by neutralizing pH, adding nutrients, and improving soil physical properties. When the same N-Viro Soil used in the turf study was applied to an adjacent area at a rate of 500 Mg/ha, soil physical properties including aggregate stability, compactability, saturated hydraulic conductivity, and water holding capacity were significantly improved 1 yr after application.

Soil Blend Ingredient

We have conducted a number of studies in which N-Viro Soil was blended with other materials to make a synthetic soil for horticultural uses. The N-Viro Soil blended 1:3 by volume with a yard waste compost produced bluegrass growth equivalent to that produced with 100 Mg/ha N-Viro Soil added as a soil amendment (Logan & Bargar, 1992). Ratios of N-Viro Soil to yard waste compost of 1:1 and 3:1 produced lower growth than the soil-only control, probably because of high soluble salts. After 3 to 6 mo, however, all plots had high quality turf stands. In companion greenhouse studies, Logan and Bargar (1992) found that sewage sludge compost was incompatible with N-Viro Soil as a blend because of the release of NH_3 from the sludge compost as pH of the blend increased to >7. In another study in which N-Viro Soil was blended with three mineral soils at ratios of N-Viro Soil to mineral soil ranging from 99:1 to 20:80, optimum germination and early growth was obtained at a ratio of 40:60. Higher ratios of N-Viro Soil reduced germination and early growth because of high soluble salts.

Landfill Cover Material

The N-Viro Soil is used as daily and intermediate cover at a number of locations, notably in New Jersey, California and Kentucky. At some locations it is blended 1:1 with soil, while other locations use it as is. It is important to maintain a solids content $>60\%$ for ease of handling with a bulldozer.

CONCLUSIONS

The ASAD Process for alkaline stabilization of municipal sewage sludge has been proven to be an effective alternative to traditional digestion methods of sludge treatment in achieving complete pathogen destruction. In addition, the product of this technology, N-Viro Soil, appears to have a wide range of beneficial uses that are based on the lime and fertilizer value, physical properties and biological activity of this material. The high degree of biological stability and low odor potential make it a potentially more marketable product than traditional sewage sludges. Because it must meet the

USEPA concentration limits for trace elements and is pathogen free, N-Viro Soil also is environmentally acceptable.

REFERENCES

American Society for Testing and Materials. 1993. Standard test methods for chemical analysis of limestone, quicklime, and hydrated lime. C25. 4.01:9–36. ASTM, Philadelphia, PA.

Bennett, G.F. 1989. Effects of cement kiln dust on the mobility of heavy metals in treatment of wastewater treatment plant sludge. Rep. to The Thomas Edison Program for Innovative Technol., Ohio Dep. Develop., Columbus, OH.

Burnham, J.C., J.F. Donovan, J. Forste, J. Gschwind, T.J. Logan, and D. Zenz. 1992a. Production and distribution of municipal sewage sludge products. p. 479–530. *In* C Lue-Hing et al. (ed.) Municipal sewage sludge management. Technomic Pub., Lancaster, PA.

Burnham, J.C., N. Hatfield, G.F. Bennett. 1992b. Use of kiln dust and quicklime for effective municipal sludge pasteurization and stabilization with the N-Viro Soil process. p. 128–141. *In* D.D. Walker et al. (ed.) Innovations and uses for lime. ASTM STP 1135. ASTM, Philadelphia, PA.

Logan, T.J. 1992a. Mine spoil reclamation with sewage sludge stabilized with cement kiln dust and flue gas desulfurization byproduct (N-Viro Soil Process). p. 220–231. *In* Proc. 9th Ann. Natl. Meet. Duluth, MN. Am. Soc. Surface Mining & Reclam., Princeton, WV.

Logan, T.J. 1992b. Chemistry and bioavailability of metals and nutrients in cement kiln dust-stabilized sewage sludge. p. 493–502. *In* Proc. the Environ. Technol. Expo., East Chicago. 24–27 February. Cahners Publ. Co., Des Plaines, IL.

Logan, T.J. 1993. Alkalinity and liming value of N-Viro Soil materials. AC93-031-006. Ann. Conf. Water Environ. Fed., Alexandria, VA.

Logan, T.J., and B. Bargar. 1992. Field and greenhouse studies on turfgrass growth with N-Viro Soil and N-Viro Soil/compost mixtures. N-Viro Energy Systems, Ltd., Toledo, OH.

Logan, T.J., B. Harrison, and M.D. Che. 1989. Agronomic effectiveness of cement kiln dust-stabilized sludge. Final Rep. The Thomas Edison Prog. for Innovative Technol., Ohio Dep. Dev., Columbus, OH.

Logan, T.J., and B.J. Harrison. 1995. Physical characteristics of alkaline stabilized sewage sludge (N-Viro Soil) and their effects on soil physical properties. J. Environ. Qual. 24:153–164.

Ma, Q.Y., S.J. Traina, T.J. Logan, and J.A. Ryan. 1993. In situ Pb immobilization by apatite. Environ. Sci. Technol. 27:1803–1810.

Pierzynski, G.M., and A.P. Schwab. 1993. Bioavailability of zinc, cadmium and lead in a metal-contaminated alluvial soil. J. Environ. Qual. 22:247–254.

Smith, K.E., C.E. Clapp, T.R. Halbach, S.A. Stark and A.M. Fulop. 1993. Nitrogen mineralization of N-Viro amended soil. p. 50. *In* Agronomy abstracts. ASA, Madison, WI.

U.S. Environmental Protection Agency. 1990. 40 CFR Part 503. National sewage sludge survey; availability of information and data, and anticipated impacts on proposed regulations; proposed rule. Fed. Reg. 55:47 210–47 283.

U.S. Environmental Protection Agency. 1993. 40 CFR Part 503. Standards for the use or disposal of sewage sludge. Fed. Reg. 58:9387–9404.

11 Issues Affecting Application of Noncomposted Organic Waste to Agricultural Land

J. H. Edwards, E. C. Burt, and R. L. Raper

USDA-ARS
National Soil Dynamics Laboratory
Auburn, Alabama

R. H. Walker

Auburn University
Auburn, Alabama

Organic wastes (sewage sludge, feedlot manures, poultry litter, crop residues) have been applied to agricultural land for decades, in many countries of the world today they are the only available fertilizers for agricultural production. Land application recycles valuable nutrients and effectively disposes of the wastes. However, when waste production is concentrated in a relatively small area such as the poultry-producing areas of Alabama, Arkansas, Georgia, North Carolina, and West Virginia and in the beef cattle feedlots of the Midwest, the potential for environmental contamination is enhanced when organic waste is overapplied to agricultural land. Although beneficial effects of poultry waste on crop production have been demonstrated repeatedly (Ketcheson & Beauchamp, 1978; Sims, 1987; Flynn et al., 1993; Wood et al., 1993), land application of such materials can promote degradation of water quality (Liebhardt et al., 1979; Ritter & Chirnside, 1984).

Composting newsprint and other organic waste is becoming increasingly recognized as a viable method for waste management in Europe and the USA (Goldstein, 1989). While composting reduces the volume of organic waste and improves its value as a soil amendment, it also involves additional handling, moving, and processing. The total amount of energy required to produce, transport, and apply compost is on the order of 34.4 kWh Mg^{-1} of municipal solid waste (MSW), and the composting step requires 50 to 70% of the total energy used in the entire process (Diaz et al., 1987). Furthermore, the resistance of newsprint to decomposition may increase operational costs, thus reducing the potential benefits of composting newsprint using traditional methods.

The demands on soil with low or marginal productivity for increased crop production are escalating in response to increased population, loss of prime cropland to urbanization, and continued degradation through soil erosion. This is especially true for soils in the Gulf and Atlantic Coastal Plain and Appalachian Plateau region which have chemical and/or physical barriers that can limit crop production. Soils of these regions have sandy loam and loamy sand surface horizons, weak soil structure, low organic matter content, and acid subsoils (Adams, 1981). Compacted soil layers often prevent crops from achieving full yield potential by retarding plant root growth and limiting the soil volume available to roots for water and nutrient uptake. Tillage pans have reduced cotton (*Gossypium hirsutum* L.) yields up to 50% and affected root growth when the soil pan strength increased to 2.5 MPa (Taylor et al., 1964).

The benefits to be derived from applying newsprint and other organic wastes to land are dependent on geographical location. In the southeastern USA, climatic conditions favor the rapid incorporation of organic wastes into soil organic matter where they can improve the physical and chemical properties of soil, thus increasing soil aggregation and water infiltration, improving soil moisture-holding capacity, soil tilth, soil nutrient-holding capacity, and reducing erosion and sediment loss by surface runoff, and ultimately increasing yields and sustainability of crop production. In drier regions of the USA, applications of organic wastes may help conserve soil water resources and reduce sediment loss caused by wind erosion.

Application of noncomposted organic wastes into agricultural soils accomplishes the composting process in situ because microorganisms eventually incorporate C and other elements into soil organic matter and valuable nutrients are returned to the soil where they originated. However, there are many interacting factors affecting the biorecycling of newsprint and other organic wastes. Some issues to be discussed are: (i) Who will benefit from the development of organic waste soil-amendment systems? (ii) How much organic waste can be safely applied to agricultural lands? (iii) What forms (composted, pelleted, or noncomposted) are best suited for land application? (iv) What is the best time of the year to apply organic wastes? (v) What methods of waste application should be used? (vi) How will crops respond to organic wastes? (vii) How will weeds respond to organic wastes? (viii) Why is there an increase in crop yield when organic waste plus poultry litter are applied to agricultural land? (ix) What mechanisms are involved in soil nutrient dynamics created by newsprint application and how can they be balanced? (x) What N sources can be used to adjust the C/N ratio of organic wastes without inducing nutrient imbalances? (xi) How does organic waste application influence microbial populations and activity? (xii) Are sediment, nutrient, and pesticide loss profiles acceptable? (xiii) Can organic waste be used as residue cover during fallow periods to reduce soil and wind erosion? (xiv) What are current and future trends in research with organic wastes? Some of these questions will be addressed in this chapter from research conducted at USDA-ARS, National Soil Dynamics Laboratory and Auburn

University, Auburn, AL on the ecological impact of application of MSW and poultry litter on soil nutrient and soil microbial dynamics.

WHO WILL BENEFIT FROM THE DEVELOPMENT OF ORGANIC WASTE SOIL-AMENDMENT SYSTEMS?

Poultry production is one of the most rapidly growing agricultural industries in the southeastern USA, and generates as a byproduct about 1 814 000 t (2 million tons) of litter from approximately 850 million broiler chickens grown annually, just in the state of Alabama alone (third in the nation) (Hinton, 1991; Mitchell et al., 1989). Much of the litter contains organic or C-based wastes, such as wood shavings or peanut (*Arachis hypogea*) hulls that are used as bedding materials, as well as poultry excreta. Application of poultry litter at recommended loading rates can supply the necessary N and P requirements for most row and hay crops (Flynn et al., 1993; Wood et al., 1993). However, when litter is removed from the poultry house, its disposal is usually paid for by the truck load, and land application is frequently carried out in close proximity to the poultry house. This repeated application has led to groundwater contamination with excessively high levels of nitrates since poultry litter is often spread on fallow and dormant lands in winter, which coincides with a high rainfall and low evapotranspiration season (Thomas, 1976, p. 20). Thus, a major challenge facing producers is the safe disposal of poultry litter.

Another problem is finding ways to manage and use the vast amounts of MSW such as waste paper and wood products, leaves and yard waste, currently going into landfills (Edwards, 1992). These materials comprise approximately 65% of landfill volume and are obvious targets for reduction goals. Waste paper accounts for 40% of this volume, but only 17% is currently being recycled (USEPA, 1989, p. 1–70). Alabama recently passed laws requiring a 25% reduction in landfill volume, and other states have mandated reductions as high as 60% by 1995. Disposal of yard waste and leaves in landfills is now banned in many states.

On average, 73% of MSW in the USA goes to landfills, 13% is recycled, and 14% is incinerated. Average cost of disposal of MSW in landfills in the USA ranged from \$42 to \$97 t^{-1} (\$46–\$107 ton^{-1}) in 1986, \$54 to \$176 t^{-1} (\$59–\$194 ton^{-1}) in 1991, and is predicted to increase from \$116 to \$199 t^{-1} (\$128–\$219 ton^{-1}) by 1996 (Greshman, 1992; Glenn, 1992). Even though we may be willing to pay the increased cost of landfill disposal of MSW, we are all NIMBY's (Not In My Back Yard) when it comes to building new landfills, incinerators or other means of disposing of the byproducts of our throw-it-away society.

The development of organic waste-amended soil-crop production systems that use poultry litter, newsprint, yard waste and/or other waste byproducts in an environmentally safe manner has potential for broad-based benefits. Poultry producers may have a system for recycling waste into valuable resources for agriculture. Municipalities may have alternate systems for managing C-based wastes that have been traditionally disposed of in land-fills. This practice will extend the life of current landfills and reduce the need to construct additional ones. Farmers may benefit from the use of waste byproducts to improve soil quality and thus crop yields and improve the potential for agricultural sustainability. Consumers may benefit from lower tax revenues for garbage disposal and possibly lower food prices as a result of the development of these organic waste-amended soil-crop systems.

HOW MUCH ORGANIC WASTE CAN BE SAFELY APPLIED TO AGRICULTURAL LANDS?

The recommended loading rates of sludge or poultry litter to agricul-tural land are determined by the concentrations of the nutrients or metals. In the case of poultry litter the content of the N, P, and/or Cu may limit the loading rate when it is applied as a nutrient source for row or hay crops. However, these nutrients may not be the limiting factor when poultry litter is used as a nutrient source to initiate microbial activity for organic wastes with high C/N ratio.

The loading rates of organic wastes with high C/N ratio ($\geq 100:1$) is dependent on many interrelated factors, i.e., the nutrient requirements of the organic waste to initiate microbial decomposition, source (organic or in-organic) to be used to satisfy the nutrient requirements of the waste, metal content of the C-based organic waste that can impact soil, plant, animal or human health. In this section we will explain several concepts that are im-portant in studying the agronomic and environmental aspects of land appli-cation of organic wastes, i.e., what factors may influence the decomposition rate of organic wastes, N and P mineralization or immobilization rates, how the C/N ratio of the C-based organic waste affects its use as a soil amendment.

Calculated loading rates for applying organic waste, such as poultry lit-ter, manures, or sludges to agricultural land are usually based on the total N content of the waste (O'Keefe et al., 1986), or on the plant-available N (PAN) content of the organic waste (King, 1984). These loading rates are based on indices that estimate soil N needed to maximize crop production and minimize the potential for enviornmental pollution from the different forms of N. When poultry litter is applied to agricultural land at rates in excess of the N requirement of the crop, leaching of NO_3–N and surface runoff is ultimately the fate of the excess or unused N (Mitchell et al., 1989).

The N and P in poultry litter, present in inorganic and organic frac-tions, are subject to volatilization, denitrification, immobilization, miner-alization, leaching, and surface runoff, as well as plant uptake. When poultry litter is applied to soil, there are no simple, rapid and reproducible methods of predicting how much N and P will be mineralized because variables such as soil type, temperature, soil moisture content, aeration, as well as species and amount of soil microorganisms are involved. The rate and extent that these processes occur determines the usefulness of poultry litter as a source of N and P for crops, as well as the potential for environmental contamination. In most states, the recommended loading rates for poultry litter is limited to the crop requirements of N or P (Payne & Donald, 1991). Application at recommended agronomic fertilizer rates has reduced nonpoint pollution hazards from overapplication of litter.

Nitrogen in poultry litter ranges from 2.0 to 4.0% on a dry weight basis, however, it is the fraction that is of most environmental concern. Seventy-five percent of the N is in the organic form, and slowly becomes mineralized or changed to an inorganic form, and thus is made plant available during the growing season. Sixty-one percent of the extractable N from poultry litter is in the form of uric acid, which can be decomposed slowly as NH_3 and CO_2 by aerobic organisms (Eno, 1966, p. 18). The other 25% is urea and ammonium-N compounds, which are already inorganic forms, and readliy available for plant and microbial absorption (Payne & Donald, 1991; Wilkinson, 1979; Hileman, 1967).

The mineralization of N in poultry litter does not necessarily coincide with the N demands of the plant. At equivalent rates, poultry litter left 42% more N in the soil system at the end of the growing season than N from a commercial inorganic source (Wood et al., 1991). Thus, poultry litter proved to be a less efficient source of N for corn [*Zea may* L.] than NH_4NO_3–N. This reduced efficiency in supplying nutrients has contributed to excessive poultry litter application to land, resulting in leaching of NO_3–N to the groundwater (Liebhardt et al., 1979, Cooper et al., 1984).

Poultry litter can be a source of P contamination to soil and water when it is applied at excessive rates. The P in poultry litter, collected from 147 poultry houses in Alabama, ranged from 0.61 to 3.9% on a dry weight basis, with an average P content of 1.6% (Payne & Donald, 1991). The P fraction is considered to be about 75% as effective as commercial fertilizer source of P during the year of application. In Arkansas, noncomposted poultry litter contained an average P content of 1.45%, of which 86% was water soluble (Hileman, 1971). However, Beegle (1988) reported the P in poultry litter was mainly in the organic fraction, and only slowly available to the crop. In some states, consideration is being given to treating poultry litter with Al, Ca and/or Fe amendments to precipitate P and reduce the risk of P contamination before it is applied to land (Moore & Miller, 1994).

The organic matter content of poultry litter ranged from 34 to 65% (Wood et al., 1991). The C/N ratio or poultry litter ranged from 9:1 to 10:1. Twenty-five percent of the organic C in poultry litter is water soluble or readily plant available (Table 11-1). Newsprint and wood wastes have a high C/N ratio $\geq 100/1$, yard waste (includes leaves and yard clippings) has a C/N of $\leq 40:1$ (Mitchell & Edwards, 1993, p. 1–4). A lower C/N ratio favors mineralization of the organic residues and release of N, while C/N ratio $\geq 50:1$ will result in the soil microorganisms immobilizing some or all of the soil N as they attempt to use the C as an energy source. Thus, there is competition between the microorganisms in the soil and plants for the available soil nutrients and an imbalance detrimental to both can result. In other words, the microorganisms that aid in the decomposition process require some N; the N will come from either the soil N pool, or some added N source. A

Table 11-1. Determination of the organic matter fraction in cellulose wastes.

	Ash	H$_2$O-soluble polysaccharide	Hemicellulose + cellulose	Protein	Lignin	Total recovery
			%			
Poultry litter	11.1	17.7	35.4	17.3	13.5	94.9
No. 1 newsprint	3.3	1.4	47.7	0.8	41.2	94.4
No. 2 newsprint	0.7	0.9	44.7	0.8	45.9	93.0
Colored insert	6.7	1.5	23.2	0.7	64.5	96.6
Cardboard	9.5	1.6	23.3	0.7	53.7	88.9
Junk mail	3.4	3.2	34.0	0.5	52.4	94.4
Computer paper	8.5	2.9	34.2	0.5	50.9	97.2
Phone book	6.3	1.8	25.1	0.5	58.4	92.7
Gin trash	4.9	4.7	47.2	9.3	30.9	96.9
Wood chips	9.9	3.3	25.9	2.2	52.4	94.3

choice might be another organic waste that is high in N. The high N and P content of poultry litter, along with high water solubility, make it an excellent substrate for the initiation of microbial activity when it is applied to agricultural land to balance the C/N ratio of newsprint, wood chips or yard waste to about 30:1 (Edwards et al., 1993a).

The effectiveness of organic mulches in conserving soil water is determined by the percentage of their surface coverage. In general, soil surface coverage of organic mulches should be greater than 30 to 50% to receive maximum benefits from the mulch. The plant residue that is produced along with 6720 kg ha^{-1} grain (6720 kg ha^{-1} of corn stover) will give a 100% coverage of the soil surface (Larson et al., 1978). A rate of 4.9 kg m^2 of ground newsprint provides a mulch that is approximately 7.5 cm thick and gives a 100% surface coverage. The rate of newsprint or other organic waste that should be applied will probably fall between 2.9 and 4.9 kg m^{-2} to achieve benefits comparable to organic mulch for water conservation. However, loading rates of MSW to agricultural land have not been determined.

Another potential environmental concern is the concentration of micronutrients Cu, Mn, and Zn contained in newsprint. If they are high enough they could become a problem with repeated application of newsprint to land (Tables 11-2 and 11-3). The metals Ba, Cd, Cr, and Pb, which are contained in newspaper printed with colored inks, usually found in advertisement inserts are also of some concern (Edwards, 1992). Red inks primarily contain Ba, Cr, and Cd compounds; blue or green inks may contain Cu compounds. In colored ad inserts with a metallic finish, the heavy metals Co, Mn, and Pb could be contained in ink. The current trend in the printing industry is to use inks with a soybean [*Glycine max* (L.) Merr.] oil base, these are less harmful to the environment than inks made with petroleum-based organic solvents. However, old sources of newsprint may still present problems from elevated metal concentrations.

The application of MSW to agricultural land is in an infancy stage and the loading rates have not yet been determined. The nutrient requirements for the decomposition of organic wastes with high C/N ratios must be es-

Table 11–2. Total elemental analyses of broiler litter, wood chips, newsprint and other waste paper.

	Ca	K	Mg	P	Cu	Fe	Mn	Zn	B	Mo
					mg kg					
Broiler litter[†]	27 055	28 501	5 662	20 203	550	2 144	632	533	69.0	5.7
Wood chip[‡]	5 378	5 043	1 882	1 888	49.5	354	200	52.5	11.7	0.9
Newsprint	1 274	89	110	41	17.3	46.8	20.7	5.1	5.6	1.7
Office paper	4 456	310	152	99	0.6	62.7	5.6	2.7	2.8	1.7
F container	1 401	115	86	43	4.6	30.0	13.6	24.9	6.4	1.2
Insert advertisement	458	204	98	99	10.9	27.3	1.2	17.6	4.7	0.8
Junk mail	4 320	92	626	181	16.4	151.0	5.1	4.5	5.2	3.5
Cardboard	682	89	96	36	1.2	16.7	23.2	5.9	23.3	0.4
Large pellets§	1 908	318	236	111	5.7	57.2	45.2	5.6	1.0	1.8
Medium pellets¶	966	79	72	48	0.9	21.2	44.1	7.3	1.3	0.5

† Collected from a poultry house in Alabama.
‡ Collected from a planing mill in Alabama.
§ Large cardboard paper pellets supplied by E-Z Fuel LTD., Sioux City, IA 51250 (telephone is 712-722-3708.
¶ Medium paper pellets supplied by Tascon, Inc, Houston, TX 77041 (telephone is 713-937-0900).

Table 11-3. Total elemental analyses of broiler litter, wood chips, newsprint and other waste paper.

	Na	Al	Ba	Si	Co	Cr	Pb	Ni	Cd
				mg/kg					
Broiler litter	6977	2573	31.6	2075	2.0	8.5	14.6	7.6	2.4
Wood chip	1369	444	15.3	734	0.4	1.1	0.4	0.9	2.4
Newsprint	640	1410	53.3	573	0	0.2	1.9	--	--
Office paper	289	2056	8.7	554	0	3.2	2.9	--	--
F container	468	998	8.5	454	0	0.2	1.8	--	--
Insert advertisement	569	1550	51.2	386	0	0	2.4	--	--
Junk mail	409	2750	11.3	145	0	3.3	9.1	--	--
Cardboard	806	542	5.1	254	0	0	0.05	--	--
Large pellets	2718	911	12.7	559	0	0.7	1.9	--	--
Medium pellets	578	3102	11.7	336	0	0.7	7.0	--	--

tablished so that the waste does not compete for nutrients with the crop. What metals or metal complexes contained in the organic waste that may induce nutrient imbalances in the soil and plant ecosystems, or their possible use as chelating agents to remove contamination by binding it into some form unusable in the plant/soil system have yet to be discovered. An accumulation of data in several interrelated areas is needed before regulatory agencies in the different states can make recommendations to set maximum loading rates of noncomposted organic waste on agricultural land.

WHAT FORMS (COMPOSTED, PELLETED, NONCOMPOSTED) ARE BEST SUITED FOR LAND APPLICATION?

The waste paper fraction of MSW, as collected by municipalities, is not suited for direct land application because it is very resistant to decomposition by soil microorganisms and can create barriers that emerging plant seedlings may not be able to penetrate. A processing step may include grinding, chopping, or shredding the paper; however, then it becomes very light and can be easily blown about. It also may require large storage areas. Application of the ground or chopped noncomposed newsprint to large land areas is not practical because specialized equipment is needed. Much of our research has been conducted with ground newsprint, minus the fire retardant, purchased in bulk from an insulation company. It was applied to the field plots by hand. When a composting step is added to the process, more energy and equipment is required for handling, moving, and applying the ground newsprint.

What is needed is a form of waste paper that can be applied to land using existing, easily modifiable, or readily available equipment. Recently, several companies began using pelletizing equipment to compress waste paper. Perhaps it can be treated to remove undesirable elements or mixed with fertilizer during the manufacturing process. Although today the cost of pelleting waste paper may prohibit its use for land application, in the near future demand from increasing experimental use may help to make it cost effectve (Adamoli, personal communication).

WHAT IS THE BEST TIME OF YEAR TO APPLY ORGANIC WASTES?

The infrastructure needed for the collection, storage, and transport steps in MSW management by municipalities is beyond the scope of this chapter. However, these are important considerations because of the discontinuous nature of MSW collections and the participatory goals that may apply. The time of application of MSW to land as it relates to agricultural production systems is discussed here.

We initiated a series of field experiments to determine, among other treatment variables, the best time (fall or spring) to apply noncomposted organic C-based waste, i.e., newsprint, yard waste, wood chips, and cotton gin trash, to agricultural land. Also, when should the C/N ratio of the noncomposted organic waste be adjusted to achieve maximum crop production benefits, and what form of N (organic or inorganic) is best suited for adjustment of C/N ratio?

When ground newsprint less than 6 mm in diameter was applied to the surface and incorporated into the top 15 cm of soil at a rate of 4.9 kg m^{-2} (43 560 lb/acre) immediately before planting in the spring, it severely stunted cotton seedlings in the first 6 wk after germination, particularly in treatments where the C/N ratio of the newsprint was not adjusted to 30:1 (Edwards et al., 1992a, b). A number of fungal organisms were found on the newsprint that require nutrients from the soil nutrient pool, organisms detected were those generally found on organic material with high C/N ratios ($\geq 100:1$).

Cotton yields were not affected if newsprint was surface applied, and C/N ratio was adjusted, 4 to 6 wk prior to planting (Edwards et al., 1993c). When newsprint C/N ratio was adjusted to $\leq 30:1$ in the fall, the following spring-planted cotton yields were increased 60% when compared to a control (standard cultural practices). With spring-applied newsprint, with the C/N ratio adjusted $\leq 30:1$, yields were increased 54% when compared to a control. When the C/N ratio of newsprint was not adjusted (C/N $\geq 150:1$), fall or spring application of newsprint decreased yields below the control.

Results show that surface-applied noncomposted ground newsprint should be composted in situ for an interval of 4 to 6 wk, or care should be taken to adjust the C/N ratio of the newsprint before crops are planted in the area. When these two considerations are met, the results were satisfactory whether the organic waste was applied in the fall or the spring of the year.

WHAT METHODS OF WASTE APPLICATION SHOULD BE USED?

We evaluated the effects of ground newsprint, with and without poultry litter, mixed with soil in narrow vertical trenches between rows of cotton, as well as on the soil surface (Edwards et al., 1992a, b). The treatments included ground newsprint applied to the soil surface; 50:50:0 (volume of soil removed from trench 10 cm wide by 61 cm deep by 18.5 m long mixed with equal volume of ground newsprint); 50:40:10 (volume of soil removed from trench 10 cm wide by 61 cm deep by 18.5 m long mixed with ground newsprint to 40% of the soil volume and 10% of the soil volume supplied as poultry litter); backfilled (volume of soil removed from trench 10 cm wide by 61 cm deep by 18.5 m long and returned 100:0:0 into trench); and control (standard cultural practices) for cotton.

Yields ranged from a low of 29 kg ha^{-1} in the plots receiving 4.9 kg m^{-2} surface-applied newsprint without poultry litter, to a high of 922 kg ha^{-1} for cotton rows adjacent to the 61-cm-deep trench backfilled with a mixture of 50:40:10. The total yield of the surface-applied ground newsprint plots was 472 kg ha^{-1}; only 6% of the yield was harvested at the first date, the other 94% was harvested at the second date. On the average, less than 30% of the cotton yield was harvested at the second date in the other treatments. The highest yielding treatments were the two that mixed excavated soil with newsprint and poultry litter (50:40:10). Plots near the 61-cm-deep trench yielded 1255 kg ha^{-1} and plots near the 122-cm-deep trench yielded 1234 kg ha^{-1}. Cotton adjacent to the 122-cm-deep trench yielded 81% at the first harvest date.

Cotton roots in the trenches backfilled with no newsprint or poultry litter and the 50:40:10 (soil/newsprint/poultry litter) treatments grew to the bottom of the 122-cm-deep trenches. Under the row (undisturbed soil), rooting depth was confined to the top 15 cm of soil similar to the control. The primary reason for the proliferation of roots in the trenches was the difference in soil bulk density between the disturbed (1.3 Mg m^{-3}) and undisturbed (1.55 Mg m^{-3}) soil.

Other experiments determined the effects of ground newsprint in a vertical trench dug uniformly from the surface to a depth of 61 cm between rows of grain sorghum [*Sorghum bicolor* (L.) Moench] and soybean (Raper et al., 1994). An implement was designed and built to disrupt the soil hardpan and incorporate ground newsprint into the trench in one pass. The implement performed well and allowed ground newsprint to be applied uniformly in a vertical trench. The trench provided a more positive rooting environment than uninterrupted soil and this was reflected in increased yields. At the end of the growing season, a trend was found toward reduced reconsolidation in trenches formed with ground newsprint as compared to trenches formed without ground newsprint.

We evaluated the effects of surface application of four organic wastes (newsprint, yard waste, woodchips, and cotton gin trash), with and without several N sources (inorganic and organic), to adjust the C/N ratio of the waste to ≤30:1, as well as provide N needed by crops. Cotton yields were higher when poultry litter, rather than NH_4NO_3, was used to adjust C/N ratios after two annual spring applications (Edwards et al., 1994b). The organic waste that produced the lowest yield was newsprint, when NH_4NO_3 was the N source used to adjust the C/N ratio, in both years. It appears that some of the phytotoxic effects of newsprint to plants were ameliorated by the addition of poultry litter as the N source. Plots receiving yard waste, cotton gin trash, and control (standard production practices) produced the highest yields, followed by woodchips and newsprint.

Surface and trench application methods have advantages and disadvantages and the desired objectives will determine which method to use. Placement of organic wastes in vertical trenches, or in the excavated channel of a subsoiler shank, uses specialized equipment with high energy requirements. Trenching is restricted by the amount of waste that can be applied in each pass. However, the adjustment of the C/N ratio of the organic waste is not as important as it is in surface application because only a small portion of the total plant root system comes in contact with the waste, and there is very little competition between the organic waste and crops for available nutrients. Of course, if the objective is to increase plant rooting depth (Edwards et al., 1992a), this can be accomplished using tillage equipment with lower energy requirements (Karlen et al., 1991). On the other hand, surface application of organic waste requires more management to provide for optimum growth of the crop. The chemical composition and nutrient requirements of the waste as well as the crop must be determined and used to balance the soil nutrient supply.

HOW WILL CROPS RESPOND TO ORGANIC WASTE?

During the first 6 wk after cotton emergence, seedlings growing where newsprint was applied to the soil surface without adjusting the C/N ratio ($\geq 100{:}1$) had higher incidence of the disease caused by *Sclerotium rolfsii* Sacc. (Sclerotium stem rot disease), and surviving plants were severely stunted by *Rhizoctonia solani* Kühn (sore shin disease)[1]. Affected plants remained stunted, were delayed in maturity, and had lower lint yields (Edwards et al., 1993a).

Greatly reduced incidence of diseases caused by *S. rolfsii* or *R. solani* were found in plots with deep vertical trench placement of newsprint, and where no poultry litter was applied. The trenches with soil, newsprint, and poultry litter were approximately 50 cm away from cotton plants and only the root system came in contact with the soil and newsprint mixture. Since primary infection sites for these organisms are located at the soil surface, no adverse affect was observed when newsprint was placed in narrow vertical trenches.

The incidence of sclerotium stem rot and sore shin disease remained low 5 or 6 wk after planting if the C/N ratio was adjusted to $\leq 30{:}1$, or if the C/N ratio of the soil surface-applied newsprint was adjusted at planting by application of commercial NH_4NO_3. *Sclerotium rolfsii* is a high O_2-demand organism growing on or near the soil surface that uses organic matter as a

---

[1] Identification of fungal pathogens completed by Dr. P.A. Backman, Professor, Dep. of Plant Pathology, Auburn Univ.

bridge to spread from plant to plant. *Rhizoctonia solani* is very active in the upper soil profile, but can cause damage at all depths. The occurrence of sore shin was primarily due to rotation or crop history of the area.

When different N sources were used to adjust the C/N ratio of four surface-applied organic wastes (newsprint, yard waste, woodchips, and cotton gin trash) to $\leq 30:1$, no N source affected rate of cotton maturity (Edwards et al., 1993c). The delay of maturity in the newsprint treatment may have been caused by an induced nutrient imbalance during early season growth. The newsprint N concentration was less than 0.1% and C/N ratio was $\geq 100:1$ before any adjustments to C/N ratio were made. The delay of maturity in the gin trash treatment was caused by excess N. The N concentration of the gin trash was 1.2% and the C/N ratio was $\leq 30:1$. When the total amount of N was balanced in all organic waste treatments, excess N was applied in the gin trash treatment because we had no way to predict the rate of availability of the N contained in the gin trash. Thus, differences in N availability were reflected in excess vegetative growth and delay in maturity in some plots when compared to the control. These observations illustrate some of the difficulties encountered in managing N rates with different organic wastes and the need to customize the application of N.

The surface application of noncomposted newsprint or yard waste, plus NH_4NO_3 as a N source, produced cotton plants that had less leaf area, used less water, and produced lower seed cotton yield than the control (standard cultural practices); the addition of wood chips or cotton gin trash with NH_4NO_3 resulted in plants that were no different than the control (Edwards et al., 1993c). The four organic wastes, when poultry litter was the N source, increased leaf area, water use, and seed cotton yield. Beneficial effects from poultry litter were greatest with newsprint, followed by yard waste, cotton gin trash, and wood chips. However, greater water use by cotton plants did not always translate into higher yield. Water-use efficiency as reflected by yield was in the order of yard waste, followed by wood chips, gin trash, and was lowest with newsprint.

Cotton yields were higher when poultry litter, rather than NH_4NO_3, was used to adjust C/N ratios of organic wastes applied to soil in the spring after two annual applications (Edwards et al., 1994b). The organic waste that produced the lowest cotton yield was newsprint, when NH_4NO_3 was used to adjust the C/N ratio to $\leq 30:1$. Yard waste, cotton gin trash, and control produced the highest yields, followed by woodchips and newsprint.

Results to date in another field experiment reveal that some organic wastes may be ideal for use when corn is the crop, while others may be unsuitable. The order of corn yields for the plots amended with organic wastes, when the C/N ratio was adjusted to 30:1, were: woodchips (8467 kg ha^{-1} [126 bu acre^{-1}]) > yard waste (8064 kg ha^{-1} [120 bu acre^{-1}]) > gin trash (7929 kg ha^{-1} [118 bu acre^{-1}]) > control (7324 kg ha^{-1} [109 bu acre^{-1}]) > newsprint (5645 kg ha^{-1} [84 bu acre^{-1}]) (Edwards et al., 1993d). The organic wastes that were more resistant to microbial decomposition were less dependent on the soil N pool for their N supply, and as a consequence had less effect on corn growth and yield. Poultry litter was a better source of

N for corn than NH_4NO_3–N when applied in conjunction with organic wastes; however, treatments using newsprint as a soil amendment adversely affected corn production. Plots containing newsprint exhibited stunted plants, and yields were significantly reduced.

HOW WILL WEEDS RESPOND TO ORGANIC WASTE?

Effective weed control with soil-applied herbicides is dependent on a sufficient concentration of herbicide being available in the soil solution. With most herbicides, organic matter has been the principal soil property influencing adsorption (Goetz et al., 1989). Simazine[2] [6-chloro-N,N'-diethyl-1,3,5-trizine-2,4-diamine] efficacy decreased more rapidly where there was a higher organic matter and water content in the surface soil (Slack et al., 1978). Potential changes in weed/crop competition are likely as production variables change. Changes in soil temperature and soil moisture can impact both crop and weed seeds and narrow the margin of crop over weed (Staniforth & Wiese, 1985).

While high levels of soil organic matter are good for crop productivity, higher rates of many soil-applied herbicides may be necessary for adequate weed control, and thus increase production costs. When soil organic matter exceeds 5% many soil-applied herbicides become ineffective for weed control and the farmer must rely on alternative methods of control. Selective postemergence-applied herbicides may be able to fill this void.

The efficacy of herbicide application was not affected by the spring application of four organic wastes to the soil surface. However, the number of large crabgrass [*Digitaria sanguinalis* (L.) Scop.] seedlings observed when herbicide was postemergence applied were significantly affected by the interaction of organic wastes and N source (Edwards et al., 1994b, c). The organic wastes significantly reduced the number of large crabgrass seedlings and the effect was more pronounced when poultry litter was the N source (Walker et al., 1993). Newsprint plus poultry litter reduced the number of large crabgrass seedlings by 65% when compared to the standard control. Fall-applied ground newsprint also showed good control of winter annual weeds such as henbit (*Lamium amplexicaule* L.), cutleaf evening primrose (*Oenothera laciniata* Hill), and bitter cress (*Cardamine hirsuta* L.). However, when ground newsprint and newsprint pellets were compared, the newsprint pellets did not significantly reduce the number of winter annual weeds when compared to the control.

---

[2]Mention of trademark, priorietary product, or vendor does not constitute a guarantee or warranty of the product by USDA or Auburn Univ. and does not imply its approval to the exclusion of the other products or vendors that may also be suitable.

WHY IS THERE AN INCREASE IN CROP YIELD WHEN ORGANIC WASTE PLUS POULTRY LITTER ARE APPLIED TO AGRICULTURAL LAND?

Soils of the Gulf and Atlantic Coastal Plain and Appalachian Plateau region have chemical and/or physical barriers that limit crop production. However, the climate of the region offers the opportunity for high annual dry matter production when complimentary soil management practices are used. In many instances, the organic matter fraction of these soils is related to their fertility, and ultimately to their ability to sustain crop production. Although soil organic matter ranges from only 5 to 10 g kg^{-1}, it has a major influence on soil physical and chemical properties. Within three crop years, soil organic matter content was increased from 5 to 15 g kg^{-1} (Wood & Edwards, 1992), and soil bulk density was decreased from 1.59 to 1.39 Mg m^{-3} in the surface 15 cm by returning crop residues to the soil surface (Edwards et al., 1992c).

We applied ground newsprint, with and without poultry litter, to the soil surface and used these plots to determine the decomposition rate of newsprint (Edwards et al., 1993a). Initial soil organic matter content was 11.9 g kg^{-1} before the addition of newsprint. When newsprint without poultry litter was applied to the soil surface at 4.9 kg m^{-2}, soil organic matter content was increased by 5.3 g kg^{-1} (11.9–17.2 g kg^{-1}) within 7 mo, however, approximately 12.4 g kg^{-1} newsprint was not decomposed. When the surface area of newsprint without poultry litter was reduced by adding soil, the organic matter content increased by 3.9 g kg^{-1} (11.9–15.8 g kg^{-1}) within 7 mo, with 14.0 g kg^{-1} of newsprint remaining to be decomposed. In contrast, when poultry litter was added to the newsprint and soil mixture, the organic matter content was increased by 12.0 g kg^{-1} (11.8–23.8 g kg^{-1}) within 7 mo, with no visible evidence of any newsprint remaining to be decomposed. The use of a combination of organic wastes illustrates that the rate at which newsprint is transformed into the stable soil organic matter fraction can be controlled by the addition of nutrients, primarily N, used by microorganisms to aid decomposition.

On the average, soil organic matter content was increased from 9.5 to 17 g kg^{-1} after two annual applications of newsprint, yard waste, or cotton gin trash, plus N to adjust the C/N ratio to 30:1 (Edwards et al., 1994b). Since woodchips contain a higher concentration of lignin, which is more resistant to decomposition, soil organic matter content was increased from 9.5 to 14 g kg^{-1} with application of woodchips. We observed an 89% increase in soil organic matter content within 18 mo along with a 10-fold increase in soil N. The increased nutrient-storage capacity is reflected in the increase in soil N levels after two annual applications of mixtures of organic wastes. We also saw a similar increase [pH dependent increase in cation exchange capacity (CEC)] in the soil nutrient-storage capacity. These are the primary reasons for the increase in yields of cotton, soybean, and corn when organic waste and poultry litter were applied to the soil surface.

WHAT MECHANISMS ARE INVOLVED IN SOIL NUTRIENT IMBALANCES CREATED BY NEWSPRINT APPLICATION?

The manifestation of a plant nutrient disorder is an indication of an improper nutrient balance in the soil, limited soil nutrient concentrations, or toxicities from excess soil nutrient levels. Any of these conditions can affect a plant's ability to absorb nutrients. Since the uptake of nutrients depends partly on the volume of roots present, any condition that affects root growth will alter nutrient uptake and will be manifested as nutrient deficiency or toxicity by the plant (Edwards et al., 1993c). A possible explanation for the phytotoxic effect of noncomposted newsprint to plants may be the high demand for both N and water by plants and microorganisms immediately after application. This condition would temporarily immobilize all applied N and induce N deficiency in plants as the soil microorganisms compete for available N.

We observed nutrient imbalances in corn seedlings when ground newsprint was surface applied in the spring and the C/N ratio was adjusted to ≤ 30:1. A study comparing three commercial inorganic N sources (urea, NH_3, NH_4NO_3) and poultry litter to adjust the C/N ratio of the newsprint gave evidence of P, Ca and S imbalances (Lu et al., 1994a). Nutrients needed by the plants (N, K, P, etc.) for optimum growth were supplied in addition to N needed to adjust the C/N ratio of ground newsprint. When compared to poultry litter, all inorganic N sources stunted corn growth during the first 4 to 6 wk after seedling emergence. When the N supply for the corn crop was from urea or NH_4NO_3, the induced nutrient imbalances were reduced compared to a newsprint control where the C/N ratio was not adjusted; however, the corn plants were stunted, delayed in maturity, and had lower grain yield than when poultry litter was the N source. Corn plants receiving anhydrous ammonia as the N source did not recover and no grain yield was obtained.

Soil solution Al was 10 times greater when anhydrous ammonia was the N source than in the control using standard cultural practices (Lu et al., 1994b). Whole corn plant (aboveground) tissue samples were collected for chemical analyses 40 d after emergence. The normal Al concentration in plant tissue is below 0.4 mg kg^{-1}; however, Al concentrations were 1.08, 0.89, and 0.77 mg kg^{-1}, respectively, where NH_3, NH_4NO_3 and urea were used to adjust C/N ratio of newsprint, and was below 0.4 mg kg^{-1} where poultry litter was used. Corn roots exhibited signs of Al toxicity, i.e., severe stunting and thickening at the root apex due to the inhibition of cell division.

The concentration of macronutrients K, Ca, Mg, and P in the soil were not affected unless the organic waste treatments contained poultry litter (Edwards et al., 1993a). The soil concentrations of P and K had increased threefold at 4 and 16 wk after application. Newsprint amendments containing poultry litter increased soil Cu 100-fold; Mn twofold; and Zn fourfold when compared to newsprint treatments where the C/N ratio was adjusted to ≤ 30:1 with NH_4NO_3. However, the concentrations of micronutrients are still well within the concentration ranges that can be tolerated by the plant.

The surface-applied newsprint did not affect the soil concentrations of Pb and Cr after 16 wk.

WHAT NITROGEN SOURCES CAN BE USED TO ADJUST THE CARBON/NITROGEN RATIO OF ORGANIC WASTE WITHOUT INDUCING SOIL NUTRIENT IMBALANCES?

Because most of the total N and P in poultry litter is in the organic form (Peperzak et al., 1959; Wood & Hall, 1991; Cummins et al., 1993), and they must be converted to the inorganic form, the rate of mineralization is governed largely by microbial mediated processes (Stevenson, 1986) and becomes the limiting step in N and P availability for uptake by microorganisms. Much research has been directed at laboratory N mineralization studies to determine quantity of N produced from the soil organic pool over a period of time (King, 1984; Bitzer & Sims, 1988; Deans et al., 1986; Fine et al., 1989). Other N mineralization studies have been conducted under field conditions to estimate the quantity of N available for plant uptake (Sims, 1986; Barbarika et al., 1985), and chemical extraction methods to estimate available N content (Chae & Tabatabai, 1986). Because the N and P contained in organic wastes are in forms not immediately available to soil biota, we must verify the quantity and quality of N and P in the soil pools and establish how essential they are for the mediation of soil microbial activity.

Incorporation of organic wastes with low C/N ratio may lead to N immobilization, but when the conditions are suitable the immobilized N may be mineralized and incorporated into microbial biomass or absorbed by the plant. Much of the added inorganic N in paper mill sludge was not recovered as plant-available N (King, 1984). Since paper mill sludge has a high C/N ratio ratio, nonrecovered N was assumed to be immobilized. When expressed as total organic N, N mineralized was 3% for soybean residue and 97% for alfalfa residue [*Medicago sativa* L.], but corn residue or sawdust resulted in a decrease in N mineralization ranging from 3 to 687% (Chae & Tabatabai, 1986).

There has been an extensive effort to develop N and P availability indices to assess the N and P supply capacities of soils, but limited information is available where one organic waste is used for supplying N and P to another manure or organic waste. Almost all research conducted with organic wastes has focused on their use as a nutrient source for plant growth and been concerned with soil NO_3 accumulation and leaching and runoff losses from repeated application (Elliot & Stevenson, 1977; Smith & Peterson, 1982). Another use for manures with high N content (C/N $\leq$ 20:1) is to combine them with organic wastes that have C/N $\geq$ 100:1. Our research shows that N and P in manures are used more efficiently as a nutrient source for the initiation of microbial activity when they are applied to land in combination with waste paper, wood chips, and yard waste (Edwards et al., 1992a, 1993a, 1994b, c).

The primary reason that inorganic N sources are inefficient when they are used to adjust the C/N ratio of waste paper is that alum is used during the processing of green logs to remove tars and resins from grinding equipment. Alum also is used to fix the cellulose fibers in the formation of the paper, the sheets of paper pulp are then put through an alkaline solution to remove or precipitate the excess alum. The inorganic N sources that contain NH_3 or NH_4 ions react with $Al_2(SO_4)_2$ to increase Al concentration in the soil solution. We observed Al toxicity (P, Ca, and S deficiencies) symptoms in cotton, soybean, and corn (Edwards et al., 1993b; Lu et al., 1994a). If inorganic sources of N are used to adjust the C/N ratio of waste paper, the Al will have to be removed from the newsprint before it is applied to land. In the pelletized waste paper used in some of our experiments, which require additional processing steps, the Al was less than 911 mg kg^{-1} compared to 2750 mg kg^{-1} in ground waste paper (Table 11-3).

HOW DOES THE APPLICATION OF ORGANIC WASTE INFLUENCE MICROBIAL POPULATIONS AND ACTIVITY?

Amending soil with newsprint and different N sources enhanced microbial populations and activity in comparison to nonamended soil (Edwards et al., 1994a). However, fungal populations were reduced by N sources in amended soil relative to the nonamended soil. There was a rate response of fungal populations to the C/N ratio, where populations increased with increasing C/N ratio. Bacterial populations were greater in soil amended with N, and populations in soil amended with poultry litter were greater than populations in soil amended with NH_4NO_3-N. There also was a rate response of bacterial populations to poultry litter, where populations decreased with increasing C/N ratio.

A newsprint medium (5 g newsprint L^{-1} distilled water) was used to detect microorganisms that could use newsprint as a sole source of nutrients. Actinomycetes were found to grow on the medium 72 h after plating, this medium and chitin agar (pH 8.5) were used to determine populations of actinomycetes in amended soil. Actinomycetes populations were greater in soil amended with poultry litter than populations in soil amended with NH_4NO_3.

Fifty bacterial colonies were taken at random from 5% tryptic soy agar plates for each treatment. These bacterial strains were identified using fatty acid methyl esters and the Microbial Identification Software[3] (Sasser, 1990). Bacterial populations after 9 wk in newsprint-amended soil had greater diversity than populations in nonamended soil. The bacterial species in soil amended with poultry litter were different from that of nonamended soil and soils amended with NH_4NO_3-N. There were more Gram positive bacterial species in soil amended with poultry litter than in soils amended with NH_4NO_3

---

[3] Mention of trademark, prorietary product, or vendor does not constitute a guarantee or warranty of the product by USDA or Auburn Univ. and does not imply its approval to the exclusion of the other products or vendors that may also be suitable.

or nonamended soil. The dominance of Gram positive organisms in poultry litter-amended soil may indicate that the soil environment is unsuitable for coliform bacteria; thus, coliform bacteria present in the poultry litter are not present in the soil (Table 11-4).

Shifts in the community structure of microbial populations were observed after addition of soil amendments (Press et al., 1994). The total numbers of Gram positive bacteria were decreased in newsprint, poultry litter, and combination-amended soils compared to nonamended soils. There was a corresponding increase in numbers of Gram negative genera isolated from newsprint-amended soil vs. nonamended soil. Many Gram negative strains of bacteria which colonize plant roots, such as *Burkholdaria* sp. and *Pseudomonas* sp. have been associated with increased plant health, yield, and biological control of many plant pathogens (Burr et al., 1978; Kloepper et al., 1988). The shift to greater numbers of Gram negative organisms may indicate a potential increase in plant growth due to some of these organisms.

Species diversity includes the number of species (richness) and the relative abundance of each species in the community (evenness) (Ludwig & Reynolds, 1988, p. 71-103). Richness indices were higher in newsprint and poultry litter-amended soils compared to NH_4NO_3 and newsprint-amended soil, indicating a greater genus diversity in waste-amended soils. Evenness was greater for newsprint/poultry litter and poultry litter-amended soils when compared to soils amended with NH_4NO_3. Newsprint alone was not greater than the nonamended soil which suggests that poultry litter was the predominant influence on the increase in evenness.

A few Enterobacteriaceae isolates were observed in soil samples but these isolates were not increased in poultry litter-amended soil compared to soil amended with NH_4NO_3. Enterobacteriaceae are commonly isolated from nonamended soils and several strains also have been identified as beneficial to plant growth (Kloepper et al., 1988). Isolates of *Salmonella* sp. were not increased by poultry litter. Only one strain of *Salmonella* sp. was isolated and this strain was found in the newsprint-amended soil (Press et al., 1994).

ARE SEDIMENT, NUTRIENT, AND PESTICIDE LOSS PROFILES ACCEPTABLE?

In long-term poultry litter vs. nonpoultry litter studies on tall fescue (*Festuca arundinacea* Schreb.) pastures, Kingery et al. (1992) found levels of soil NO_3-N ≥ 40 mg kg^{-1} at or near bedrock under pastures where litter was applied. They also found an increase of approximately 530% in extractable P under pastures where litter was applied to a depth of 60 cm as compared to pastures where litter was not applied.

An assessment of the extent of groundwater pollution in an area of intense poultry production was conducted by the USDA-SCS, Alabama Department of Environmental Management, and the Geological Survey of Alabama

Table 11–4. Bacterial taxa isolated from soil amended with newsprint.

Soil amended with organic N	Soil amended with inorganic N	Unamended soil
Arthrobacter aurescens	*Arthrobacter globiformis*	*Bacillus brevis*
A. crystallopoietes	*A. ureafaciens*	*B. megaterium*
A. globiformis	*Aureobacterium liquefaciens*	*B. psychophilus*
A. mysorens	*A. testaceum*	*B. pumilus*
A. protophomiae/ramosus	*Bacillus amyloliquefaciens*	*B. thuringiensis*
A. uratoxydans	*B. coagulans*	*Bacillus* spp.
Arthrobacter spp.	*B. megaterium*	*Cytophaga johnsonae*
Aureobacterium liquefaciens	*B. reticulum*	*Micrococcus kristinae*
A. saperdae	*Bacillus* spp.	*Streptomyces aureofaciens*
Bacillus circulans	*Clavibacter michiganense* pv *inisidiosum*	*S. cyaneus*
B. coagulans	*Curtobacterium* spp.	*S. halstedii*
B. laterospous	*Flavobacterium ferrugineum*	*S. violaceusniger*
B. megaterium	*Flavobacterium* spp.	*Streptoverticillium* spp.
B. subtilis	*Listonella* spp.	
B. coagulans	*Micrococcus* spp.	
Bacillus spp.	*Methylobacterium radiotolerans*	
Cellulomonas cellulans	*Microbacterium imperiale*	
C. michiganense pv *insidiosum*	*Phyllobacterium rubiacearum*	
C. michiganense pv *nebraskense*	*Pseudomonas radiora*	
Clavibacter michiganense pv *tessellarius*	*P. sacchaophila*	
Corynebacterium aquaticum	*P. cepacia*	
Corynebacterium bovis	*P. gladioli*	
Curtobacterium flaccumfaciens	*Sphingobacterium spiritivorum*	
Curtobacterium spp.	*Streptomyces antibioticus*	
Flavobacterium esteroarmaticum	*S. cyaneus*	
Listeria monocytogenes	*S. glaucescens*	
Microbacterium laevaniformans	*S. lavendulae*	
Micrococcus kristinae	*S. rochei*	
M. luteus	*S. violaceusniger*	
M. varians	*Streptomyces* spp.	
Micrococcus spp.	*Streptoverticillium* spp.	
Ochrobactrum anthropi	*Xanthobacter flavus*	
Ochrobactrum spp.	*Xanthomonas campestris* pv *vasuclorum*	
Staphylococcus schleiferi	*X. campestris* pv *vitans*	
Streptomyces antibiotics		

in 1988[4]. Concentrations of NO_3-N in 30 groundwater wells from the Sand Mountain area of northern Alabama exceeded the national primary drinking water standard of 10 mg L^{-1}. Fecal coliform or streptococci were detected in 28 out of 30 wells tested.

In our studies, the concentration of NO_3-N was ≤ 5 mg kg^{-1} at 2-m-soil depth after 8 mo when newsprint plus poultry litter were applied to the soil surface (Edwards et al., 1992a, b). When NH_4NO_3 was the N source, soil NO_3-N was ≥ 9 mg kg^{-1} at 2-m-soil depth. Total N in the surface 15 cm was increased from ≤ 0.1 mg kg^{-1} to ≥ 1 mg kg^{-1} when newsprint was applied with poultry litter. Utilization of organic wastes with C/N ratio $\geq 100:1$ in combination with poultry litter may mean the loading rate of poultry litter on agricultural land can be increased while protecting the environment and assuring a sustainable agricultural production system. Influence of application of organic waste plus poultry litter on herbicides and sediment runoff have not been determined.

CAN ORGANIC WASTE BE USED AS RESIDUE COVER DURING FALLOW PERIODS TO REDUCE SOIL AND WIND EROSION?

The 1985 Farm Bill requires all farmland that is classified as highly erodible by the SCS (now NRCS) to reduce erosion to a manageable level by 1995. However, in many agricultural areas in the USA not enough plant residue is produced during the growing season to meet these federal requirements. To achieve this goal, most farmers must plant cover crops during fallow periods as well as leave crop residues on the soil surface.

Limited rainfall and wind erosion are factors which limit crop production in the Southern Plains region of the USA (Bilbro & Fryrear, 1991). Methods used for conserving soil moisture in these regions include contouring, furrow diking, land leveling, land terracing, and application of organic mulches. Organic mulches from crop residues have been used with yield increases of 25% by the 3rd yr and as much as 160% by the 5th yr, an increase in water in the soil profile was the primary reason for yield increases (Fryrear, 1981). However, 22.4 ha of cotton plant residue is required to give 30% surface coverage to 1 ha of land.

One way that organic waste could be used would be to apply it as a surface mulch. Surface mulches of organic waste would protect the soil from direct sunlight, and as a consequence reduce the loss of water due to evaporation, reduce sediment loss by wind, increase water infiltration from rainfall, decrease storm water runoff, increase total available water in the soil profile, and lower surface soil temperature during summer months. Also, as these organic mulches decay, organic matter and nutrient content of the soil would increase.

---

[4]Sand Mountain Groundwater Quality Assessment, first quarterly report, January 1988.

Our research showed soybean yields were increased by application of three organic wastes (newsprint, wood products, and yard waste) in the fall as a residue cover when compared to fallow conditions without residue cover (Edwards, 1994). There was no difference in soybean yields when fall-applied organic waste was compared to a wheat (*Triticum aestivum* L.) cover crop. Soybean yields following application of organic waste residue for cover were approximately in the order: wheat cover, yard waste and wood chips produced yields that were about the same, followed by newsprint and fallow. The N sources used to adjust the C/N ratio ($\leq 30:1$) did not influence soybean yields.

Wind tunnel studies at the USDA-ARS Station at Big Spring, Texas, showed that noncomposed paper pellets applied at a rate of 11 000 kg ha^{-1} reduced wind erosion by 95%. This rate of paper pellets created a ground cover of 30% (Fryrear, personal communication).

WHAT ARE THE CURRENT AND FUTURE RESEARCH TRENDS IN LAND APPLICATION OF ORGANIC WASTE?

With the current restrictions on landfills and landfill space, land application of MSW may be a disposal option, but at the present time it is cost prohibitive for wide use in agriculture. While land application of organic waste and MSW offer the potential to reduce solid wastes in landfills, and reduce fertilizer and herbicide inputs in crop production systems, the environmental impact on delicate soil and plant ecosystems remains largely unknown. We need to establish the loading rates of waste paper, in combination with different forms of N, to maximize crop production and minimize the potential for environmental pollution.

Intensive cropping systems have long been used to remove excess nutrients that may leach to groundwater. Cereal winter cover crops, for example, are generally reported to remove 60% of the mass of N subject to leaching in lysimeter and field studies. The more intensive the cropping system, i.e., the more biomass produced, the greater the effectiveness in removing excess nutrients that can affect water quality. How are soil nutrient dynamics influenced by application of organic waste and poultry litter in intensive cropping systems? Understanding how soil nutrient dynamics are affected will help in developing decision/predictive models for environmentally and economically sound disposal of these wastes. Also, a basic understanding of the N and P dynamics and the pathways of N and P movement in the plant–soil system is needed to maximize nutrient efficiency with the application of MSW.

Reducing production costs and/or improving yields are needed to sustain soybean production in the southeastern USA. Our research has identified cultural practices that have potential to accomplish a reduction in production cost of soybean, i.e., drill planting soybean early has improved yields and reduced the need for herbicides. Also, incorporating ground newsprint into the soil in the spring has been shown to suppress summer weeds such as large crabgrass and sicklepod [*Cassia obtusifolia* L.], while fall applications have suppressed cutleaf evening primrose, henbit and bitter cress. Combining early-plant close-row with fall application of newsprint plus no-till planting has the potential of helping to sustain soybean production in the southeast. We must ascertain if the combination of drill planting soybean plus waste paper application in the fall will reduce the need for herbicide applications in soybean production systems.

Disposing of organic wastes on agricultural soils shows potential for mitigating the effects of some weeds and thus hopefully reducing the amount of herbicides needed for crop production. However, more research is needed to identify the mechanisms involved; it is unclear whether they are physical and/or chemical in nature. Other pest-averting potentials for this type of waste utilization may be discovered.

The water extract of poultry litter exhibited binding or chelating activity on some metal cation Cu, Zn, Mg, and Al. The stability of the divalent metal-complexes decreased in the order Cu > Zn > Mg. The formation of stable metal-complexes involving carboxyl electrovalent, hydroxyl and/or amino coordinate bonding (Tan et al., 1971) makes poultry litter an excellent substrate for chelating metals that may be contained in newsprint and other organic wastes.

Another research area for the possible use of organic wastes is as a C source for the bioremediation of petroleum-contaminated sites. Before the advent of current waste management laws, landfills were commonly used as disposal sites for petroleum-contaminated soils. However, landfill operators are increasingly concerned with any waste material that can have environmental problems associated with disposal. An important option may be the in situ bioremediation of petroleum-contaminated soil by utilizing MSW with high C/N ratio and/or poultry litter to accelerate the decomposition of petroleum hydrocarbons in contaminated soil.

SUMMARY

Application of organic or cellulose waste to agricultural land creates dynamic soil ecosystems. The incorporation of C and N into the soil biomass and ultimately into the soil organic matter fraction is the desired pathway for the C and N contained in these wastes. This pathway leads to improvements in soil chemical, physical and biological properties and has beneficial effects on soil fertility and crop production. However, adverse soil conditions such as nutrient imbalances that can affect plant growth, yields and quality of the environment can be created by shifts in microbial populations.

In this chapter, management practices needed for developing disposal systems using organic wastes as resources were reviewed using information from the literature and studies conducted by USDA-ARS and Auburn University. The most significant results that we have obtained to date show that the efficiency of such wastes is improved when they are applied in combinations that provide a C/N ratio of ≤30:1. Disposing of organic wastes on agricultural soils shows potential for mitigating the effects of some weeds, thus hopefully reducing the amount of herbicides needed for crop production. The research is ongoing, with several issues to be resolved requiring a long-term commitment of time.

ACKNOWLEDGEMENTS

Contribution of USDA-ARS, National Soil Dynamics Laboratory, P.O. Box 3439, Auburn, AL 36831-3439; Department of Agronomy and Soils, 202 Funchess Hall, Auburn University, AL 36849-5412; and the Alabama Agriculture Experiment Station. Journal Series no. 3-933687.

REFERENCES

Adams, F. 1981. Alleviating chemical toxicities: liming acid soils. p. 269–301. *In* G.F. Arkin and H.M. Taylor (ed.) Modifying the root environment to reduce crop stress. Monogr. no. 4. ASAE, St. Joseph, MI.

Barbarika, A., Jr., L.J. Sikora, and D. Colacicco. 1985. Factors affecting the mineralization of nitrogen in sewage sludge applied to soils. Soil Sci. Soc. Am. J. 49:1403–1406.

Beegle, D.B. 1988. Fertilizer value of poultry manure and commercial fertilizers. p. 120–124. *In* Proc. Natl. Poultry Wastes Manage. Symp., Columbus, OH. April. Ohio State Univ., Columbus, OH.

Bilbro, J.D., and D.W. Fryrear. 1991. Pearl millet versus gin trash mulches for increasing soil water and cotton yields in a semiarid region. J. Soil Water Conserv. 46:66–69.

Bitzer, C.C., and J.T. Sims. 1988. Estimating the availability of N in poultry manure through laboratory and field studies. J. Environ. Qual. 17:47–54.

Burr, T.J., M.N. Schroth, and T. Suslow. 1978. Increased potato yields by treatment of seed-pieces with specific strains of *Pseudomonas fluorescens* and *P. putida*. Phytopathology 68:1377–1383.

Chae, Y.M., and M.A. Tabatabai. 1986. Mineralization of nitrogen in soils amended with organic wastes. J. Environ. Qual. 15:193–198.

Cooper, J.R., R.B. Reneav, Jr., W. Kroontje, and G.D. Jones. 1984. Distribution of nitrogenous compounds in a Rhodic Paleudult following heavy manure application. J. Environ. Qual. 13:189–193.

Cummins, C.G., C.W. Wood, and D.P. Delaney. 1993. Co-composted poultry mortalities and poultry litter: Composition and potential value as a fertilizer. J. Sustain. Agric. 4:7–18.

Deans, J.R., J.A.E. Molina, and C.E. Clapp. 1986. Models for predicting potentially mineralizable nitrogen and decomposition rate constants. Soil Sci. Soc. Am. J. 50:323–326.

Diaz, L.F., C.G. Golueke, and G.M. Savage. 1987. Energy balance in compost production and use. p. 6–19. *In* M. De Bertoldi et al. (ed.) Compost: Production, quality and use. Elsevier Appl. Sci., London and New York.

Edwards, J.H. 1992. Recycling newsprint in agriculture. Biocycle 33(1):71–72.

Edwards, J.H. 1994. Direct land application of waste paper. Biocycle 35(4):69–72.

Edwards, J.H., E.C. Burt, R.L. Raper, and D.T. Hill. 1992a. Effects of deep placement of nutrients, broiler litter, and newsprint on cotton yield and rooting depth. Proc. Beltwide Cotton Conf. 3:1143–1146.

Edwards, J.H., E.C. Burt, R.L. Raper, and D.T. Hill. 1992b. Recycling paper and poultry byproducts in agriculture. Highlights Agric. Res. 39(1):7.

Edwards, J.H., E.C. Burt, R.L. Raper, and D.T. Hill. 1993a. Recycling newsprint on agricultural land with the aid of poultry litter. Compost Sci. Utiliz. 1:79–92.

Edwards, J.H., C.M. Press, W.F. Mahaffee, and J.W. Kloepper. 1994a. Direct application of municipal solid waste for modification of soil biota and crop production. p. 91–93. *In* C.E. Pankhurst (ed.) Proc. Workshop Manage. Soil Biota in Sustainable Farming Systems, Adelaide, South Australia. 15–18 March. CSIRO, Australia.

Edwards, J.H., R.H. Walker, and J.S. Bannon. 1994b. Effects of repeated application of noncomposted organic wastes on cotton yield. Proc. Beltwide Cotton Conf. 3:1564–1567.

Edwards, J.H., R.H. Walker, N. Lu, and J.S. Bannon. 1993b. Applying organics to agricultural land. Biocycle 34(10):48–50.

Edwards, J.H., R.H. Walker, C.C. Mitchell, and J.S. Bannon. 1993c. Effects of soil-applied noncomposted organic wastes on upland cotton. Proc. Beltwide Cotton Conf. 3:1354–1356.

Edwards, J.H., R.H. Walker, and R. Rawls. 1993d. Effects of noncomposted organic waste in a corn production system. p. 16. In South. Branch Agronomy abstracts. ASA, Madison, WI.

Edwards, J.H., R.H. Walker, and W.D. Webster. 1994c. Effects of noncomposted organic waste as residues on cotton yields. Proc. Beltwide Cotton Conf. 3:1561–1563.

Edwards, J.H., C.W. Wood, D.L. Thurlow, and M.E. Ruf. 1992c. Long-term tillage and crop rotation effects on the fertility status of a Hapludult soil. Soil Sci. Soc. Am. J. 56:1577–1582.

Elliot, L.F., and F.J. Stevenson. 1977. Soils for management of organic wastes and waste waters. ASA, CSSA, SSSA, Madison, WI.

Eno, C.F. 1966. Chicken manure-its production, value, preservation and disposition. Univ. Florida. Agric. Exp. Stn. Circ. S-140.

Fine, P., U. Mingelgrin, and A. Feigin. 1989. Incubation studies of the fate of organic nitrogen in soils amended with activated sludge. Soil Sci. Soc. Am. J. 53:444–450.

Flynn, R.P., C.W. Wood, and J.T. Touchton. 1993. Nitrogen recovery from broiler litter in a wheat-millet production system. Bioresourc. Technol. 44:165–173.

Fryrear, D.W. 1981. Management of blank rows in dryland skip-row cotton. Trans. ASAE 24:988–990.

Glenn, J. 1992. The state of garbage in America: 1992 nationwide survey. Biocycle 33(4):46–55.

Goetz, A.J., R.H. Walker, G. Wehtje, and B.F. Hajek. 1989. Sorption and mobility of Chlorimuron in Alabama soils. Weed Sci. 37:428–433.

Goldstein, N. 1989. Solid waste composting in the U.S. Biocycle 30(4):32–37.

Greshman, H. 1992. Municipal waste costs going up. Biocycle 33(3):13.

Hileman, L.H. 1967. The fertilizer value of broiler litter. Arkansas Agric. Exp. Stn. Rep. Ser. 158.

Hileman, L.H. 1971. Effect of rate of poultry manure application on selected soil chemical properties. p. 247–248. In Livestock waste management and pollution abatement. Proc. Int. Symp. Livestock Wastes. Columbus, OH. 19–22 April. ASAE, St. Joseph, MI.

Hinton, S.A. 1991. Poultry review. In Alabama agricultural statistics. Alabama Agric. Stat. Serv. Bull. 34. Alabama Dep. Agric. Industry, Montgomery, AL.

Karlen, D.L., J.H. Edwards, W.J. Busscher, and D.W. Reeves. 1991. Grain sorghum response to silt-tillage on Norfolk loamy sand. J. Prod. Agric. 4:80–85.

Ketcheson, J.W., and E.G. Beauchamp. 1978. Effects on corn stover, manure and nitrogen on soil properties and crop yield. Agron. J. 70:792–797.

King, L.D. 1984. Availability of nitrogen in municipal, industrial and animal wastes. J. Environ. Qual. 13:609–612.

Kingery, W.L., C.W. Wood, D.P. Delaney, J.C. Williams, and G.L. Mullins. 1992. Impact of long-term land application of broiler litter on environmentally related soil properties. J. Environ. Qual. 23:139–147.

Kloepper, J.W., D.J. Hume, F.M. Scher, D. Singleton, B. Tipping, M. Laliberte, K. Frauley, T. Kutchaw, C. Simonson, R. Lifshitz, I. Zaleska, and L. Lee. 1988. Plant growth-promoting rhizobacteria on canola (rapeseed). Plant Dis. 72:42–46.

Larson, W.E., R.F. Holt, and C.W. Carlson. 1978. Residues for soil conservation. p. 1–15. In W.R. Oschwald (ed.) Crop residue management systems. ASA Spec. Publ. 31. ASA, CSSA, and SSSA, Madison, WI.

Liebhardt, W.C., C. Golt, and J. Tupin. 1979. Nitrate and ammonium concentrations of ground water resulting from poultry manure applications. J. Environ. Qual. 8:211–215.

Lu, N., J.H. Edwards, R.H. Walker, and J.S. Bannon. 1994a. Organics wastes and nitrogen sources interaction on corn growth and yield. p. 431–438. In A.B. Bottcher et al. (ed.) Proc. 2nd Conf. Environmentally Sound Agriculture, Orlando, FL. 20–22 April. ASAE, St. Joseph, MI.

Lu, N., J.H. Edwards, and R.H. Walker. 1995b. Organics wastes and nitrogen sources interaction on soil solution ionic activity. Compost Sci. Utiliz. 3(1):6–18.

Ludwig, J.A., and J.F. Reynolds. 1988. Statistical ecology: A primer on methods and computing. John Wiley & Sons, New York.

Mitchell, C.C., and J.H. Edwards. 1993. Organic by-products on agricultural soil. Alabama Coop. Ext. Serv., S-04-93.

Mitchell, C.C., J.O. Donald, and J. Martin. 1989. The value and use of poultry waste as fertilizer. Poultry by-product management–agronomy. Alabama Coop. Ext. Serv., ANR-244.

Moore, P.A., Jr., and D.M. Miller. 1994. Decreasing phosphorus solubility in poultry litter with aluminum, calcium, and iron amendments. J. Environ. Qual. 23:325–330.

O'Keefe, B.E., J. Axley, and J.J. Meisinger. 1986. Evaluation of nitrogen availability indexes for a sludge compost amended soil. J. Environ. Qual. 15:121–128.

Payne, V.W.E., and J.O. Donald. 1991. Poultry-waste management and environmental protection manual. Alabama Coop. Ext. Serv. ANR-580.

Peperzak, P, A.G. Caldwell, R.R. Hunziker, and C.A. Black. 1959. Phosphorus fractions in manures. Soil Sci. 87:293–302.

Press, C.M., W.F. Mahaffee, J.H. Edwards, and J.W. Kloepper. 1994. Population shifts and species diversity of bacteria associated with newspaper and chicken litter amended soils. Compost Sci. Utiliz. 2:(In press.)

Raper, R.L., T.R. Way, E.C. Burt, D.T. Hill, J.H. Edwards, D.W. Reeves, and A.A. Trotman. 1994. Hindering hardpan reformation with cellulose waste. p. 251–259. In D.L. Karlen et al. (ed.) Agricultural utilization of urban and industrial by-products. ASA Spec. Publ. 58. ASA, CSSA, and SSSA, Madison, WI.

Ritter, W.F., and A.E.M. Chirnside. 1984. Impact of land use on ground water quality in southern Delaware. Ground Water 22:38–47.

Sasser, M. 1990. Identification of bacteria through fatty acid analysis. p. 199–204. In Z. Klement et al. (ed.) Methods in phytobacteriology. Akademiai Kiado, Budapest, Hungary.

Sims, J.T. 1986. Agronomic evaluation of poultry manure amended soil: Temperature and moisture effects. J. Environ. Qual. 15:59–63.

Sims, J.T. 1987. Agronomic evaluation of poultry manure as a nitrogen source for conventional and no-tillage corn. Agron. J. 79:563–570.

Slack, C.H., R.L. Blevins, and C.E. Rieck. 1978. Effect of soil pH and tillage on persistence of Simazine. Weed Sci. 26:145–148.

Smith, S.J., and J.R. Peterson. 1982. Recycling of nitrogen through land application of agricultural, food processing, and municipal wastes. p. 791–832. In F.J. Stevenson (ed.) Nitrogen in agricultural soils. Agron. Monogr. 22. ASA, CSSA, SSSA, Madison, WI.

Staniforth, D.W., and A.F. Wiese. 1985. Weed biology and its relationship to weed control in limited-tillage systems. p. 15–25. In A.F. Wiese (ed.) Weed control in limited-tillage systems. Weed Sci. Soc. Am., Champaign, IL.

Stevenson, F.J. 1986. Cycles of soil. John Wiley & Sons, New York.

Tan, K.H., R.A. Leonard, A.R. Bertrand, and S.R. Wilkinson. 1971. The metal complexing capacity and the nature of the chelating ligands of water extract of poultry litter. Soil Sci. Soc. Am. Proc. 35:265–269.

Taylor, H.M., L.E. Locke, and J.E. Box. 1964. Pans of the Southern Great Plains soils. III. Their effects on yield of cotton and grain sorghum. Agron. J. 56:542–545.

Thomas, G.W. 1976. Development of environmental guidelines for fertilizer use. Southern Coop. Res. Bull. 211.

U.S. Environmental Protection Agency. 1989. The solid waste dilemma: An agenda for action. U.S. Gov. Print. Office, Washington, DC.

Walker, R.H., J.H. Edwards, and C.C. Mitchell. 1993. Effects of noncomposted waste materials on weed control in crops. Highlights Agric. Res. 40:3.

Wilkinson, S.R. 1979. Plant nutrient and economic value of animal manures. J. Anim. Sci. 48:121–133.

Wood, C.W., C.D. Cotton, and J.H. Edwards. 1991. Broiler litter as a nitrogen source for strip and conventional tillage corn. p. 568–575. In A.B. Bottcher et al. (ed.) Proc. Conf. Environmentally Sound Agriculture, Orlando, FL. 11–18 April. IFAS, Univ. Florida, Gainesville, FL.

Wood, C.W., and J.H. Edwards. 1992. Agroecosystem management effects on soil carbon and nitrogen. Agric. Ecosyst. Environ. 39:123–138.

Wood, C.W., and B.M. Hall. 1991. Impact of drying method on broiler litter analyses. Commun. Soil Sci. Plant Anal. 22:1677–1688.

Wood, C.W., H.A. Torbert, and D.P. Delaney. 1993. Poultry litter as a fertilizer for bermudagrass: Effects on yield and quality. J. Sustain. Agric. 3(2):21–36.

12 Hindering Hardpan Reformation with Cellulose Waste

R. L. Raper, T. R. Way, E. C. Burt, J. H. Edwards, and D. W. Reeves

USDA-ARS
National Soil Dynamics Laboratory
Auburn, Alabama

D. T. Hill

Auburn University
Auburn, Alabama

A. A. Trotman

Tuskegee University
Tuskegee, Alabama

Disposal of solid wastes in the USA is becoming increasingly challenging. Our 5500 landfills are filling up at an alarming rate and few new ones are being approved (Rathje, 1991). The nation's annual trash bill now exceeds $15 billion and new ways of handling waste materials are sorely needed.

According to Rathje (1991), misperceptions must be dealt with before a reduction in trash can be achieved. He reports that most people believe that landfills are filled with fast-food packaging, polystyrene foam and disposable diapers. His research has showed that these components add up to only about 2% of the total solid waste.

The largest component of landfill waste is paper. Waste paper now occupies some 50% of the waste stream, and newspapers themselves occupy between 10 to 15%, even though many are now being recycled or exported. Rathje (1991) reports that paper does not decompose as rapidly as once believed. His research has found legible copies that have been buried for 40 yr. Alternative methods for disposal of waste paper should be investigated before our expensive landfill space is filled to capacity.

An alternative paper disposal method that has been investigated at the USDA-ARS National Soil Dynamics Laboratory is to use shredded newspaper as a soil amendment for crop production (Burt et al., 1992; Edwards et al., 1992a, b, 1994). Experiments have been conducted on both surface incorporation of the waste paper and on burying this material in vertical trenches. Preliminary results have indicated that increased cotton (*Gossypium hirsutum* L.) yields have resulted when the waste paper was applied in a vertical trench between plant rows. The trenches alleviated the shallow rooting problems associated with a hardpan and the waste paper may decrease the rate at which the hardpan reconsolidates. The cellulose waste offers the benefit of keeping the hardpan from reforming and offers a disposal alternative to landfills.

Vertical mulching of agricultural crop residue was investigated as a method of alleviating heavy straw concentrations left on the soil surface after large yields (Spain & McCune, 1956). This heavy straw concentration was interfering with planting operations and preventing timely field operations. Curley et al. (1958) investigated burial of organic materials, including hay and corn cobs, as a method of increasing water availability and deepening the plant root zone. They found that in the 1st yr of the experiment, large differences in crop yield were achieved due to the benefits afforded by vertical mulching. In later years, however, yields were variable due to heavy winter rains which provided adequate moisture for all plots. A Canadian experiment (Clark & Hore, 1965) found that vertical mulching had little effect on crop yields, soil moisture distribution, unit draft or watertable levels.

Saxton (1980) hypothesized that one reason for the failure of the vertical mulching system developed in the 1950s is that the experimental plots were plowed following the vertical mulching treatment, which sealed the trench surfaces. Saxton (1980) proposed a system by which the mulch was allowed to penetrate the soil surface and intercept free water. The trench would then have a longer lifespan, infiltration would be increased, and the problems associated with surface sealing would be alleviated.

An implement was developed for this application to vertically mulch residue cover in the Pacific Northwest of the USA (Hyde et al., 1986, 1989). Several methods were tested to establish the vertical trenches necessary for the vertical mulch operation. These included vertical shanks, parabolic shanks, and rotary slot cutting devices. They determined that a vertical shank with a 22° lift angle minimized draft requirements and formed the best slot when combined with a pair of preceding coulters that cut to about one-half the depth of the slot. The slot depth was 26 cm. A machinery system similar to the one developed in the Pacific Northwest region was developed to handle corn (*Zea mays* L.) residue (Edwards & Rumsey, 1989).

Despite successful attempts by researchers to incorporate crop residue into a vertical trench, commercialization of this process has not been successful. Apparently, the potential yield increases have been insufficient to overcome the significant machinery and tillage energy costs. If waste cellulose materials are applied, these machinery and energy costs may be offset by potential payments from municipalities to farmers for disposal of a portion of their solid waste. This would be in addition to any potential yield increases that farmers might achieve. According to Burt et al. (1992) waste disposal costs for municipal landfills very from $44 to $154/Mg, depending on location. They reported that if waste could be applied to agricultural soil at a rate of 27 Mg/ha with a 1-m spacing between trenches, landfill costs of $1188 to $4158/ha could be avoided.

To facilitate experiments of this nature, an implement was designed and built at the USDA-ARS National Soil Dynamics Laboratory to disrupt the hardpan and incorporate cellulose material into the trench in one pass. This implement was used in an experiment to determine the response of grain sorghum (*Sorghum bicolor* [L.] Moench) to the tillage/disposal system. Various C/N ratios were examined by mixing poultry litter or liquid ammonium nitrate with the cellulose.

The objectives of this experiment were to:

1. Determine the effectiveness of an implement designed to apply shredded newspaper in a vertical trench in soil.
2. Determine the effect of shredded newspaper applied in a vertical trench between crop rows on crop response.
3. Determine the effect of shredded newspaper applied in a vertical trench between crop rows on hardpan reconsolidation.

METHODS AND MATERIALS

An experiment to investigate vertical trenching of shredded newspaper was conducted near Shorter, AL, at the Auburn University E.V. Smith Experiment Staton. The shredded newspaper was obtained from CEL-PAK[1] (Decatur, AL) who manufactured it for insulation. This material was passed through a 9.5-mm size screen before being packaged into 11.4-kg bundles. A C/N analysis of the shredded newspaper showed it to contain 48% C and less than 0.5% N.

This experiment used an implement developed at the USDA-ARS National Soil Dynamics Laboratory for digging a trench and applying the shredded newspaper in the trench. This implement consisted of a 15-cm wide shank which was used to subsoil to a depth of 61 cm. A large hopper mounted above the shank held the shredded newspaper that was to be applied into the trench. The shredded newspaper was dispensed directly behind the shank to allow some of the cellulose material to flow to the bottom of the trench. Excavations performed during initial trials showed that a uniform mixture was being obtained along the depth of the trench.

Nitrogen was mixed with the cellulose material to adjust its C/N ratio to an acceptable level for plant growth. The purpose of the added N was to optimize yields and minimize the effects of the large quantities of C being placed into the soil. Two different methods were used for this adjustment. The simplest method consisted of mixing a 32% ammonium nitrate solution with the shredded newspaper prior to applying it in the trench. The C/N ratio was adjusted to 25:1 using this technique.

---

[1] The use of tradenames or company names does not imply endorsement by USDA-ARS, Auburn University, or Tuskegee University.

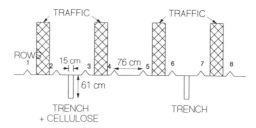

Fig. 12-1. Position of trenches relative to rows.

The second method consisted of mixing poultry litter with the shredded newspaper to achieve this same C/N ratio. A chemical analysis of the poultry litter was performed a few days prior to the experiment and it was found to have 22% C and 3% N. A third C/N ratio was achieved by applying the shredded newspaper directly into the trench without mixing it with any N source. This third ratio was approximately 125:1.

An arrangement of rows and trenches provided a plot area which could be used to determine both C/N effects and trench effects. Eight-row plots with a row spacing of 0.76 m were used with trenches placed halfway between Rows 2 and 3 and Rows 6 and 7 (Fig. 12-1). Each plot was 18.3 m in length with a 6-m buffer zone between plots. The experimental layout consisted of placing a trench with the cellulose material mixture on one side of the plot (between Rows 2 and 3) and placing a trench with no cellulose material on the other side of the plot (between Rows 6 and 7). The trench between Rows 6 and 7 with no cellulose material was constructed with the cellulose disposal implement, but with nothing applied in the trench area. It was assumed that Rows 2 and 3 would benefit from the trench with the cellulose material, and Rows 6 and 7 would benefit from the trench alone. Rows 4 and 5 would receive no trenching benefit and should provide a useful control. Rows 1 and 8 were used as buffer rows.

Two different experiments were conducted and each contained four replications. A Cahaba-Wickham-Bassfield sandy loam soil (fine-loamy, siliceous, thermic Typic Hapludults–fine-loamy, mixed, thermic Typic Hapludults–coarse-loamy, siliceous, thermic Typic Hapludults) with a predominant hardpan which is prone to reconsolidation was used to grow grain sorghum. A similar experiment was performed in a Norfolk sandy loam soil (fine-loamy, siliceous, thermic Typic Kandindults) with soybean [*Glycine max* (L.) Merr.]. The soybean experiment received no poultry litter treatment.

The trenches were formed and the shredded newspaper applications were made in June 1993. Because very little rain fell for the rest of the summer, a small amount of water was used to irrigate both sets of plots early in the growing season to prevent loss of data. In the grain sorghum experiment, Rows 4 and 5 were noted to contain plants with reduced height and immature heads as compared to those plants in Rows 2 and 3 or Rows 6 and 7. At the conclusion of the growing season, the center 9.1 m of each plot along the plot length was harvested for crop yield. These harvests placed Rows 2 and 3 together, as well as Rows 4 and 5, and Rows 6 and 7.

The soil condition at the end of the growing season was determined by recording the force required to push a soil cone penetrometer (ASAE, 1993) vertically into the soil at depth increments of approximately every 3 mm. The soil cone penetrometer is an indicator of soil strength and simulates the forces that roots would encounter when growing in a particular soil condition. The forces recorded are then divided by the cross-sectional base area of the cone to calculate cone index. Several profiles were taken in the first replication of the experiment across the plot area both inside the trench and in nearby untrenched areas in October 1993, to investigate the effect of trenching on soil condition. A series of 17 measurements were made across the plot in all three different C/N treatments. The soil condition when these measurements were taken was extremely dry which gave large cone index values. Another set of cone index measurements were taken in all treatments in all replications in a saturated soil condition in November 1993. A soil cone penetrometer was inserted both in the trench that contained cellulose and in the trench that did not contain any cellulose at five separate locations within each plot.

Within each plot, two ceramic cup lysimeters with $0.4-\mu$ pore size (Soil Moisture Equipment, Inc., Santa Barbara, CA) were installed midway between the plant row and the trench at depths of 0.3 and 0.6 m. Water samples were taken monthly to determine if any N leaching could be detected, particularly in the plots that received the added amounts of N. No data from these lysimeters was reported in this paper.

Grain yield data were analyzed using C/N ratio and row positions as the independent variables. Analyses of variance and multiple comparisons were made using SAS statistical software (Cary, NC).

RESULTS

Grain yield results from the grain sorghum and soybean plots were similar. The most interesting result is the benefit of the rows that were adjacent to the trenches. Grain yields of grain sorghum and soybean in rows adjacent to trenches (Rows 2, 3, 6, and 7) were significantly greater than those of rows not adjacent to a trench (Rows 4 and 5) (Fig. 12-2 and 12-3). The improved rooting depth afforded by the trench allowed plants adjacent to this area to take advantage of available moisture during a growing season which received little rainfall.

In both the grain sorghum and soybean plots, grain yields were reduced slightly in those plots with shredded newspaper in their trenches. The yield reduction was not statistically significant.

No significant effect of adjusting the C/N ratio of cellulose was found (Fig. 12-4 and 12-5). On the contrary, the crop yield tended to be reduced when these ratios were adjusted, although the differences were not statistically significant.

An effort was made to excavate and visually inspect the trenches for rooting activity near the end of the growing season. Only a small amount

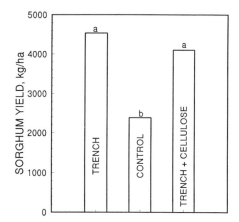

Fig. 12–2. Grain sorghum yield in rows adjacent to trenches (Rows 2 and 3 for cellulose plus trench and Rows 6 and 7 for trench alone) and in control rows (Rows 4 and 5). Means with the same letter are not significantly different at $\alpha = 0.05$.

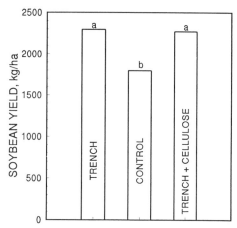

Fig. 12–3. Soybean yields in rows adjacent to trenches (Rows 2 and 3 for cellulose plus trench and Rows 6 and 7 for trench alone) and in control rows (Rows 4 and 5). Means with the same letter are not significantly different at $\alpha = 0.05$.

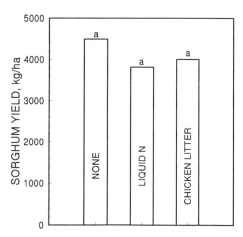

Fig. 12–4. Grain sorghum yields in Rows 2 and 3 adjacent to trenches filled with cellulose material that had C/N ratios balanced to 25:1. Means with the same letter are not significantly different at $\alpha = 0.05$.

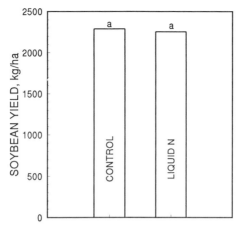

Fig. 12-5. Soybean yields in Rows 2 and 3 adjacent to trenches filled with cellulose material that had C/N ratios balanced to 25:1. Means with the same letter are not significantly different at $\alpha = 0.05$.

of cellulose remained visible in the trenched area, but rooting activity was abundant to the bottom of the trench. Cone index measurements were taken across the plot width to determine the effect of the trench. The results show that the effect of the trench is much broader than the 15-cm width of the tillage tool (Fig. 12-6). This is evident in both trenches, with and without cellulose.

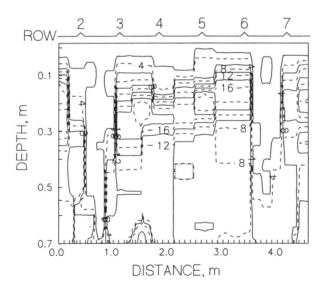

Fig. 12-6. Cone index (MPa) profile taken at end of growing season across the grain sorghum experiment with the cellulose trench between Rows 2 and 3 and the non-cellulose trench between Rows 6 and 7. The C/N treatment for this plot was ammonium nitrate.

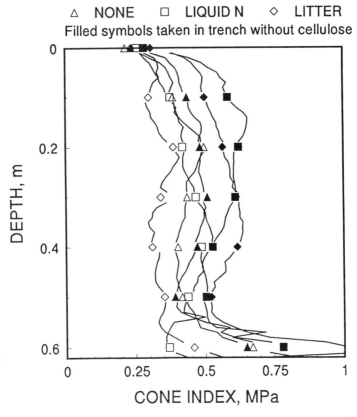

Fig. 12-7. Cone index measurements taken at end of growing season in the center of the trenches of the grain sorghum experiment.

A trend was noted while examining the cone index profiles: trenches without any cellulose had greater cone index values than trenches in which cellulose had been applied (Fig. 12-7). This trend may indicate that reconsolidation is already beginning to occur, even though it had only been 5 mo since the trenches were created.

SUMMARY AND CONCLUSIONS

Experiments were conducted to determine the effect of inserting cellulose waste in a trench uniformly between the surface and down to a depth of 61 cm between rows in a grain sorghum and soybean experiment. An implement was designed and built to disrupt the hardpan and incorporate these cellulose materials into the trench in one pass. The cellulose application implement performed well and enabled shredded newspaper to be applied uniformly in a vertical trench.

The vertical trench provided a dramatic positive rooting environment improvement over soil that was not subsoiled and this benefit was reflected in grain sorghum and soybean yields. At the end of the growing season, a trend was found toward reduced reconsolidation in trenches formed with shredded newspaper applied in the trenches as compared to trenches formed without shredded newspaper applied in the trenches. Decreasing the C/N ratio of the shredded newspaper to 25:1 also was found to have no effect on grain sorghum or soybean yields.

ACKNOWLEDGMENTS

The authors would like to thank B.H. Washington, Design Engineer at the USDA-ARS National Soil Dynamics Laboratory in Auburn, Alabama, for his contribution in designing the combination trenching and cellulose disposal implement. This manuscript is designated Journal Series no. 2-933676 by the Alabama Agricultural Experiment Station.

REFERENCES

American Society of Agricultural Engineers. 1993. Soil cone penetrometer. ASAE, St. Joseph, MI.

Burt, E.C., J.H. Edwards, R.L. Raper, and D.T. Hill. 1992. Cellulose as a soil conditioner. ASAE Paper no. 921560. ASAE, St. Joseph, MI.

Clark, D.E., and F.R. Hore. 1965. Vertical and horizontal mulch on Haldimand clay. Trans. ASAE 8:592–593.

Curley, R.G., B.A. Krantz, J.A. Vomicil, and W.J. Chancellor. 1958. Vertical mulching; one possible solution to soil compaction problems. Univ. California Agric. Ext. Stn. Bull.

Edwards, B.A., and J.W. Rumsey. 1989. Development of a vertical mulching implement. ASAE Paper no. 891503. ASAE, St. Joseph, MI.

Edwards, J.H., E.C. Burt, R.L. Raper, and D.T. Hill. 1992a. Effects of deep placement of nutrients, broiler litter, and newsprint on cotton yield and rooting depth. p. 1143–1146. In Proc. of the Beltwide Cotton Conf., Vol. 3, Nashville, TN. 6–10 January. Natl. Cotton Council, Memphis, TN.

Edwards, J.H., E.C. Burt, R.L. Raper, and D.T. Hill. 1992b. Recycling newsprint on agricultural land with the aid of broiler litter. Compost Sci. Utiliz. 1:79–92. 1993.

Edwards, J.H., R.H. Walker, E.C. Burt, and R.L. Raper. 1995. Issues affecting application of noncomposted organic waste to agricultural land. p. 225–249. In D.L. Karlen et al. (ed.) Agricultural utilization of urban and industrial by-products. ASA Spec. Publ. 58. ASA, CSSA, and SSSA, Madison, WI.

Hyde, G.M., J.E. George, K.E. Saxton, and J.B. Simpson. 1986. A slot-mulch implement design. Trans. ASAE 29:20–25.

Hyde, G.M., J.E. George, K.E. Saxton, and J.B. Simpson. 1989. Slotmulch insertion machine: Design and performance. ASAE Paper no. 891502. ASAE, St. Joseph, MI.

Rathje, W.L. 1991. Once and future landfills. National Geographic, May, p. 116–134.

Saxton, K.E. 1980. Slot mulch tillage for water management and reduced erosion. p. 42–49. In Proc. Conf. on Crop Production with Conservation in the 80's. Vol. 3, Chicago, IL. 1–2 December. ASAE, St. Joseph, MI.

Spain, J.M., and D.L. McCune. 1956. Something new in subsoiling. Agron. J. 48:192–193.

13 Revegetating Bentonite Mine Spoils with Sawmill By-Products and Gypsum

Gerald E. Schuman

USDA-ARS
High Plains Grasslands Research Station
Cheyenne, Wyoming

Lands disturbed by, or abandoned after, bentonite mining in the Northern Great Plains are difficult to reclaim because of the adverse chemical and physical properties of the spoil material, the limited inherent topsoil, the arid/semiarid climate of the area, and to some extent, the mining methods utilized. Reclamation technology utilizing sawmill by-products and gypsum was developed to enhance revegetation of these abandoned lands. This technology and research leading to its development will be presented in this chapter.

Although large-scale bentonite mining began in the region in the 1930s, few areas were reclaimed until after enactment of reclamation laws in the early 1970s. Ninety percent of the US's supply of bentonite is mined in the Northern Great Plains states of Montana, South Dakota, and Wyoming (Ampian, 1980). The National Academy of Sciences (1974) reported that more land in Montana was disturbed in 1973 by bentonite mining than by coal mining, and that more abandoned spoils had accumulated over the years from bentonite than from coal mining. Although these statements refer to Montana, they are indicative of the Northern Great Plains Region as well.

Natural revegetation and man-assisted reclamation seedings of nontopsoiled, nonamended bentonite spoils have resulted in no- to poor-plant establishment (Dollhopf & Bauman, 1981; Sieg et al. 1983); therefore, spoil modification is essential for successful revegetation of these sodic, saline, high clay content spoils. Reclamation of abandoned bentonite spoils has met with limited success (Hemmer et al., 1977; Bjugstad et al., 1981; Dollhopf & Bauman, 1981) unless organic amendments were utilized. Schuman and Sedbrook (1984) reported on a 1979 to 1983 study which demonstrated the effectiveness of sawmill wastes (sawdust, woodchips, and bark) in improving the physical characteristics of the spoil, greatly enhancing water infiltration, and promoting vegetation establishment.

Dollhopf and Bauman (1981) evaluated the use of woodchips (1660 m^3 ha^{-1}) and manure (224 Mg ha^{-1}) in conjunction with 4500 kg ha^{-1} of straw mulch in a 1980 study. They obtained good 1st yr plant densities with woodchips; however, the manure only resulted in about 23% of the density obtained with the woodchips. They also evaluated the effectiveness of the inorganic amendments gypsum, calcium chloride, and sulfuric acid, but these amendments resulted in poor initial seedling establishment. Dollhopf and Bauman (1981) also evaluated the effect of these inorganic amendments on the hydraulic conductivity of the spoil and found that, although they increased the hydraulic conductivity it was still below the "very slow" permeability class, 0.125 cm h^{-1} (Kohnke, 1968). The authors concluded that the poor plant establishment observed was because these chemical amendments did not alter the physical condition of the spoil rapidly enough to enable successful seedling emergence, but that over time these amendments may prove beneficial.

The findings of Dollhopf and Bauman (1981) were not unexpected since inorganic amendments used alone may require considerable time (perhaps years) to improve spoil physical characteristics through the replacement and leaching of Na under natural precipitation conditions. Dissolution of the gypsum or calcium chloride, and replacement of the Na might have occurred. However, leaching probably did not occur because of a lack of time and/or the alteration of the spoil in a way that improved infiltration, allowing the Na to be leached from the spoil profile (U.S. Salinity Lab. Staff, 1954; Prather et al., 1978; Hira et al., 1981). Conversely, the dissolution of the inorganic amendments can result in a large increase in the salinity at levels detrimental to seedling development but not detrimental to growth and development of established plants (U.S. Salinity Lab. Staff, 1954; Heather & Hegarty, 1979). It is noteworthy, however, that Dollhopf and Bauman (1981) demonstrated a very good initial plant density on the woodchip amendment and topsoiled treatments which had an immediate affect on the improvement of spoil physical characteristics even under the same precipitation conditions with which poor results were obtained by the use of inorganic amendments.

These research findings demonstrated the importance of a bentonite spoil amendment program to achieve rapid and successful reclamation. To be successful, the spoil amendment program must immediately increase water infiltration for vegetation establishment and Na leaching to ensure the effectiveness of any inorganic amendment used to ameliorate the sodicity problems. If both of these requirements of improved infiltration and Na replacement/leaching are not met, long-term reclamation may not be achieved or may require considerable time.

The placement of topsoil over bentonite spoil is not seen as a viable option for abandoned bentonite mined land reclamation. No topsoil salvage was accomplished during the mining and in most cases topsoil borrowing is not recommended. Topsoil is a limited resource in bentonite mining areas, and the disturbance of native rangeland results in significant additional land area requiring revegetation and postreclamation management (Richmond, 1991). Because existing soils are very shallow, only a limited depth of topsoil would be available to cover the characteristically sodic spoil materials. A shallow covering may enable initial revegetation, but experience has shown that, unless sodic spoils are buried sufficiently, sodication of the soil will occur within a short time (Merrill et al., 1980; Dollhopf et al., 1985). Therefore, the alternative practice of topsoiling has been considered to be both technically and environmentally unsound.

The purpose of this paper is to discuss the role of organic amendments, for achieving immediate improvement in the physical properties of bentonite spoil, and inorganic amendments for long-term amelioration of the sodic properties of the spoil. Field research data will be used to demonstrate these relationships.

PRELIMINARY STUDY OF ORGANIC AMENDMENTS: EFFECTS ON SPOIL PHYSICAL PROPERTIES AND VEGETATION

Field studies were initiated in 1979 on 1.5 ha of leveled abandoned bentonite spoils near Upton in northeastern Wyoming (Schuman & Sedbrook, 1984). These spoils were typical of the abandoned spoils associated with the Mowry shale formation (Table 13-1). The study involved an initial evaluation of the feasibilty and effectiveness of Ponderosa pine (*Pinus ponderosa* Laws) wood residues (woodchips, bark, and sawdust) as an amendment to improve the physical characteristics of the spoil and thereby aid revegetation. Wood residue rates of 0-, 112-, and 224-Mg ha^{-1}, N fertilizer at 5.0 kg N Mg^{-1} wood residue and P fertilizer at 90 kg P ha^{-1} were incorporat-

Table 13-1. Physical and chemical characteristics of pretreatment bentonite spoil samples, Upton, WY, 1981 (Smith, 1984).

Parameter	Mean and standard error†
Particle-size separates (%)	
Sand	10.8 ± 0.8
Silt	29.6 ± 0.8
Clay	59.6 ± 1.1
Saturation percentage (%)	80.9 ± 1.7
NO_3-N (mg kg^{-1})	7.7 ± 0.4
NH_4-N (mg kg^{-1})	2.6 ± 0.1
TKN (mg kg^{-1})	751.1 ± 5.8
P (mg kg^{-1})	8.1 ± 0.3
C (mg kg^{-1})	10.0 ± 1.0
pH	6.8 ± 0.1
Electrical conductivity (dS m^{-1})	13.4 ± 1.1
Water-soluble cations	
Ca (mg kg^{-1})	187.9 ± 9.2
Mg (mg kg^{-1})	73.6 ± 4.2
Na (mg kg^{-1}	3613.7 ± 101.3
K (mg kg^{-1})	32.0 ± 0.8
Sodium adsorption ratio	63.1 ± 1.2

† Particle-size separates obtained from five observations. All other parameters are a mean of 144 samples.

Table 13-2. The effect of wood residue amendment of bentonite spoils on the soil–water content, Upton, WY, 1980 and 1982 average (Schuman & Sedbrook, 1984).

	Spoil depth, cm		
	0–20	20–40	40–60
Wood residue treatment	Water content		
——Mg ha^{-1}——	g kg^{-1}		
0	115a†	138a	139a
112	212b	166b	143a
224	232b	180b	155a

† Means among wood residue levels within a spoil depth followed by the same letter are not significantly different, $P \leq 0.05$.

ed into the leveled spoil to a depth of 30 cm. Plots were drill-seeded in mid-June 1979, with a mixture of wheatgrasses, forbs, and shrubs.

The addition of 112 and 224 Mg ha^{-1} of wood residue nearly doubled the soil–water content in the 0- to 20-cm spoil depth (Table 13-2). Soil–water at the 20- to 40-cm depth also was slightly greater in the amended plots compared to the control. The decreased effect of wood residue on soil–water at the deeper spoil depth is related to the fact that amendment incorporation only occurred to 30 cm; therefore, percolation below that depth was limited. The observed increase in soil–water storage allowed for good seed germination, seedling establishment, and forage production. Seeded species production was significantly improved by the addition of the wood residue amendment (Table 13-3). In 1980, no statistically significant differences in forage production across wood residue rates were evident, partially due to the variability in the data. However, the 112- and 224-Mg ha^{-1} wood residue treated plots averaged 386 kg ha^{-1} compared to 17 kg ha^{-1} for the control. In other years, the forage production of the amended plots was significantly greater than the control, and responded to the 28% above average precipitation in 1982. This preliminary study demonstrated the effectiveness of the wood residue amendment in enabling establishment of vegetation and subsequent forage production from abandoned bentonite mined lands.

Table 13-3. Seeded species aboveground biomass on bentonite mine spoils as affected by wood residue amendment, 1980 to 1983, Upton, WY (Schuman & Sedbrook, 1984).

	Wood residue rates (Mg ha^{-1})		
Year	0	112	224
	kg ha^{-1}		
1980	17a†	381a	392a
1981	15a	703b	554b
1982	12a	1006b	1332b
1983	3a	760b	1202b

† Means within a year followed by the same letter are not significantly differnet, $P \leq 0.05$.

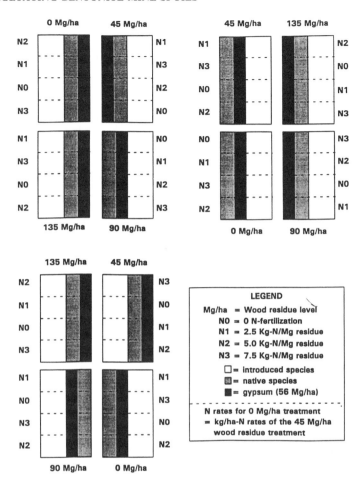

Fig. 13-1. Field plot design for evaluating the effectiveness of wood residue, N fertilizer, and gypsum amendments on ameliorating the physical and chemical properties of bentonite spoils.

ORGANIC AMENDMENT EFFECTS ON PLANT ESTABLISHMENT AND PRODUCTION AND BENTONITE SPOIL CHEMISTRY

Study Design

To better understand the long-term stability and management of these lands, information was needed that demonstrated the effects of wood residue amendments on the spoil material chemistry. Therefore, in 1981 a second study was established about 1 km from the preliminary study site. This study was conducted to refine the wood residue and N fertilizer requirements to ensure the most efficient amendment combination for revegetation, and to evaluate the treatment effects on the spoil chemistry. The study was a split-plot design with a split-block within the wood residue treatments (Fig. 13-1). Four wood residues rates (0, 45, 90, and 135 Mg ha^{-1}), four N fertilizer

Table 13-4. Perrenial grass production (averaged across species mixtures and N fertilizer treatments) in response to wood residue rate, Upton, WY 1983 to 1986 (Smith, 1984; Belden, 1987; unpublished data).

| | Wood residue rates (Mg ha^{-1}) | | | |
Year	0	45	90	135
		kg ha^{-1}		
1983	59a†	669b	1748c	2550d
1984	80a	361a	1220b	1956c
1985	10a	55a	148a	448a
1986	15a	116a	324a	886b

† Means among wood residue rate followed by the same letter are not significantly different, $P \leq 0.05$.

rates (0-, 2.5-, 5.0-, and 7.5-kg N Mg^{-1} wood residue) and two seed mixtures (native and introduced species) were established. Nitrogen was applied at the rate of 0, 112, 224, and 336 kg N ha^{-1} on the 0-wood residue treatment. These N rates were equal to those applied on the 45 Mg ha^{-1} wood treatment on a per hectare basis. Phosphorus was applied uniformly at the rate of 90 kg P ha^{-1}. Smith et al. (1985) provided complete details on the study design and implementation.

Plant Responses

Plant response to wood residue amendment in the Smith et al. (1985) study was similar to that observed in the preliminary study. Seedling density was significantly improved by wood residue amendments because of its positive effect on soil–water content, crusting, and bulk density. Seedling density was greater for the three wood residue treatments (41, 60, and 70 plants m^{-2}) than for the no-wood residue treatment (14 plants m^{-2}). However, there were no significant differences in seedling density between the 90 and 135 Mg ha^{-1} wood residue levels. Perennial grass production increased as wood residue rates increased, with maximum production occurring at the 135 Mg ha^{-1} wood residue levels (Table 13–4). Production in 1985 was extremely low compared to previous years because of an extreme drought that year. Precipitation from August 1984 through June 1985 (17 cm) was approximately 60% of normal for the period. However, normal precipitation occurred in late summer 1985 and throughout the 1986 growing season, resulting in the improved production observed in 1986. Visual observation of the plant community in 1986 indicated that some plant mortality had occurred as a result of the drought; therefore, predrought production levels may not be reached again for several years. Perennial grass production also responded to N fertilizer rates in 1983 and 1984 with peak biomass occurring at the 2.5 and 5.0 kg N Mg^{-1} of wood residue rates, respectively. In contrast, N had no appreciable effect on production in 1985 and 1986. The limited N response in these years was probably the result of the drought.

With the exception of tall wheatgrass (*Thinopyrum ponticum* (Podp.) Barkw. D.R. Dewey], all successfully established grass species in the initial

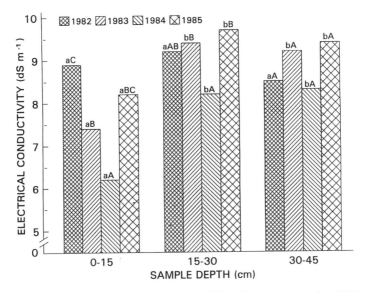

Fig. 13-2. Mean EC averaged across wood residue and N fertilizer treatments, for 1982 to 1985, at three spoil depths. Means with the same letters among years (lowercase) or within years (uppercase), are not significantly different, $P \leq 0.05$ (Belden, 1987).

growing season were rhizomatous (see Smith et al., 1986, for details of individual species response to the amendments). This suggests that sod-forming grasses are generally better suited than bunchgrasses for revegetation of bentonite spoils. The predominance of sod-forming grasses on clay soils in the region support this observation (Weaver & Albertson, 1956). Rhizomes have been noted to exhibit physical resistance to breakage and the capacity for regrowth and/or increased production from rhizomes if breakage occurs in a high clay soil (White & Lewis, 1969). Findings have demonstrated that plant species potentially useful and/or successful in revegetation of wood residue amended, abandoned bentonite spoils should have at least some of the following characteristics: sod-forming morpohlogy, drought and salt tolerance, adaptation to clay texture, and adaptation to a shallow, poorly drained spoil/soil environment (Smith et al., 1986).

Spoil Responses

Edaphic responses to wood residue amendment were exhibited in three ways: increased water infiltration and storage, decreased salinity due to leaching, and increased sodicity. The improved water storage as a result of wood residue amendment has been adequately discussed, using results of the earlier studies by Schuman and Sedbrook (1984). However, the increased water movement into the spoil in the second study demonstrated significant leaching of the soluble salts from the surface 15 cm of the spoil from 1981 to 1984 (Fig. 13-2). In 1985, the severe drought resulted in a significant upward migration of salts as a result of upward water movement in response

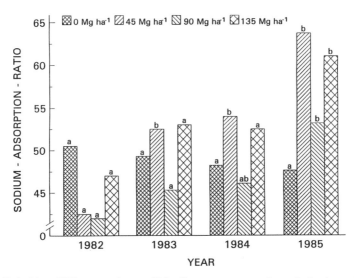

Fig. 13–3. Mean SAR averaged across N fertilizer treatments and sample depths, at four wood residue levels for 1982 to 1985. Means among years with the same lowercase letter are not significantly differnet, $P \leq 0.05$ (Belden, 1987).

to high evapotranspiration demands. Although upward salt migration occurred during the drought year, the electrical conductivity (EC) of the 0- to 15-cm spoil depth in 1985, nonetheless, did not exceed the 1982 levels observed before leaching occurred.

Leaching of soluble salts is desirable and necessary to encourage soil development and long-term stability of these lands. However, the leaching process resulted in an increase in spoil sodium-absorption-ratio (SAR) in the residue amended plots over time (Fig. 13–3). The pool of soluble Na (91% of soluble cations) was so large that leaching occurred, the relative proportion of Na in the system compared to Ca and Mg became greater, thereby increasing the SAR. This observed increase in SAR can have significant long-term effects on plant nutrition, spoil physical qualities, and subsequent maintenance of the vegetation community. The increasing SAR indicates that chemical amendments are necessary in addition to the wood residue, to ensure long-term reclamation and revegetation success. Such chemical amendments may include such products as gypsum, phosphogypsum, and calcium chloride.

Wood Residue Composition

Sustained long-term success of these reclaimed lands depends on the continued improvement and development of a "soil." In addition to amelioration of the saline-sodic condition and improved structure, "the new soil" must develop active microbial functions to ensure nutrient availability through sustained nutrient cycles. Evaluation of wood residue decomposition indicates that microbial functions have begun to develop. Wood residue decom-

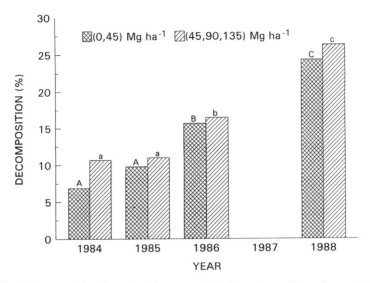

Fig. 13–4. Decomposition of wood residues amended to bentonite spoil for a 5-yr period. Data analyzed separately for 0- and 45-; and 45-, 90-, and 135-Mg ha^{-1} wood residue treatments. Means within the 0- and 45-; and 45-, 90-, and 135 Mg ha^{-1} treatments with the same upper- and lowercase letter, respectively, are not significantly different, $P \leq 0.05$ (Schuman & Belden, 1991).

position after 1, 2, 3, and 5 yr was 10.7, 11.0, 16.5, and 26.3%, respectively (Fig. 13–4). The single N addition in 1981 had a pronounced effect on decomposition during that 5-yr period (Fig. 13–5). Whitford et al. (1989) conclud-

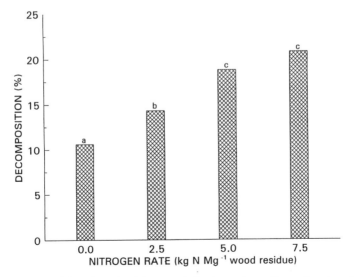

Fig. 13–5. Decomposition of wood residue amended to bentonite spoils as a function of N fertilizer application rate (averaged across all wood residue rates and years). Means with the same letter are not significantly different, $P \leq 0.05$ (Schuman & Belden, 1991).

ed that in semiarid rangelands where moisture availability affects N immobilization and mineralization, high C/N ratio amendments (such as wood residue) can be beneficial. They suggested that more resistant sources of organic mulches are superior to readily decomposed material because they provide a slow release of organic particles that serve as energy sources for the microflora. The wood residues have enabled successful revegetation of these lands, which adds readily decomposed material in root and litter turnover. These two sources of organic materials (plant material and wood residue) help to ensure long-term success through sustainable nutrient cycles.

AMELIORATION OF SODIC SPOIL WITH GYPSUM

An inorganic amendment study was designed to evaluate the feasibility and effectiveness of amending previously revegetated lands with gypsum. Gypsum was surface-applied, in April 1987, at the rate of 56 Mg ha^{-1} to approximately 40% of each of the native seed mixture plots of the 1981 study (Fig. 13-1). This level of gypsum amendment was calculated to reduce the exchangeable-sodium percentage (ESP) of the sodic spoil to 15. By using the existing 1981 study plots, we could evaluate the effects of gypsum on previously established vegetation, and obtain long-term baseline spoil data with which to determine the effectiveness of gypsum in alleviating the increasing sodicity problem.

Spoil samples were collected in October 1986 and in May 1988–1990 to evaluate the effects of gypsum. Gypsum amendment significantly increased EC at all spoil depths (Fig. 13-6). Such increases in salinity could result in reduction of germination and seedling establishment if applied during initial reclamation. Indeed, if gypsum had been applied during initial reclamation of this site, the EC would have been even higher in these spoils, since by 1987, leaching from the surface spoil depth had occurred over several years to an EC that was considerably below the initial EC of 13.4 dS m^{-1} in 1981 (Table 13-1). However, the observed increase would have been less if the gypsum had been incorporated into a larger volume of spoil. Such a large increase in salinity may have influenced seedling establishment in Dollhopf and Bauman's (1981) inorganic amendments study, since they observed fair-to-good seedling establishment when using deep chiseling with irrigation or woodchip amendment treatments. Even though the EC increased in 1988, it decreased significantly over the next 2 yr. This decrease is conaidered to be the result of leaching (Schuman & Meining, 1993). However, analysis of vegetation production data suggested no detrimental effects of the increased EC on the established plants. Perennial grass biomass averaged 175 and 317 kg ha^{-1} over the years 1988 to 1990 for the 0- and 56-Mg ha^{-1} gypsum treatments, respectively, indicating a significant enhancement of productivity by gypsum amendment (Schuman et al., 1994). Perennial grass biomass increased from 211 and 202 kg ha^{-1} in 1988 and 1989, to 325 kg ha^{-1} in 1990 (Schuman et al., 1994).

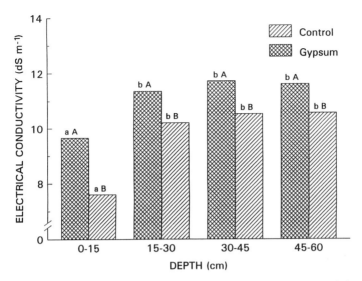

Fig. 13-6. The effect of gypsum amendment on the EC of wood residue-amended bentonite spoil at four spoil depths, 1988 to 1990. Means with the same letter within a treatment (lowercase) or within a depth (uppercase) are not significantly different, $P \leq 0.10$ (Schuman & Meining, 1993).

Gypsum treatment significantly reduced the ESP of the 60-cm spoil profile (Fig. 13-7). These changes began to become evident within the first 13 mo after treatment. The gypsum amendment also significantly increased the spoil–water storage in the 60-cm spoil profile (Fig. 13-8). The benefits ex-

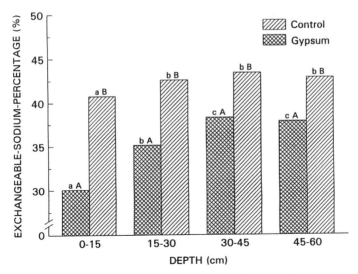

Fig. 13-7. The effect of gypsum amendment on the ESP of wood residue-amended bentonite spoil at four depths, 1988 to 1990. Means with the same letter within a treatment (lowercase) or within a depth (uppercase) are not significantly different, $P \leq 0.10$ (Schuman & Meining, 1993).

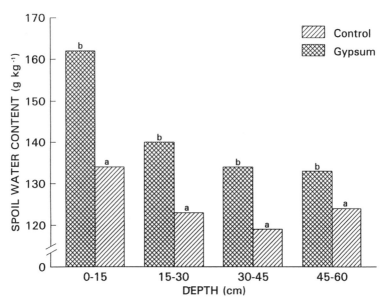

Fig. 13-8. Response of spoil-water content of revegetated saline-sodic bentonite mine spoil to gypsum amendment, 1988 to 1990. Means within a spoil depth with the same lowercase letter are not significantly different, $P \leq 0.10$ (Schuman & Meining, 1993).

hibited during the 3 yr documented by this phase of the study should be further enhanced with time.

SUMMARY

This paper has reviewed and summarized the findings of research leading to the development of a reclamation technology for abandoned bentonite mine spoils. Wood residue amendment resulted in immediate improvement of the physical characteristics of bentonite spoils, enabling improved water infiltration, leaching of salts, and the concurrent establishment of a desirable productive plant community. These studies also have pointed out the need for an inorganic amendment, such as gypsum, to be incorporated with the wood residue to replace Na in the system with Ca and prevent further sodication of the spoil. It is important to understand that inorganic amendments are complimentary to the wood residue (or other recalcitrant organic amendment that readily improves physical qualities) and are not substitutes for organic amendments. Nitrogen fertilizer also should be incorporated with the wood residue to provide adequate N for both plant growth and wood residue decomposition. Wood residue decomposition data indicated that some level of microbial function has developed in these spoils, helping to promote the long-term success of the reclamation.

This technology has been readily adaptable to large-scale reclamation of abandoned bentonite spoils. The Wyoming Abandoned Mine Land Program has reclaimed over 6000 ha of these spoils since 1985 using the technology. Average cost per hectare was $12 000, with amendments accounting for only 21.3% of the total cost (Richmond, 1991). Reclaiming these abandoned lands has eliminated a significant environmental problem (on- and off-site) and aesthetic issues in the region. Other states in the region are also developing programs to address abandoned bentonite spoil reclamation. Much of the information and technology developed from this research is directly applicable to contemporary mining and reclamation of bentonite mined lands even when topsoil salvage is practiced, especially since topsoil quantities are generally quite limited.

Although the research reviewed covered a period of 12 yr, further investigations of lands reclaimed using this technology is desirable to evaluate long-term stability under normal, postmining land uses. This would enable incorporation of new findings into the technology and reduce the chance of failure or deterioration of these lands based upon relatively short-term information.

REFERENCES

Ampian, S.G. 1980. Clays. p. 183–197. *In* Mineral facts and problems. U.S. Dep. Interior. Bureau of Mines Bull. 671. U.S. Gov. Print. Office, Washington, DC.

Belden, S.E. 1987. Edaphic responses to wood residue and nitrogen amendment of abandoned bentonite spoil. M.S. thesis. University of Wyoming, Laramie.

Bjugstad, A.J., T. Yamamoto, and D.W. Uresk. 1981. Shrub establishment on coal and bentonite clay mine spoils. p. 104–122. *In* Shrub establishment on disturbed arid and semiarid lands. Wyoming Game Fish Dep., Cheyenne, WY.

Dollhopf, D.J., and B.J. Bauman. 1981. Bentonite mine land reclamation in the Northern Great Plains. Montana Agric. Exp. Stn. Res. Rep. 179.

Dollhopf, D.J., R.C. Postle, R.B. Rennick, and S.A. Young. 1985. Chemical amendment and irrigation effects on sodium migration and vegetation characteristics in 7-years old sodic minesoils. Montana Agric. Exp. Stn. Res. Rep. SR-17.

Heather, A.R., and T.W. Hegarty. 1979. Sensitivity of germination and seedling radicle growth to moisture stress in some vegetable crop species. Ann. Bot. 43:241–243.

Hemmer, D.S., S. Johnson, and R. Beck. 1977. Bentonite mine related reclamation problems in the Northwestern States. Montana Dep. Lands, Old West Regional Grant no. 10570164.

Hira, G.S., M.S. Bajwa, and N.T. Singh. 1981. Prediction of water requirement for gypsum dissolution in sodic soils. Soil Sci. 131:353–358.

Kohnke, H. 1968. Soil physics. McGraw-Hill Book Co., New York.

Merrill, S.D., F.M. Sandoval, J.F. Power, and E.J. Doering. 1980. Salinity and sodicity factors affecting suitability of materials for mined lands reclamation. p. 3-1–3-25. *In* Adequate reclamation of mined lands?-A symposium. Billings, MT. 26–27 March. Soil Water Conserv. Soc. Am., Ankeny, IA.

National Academy of Sciences. 1974. Rehabilitation potential of western coal lands. Ballinger Publ. Co., Cambridge, MA.

Prather, B.J., J.O. Goertzen, J.D. Rhoades, and H. Frenkel. 1978. Efficient amendment use in sodic soil reclamation. Soil Sci. Soc Am. J. 42:782–786

Richmond, T.C. 1991. Reclamation of abandoned bentonite mines in Wyoming with wood-waste and calcium amendments to mitigate sodic soils. p. 455–478. *In* W.R. Oaks and J. Bowden (ed.) Proc. Natl. Meet. Am. Soc. Surface Mining Reclamation, Durango, CO. 14–17 May. Am. Soc. Surface Mining and Reclamation, Princeton, WV.

Schuman, G.E., and S.E. Belden. 1991. Decomposition of wood-residue amendments in revegetated bentonite mine spoils. Soil Sci. Soc. Am. J. 55:76–80.

Schuman, G.E., E.J. DePuit, and K.M. Roadifer. 1994. Plant responses to gypsum amendment of sodic bentonite mine spoil. J. Range Manage. 47:206–209.

Schuman, G.E., and J.L. Meining. 1993. Short-term effects of surface applied gypsum on revegetated sodic bentonite spoils. Soil Sci. Soc. Am. J. 57:1083–1088.

Schuman, G.E., and T.A. Sedbrook. 1984. Sawmill wood residue for reclaiming bentonite spoils. For. Prod. J. 34:65–68.

Sieg, C.H., D.W. Uresk, and R.M. Hansen. 1983. Plant-soil relationships on bentonite mine spoils and sagebrush-grassland in the northern high plains. J. Range Manage. 36:289–294.

Smith, J.A. 1984. Wood residue and fertilizer amendments for reclamation of orphan bentonite mine spoils. M.S. thesis, Univ. Wyoming, Laramie.

Smith, J.A., G.E. Schuman, E.J. DePuit, and T.A. Sedbrook. 1985. Wood residue and fertilizer amendment of bentonite mine spoils: I. Spoil and general vegetation responses. J. Environ. Qual. 14:575–580.

Smith, J.A., E.J. DePuit, and G.E. Schuman. 1986. Wood residue and fertilizer amendment of bentonite mine spoils: II. Plant species responses. J. Environ. Qual. 15:427–435.

U.S. Salinity Laboratory Staff. 1954. Plant response and crop selection for saline and alkali soils. p. 55–68. *In* L.A. Richards (ed.) Diagnosis and improvement of saline and alkali soils.. USDA Handb. no. 60. U.S. Gov. Print. Office, Washington, DC.

Weaver, J.E., and F.W. Albertson. 1956. Grasslands of the Great Plains. Johnson Publ. Co., Lincoln, NE.

White, E.M., and J.K. Lewis. 1969. Ecological effect of a clay soil's structure on some native grass roots. J. Range Manage. 22:401–404.

Whitford, W.G., E.F. Aldon, D.W. Freckman, Y. Steinberger, and L.W. Parker. 1989. Effects of organic amendments on soil biota on a degraded rangeland. J. Range Manage. 42:56–60.

14 Application of Municipal Sewage Sludge to Forest and Degraded Land

D. H. Marx, C. R. Berry, and P. P. Kormanik

USDA-Forest Service
Institute of Tree Root Biology
Athens, Georgia

Nearly 8 million dry tons (7 256 000 t) of municipal sewage sludge are produced each year in the USA by the more than 15 000 publicly owned treatment plants and the tonnage is increasing. Municipalities are faced with increasing cost of incineration, decreasing land for and increasing costs of landfills, and environmental concerns over ocean dumping. Under the circumstances, the attractiveness of sludge application for improvement of forest and degraded lands is obvious. About 25% of the municipal sludge being produced each year is being applied to land for its fertilizer and organic matter improvement value. For two decades, researchers in the USA have been studying the feasibility of land application of municipal sewage sludges. Research, large-scale practical projects, and commercial ventures have shown that stabilized sludge is an excellent organic amendment and chemical fertilizer for various plants. One disadvantage, however, is that sludge contains every element or compound found in wastes from domestic and industrial sources. Because of this, USEPA and certain states have developed guidelines and regulations governing sludge application on human food crops. No federal guidelines have been issued for use on forests or animal food crops (Bastian et al., 1982; Sopper, 1992).

Forests occupy some 298 million of the 916 million hectares of land in the USA (Powell et al., 1993). Commercial timberlands occupy nearly 200 million hectares. The remaining forest is in parks, wilderness areas, and otherwise noncommercial land. Most of the commercial timberland is in the northern and southern regions, and about 60% is in private nonindustrial ownerships (Table 14–1).

One of the major ecological benefits of forests is their ability to absorb impure water, filter it, and release high-quality water to streams, rivers, and reservoirs. This water cycle is mediated by certain properties of the canopies, but mostly by the forest floor and soil. Most forest soils have high infiltration rates with negligible overland flow. Unlike other soils, forest soils

Table 14-1. Commercial timberlands in the USA by ownership group and region (Powell et al., 1993).

Region	All ownerships†	National forests	Other public	Forest industry	Nonindustrial private
	ha (thousands)				
North	63 885	3 864	8 405	6 558	45 058
South	80 691	4 678	3 621	15 800	56 592
Rocky Mountains	25 355	14 738	2 424	1 181	7 013
Pacific Coast	28 267	10 996	4 509	4 985	7 777
USA	198 198	34 276	18 959	28 524	116 440

† Excludes over 100 million hectares of National Parks, wilderness and other noncommercial forest lands.

have a well-defined floor of organic litter, large porous channels caused by roots and animal activity, high amounts of decomposing organic matter, and an accumulation of woody debris on the surface. Associated with these features is a perennial root system that supports diverse macro- and microorganisms in the forest floor, soil, and rhizosphere (i.e., fine root surfaces). These organisms disintegrate and decompose organic matter and release nutrient elements for eventual absorption by forest vegetation. Nutrient and water cycles in forests are synchronous and interdependent.

The large forest land base in the USA and the physical, chemical and biological characteristics of forest soils make forests suitable for municipal sludge disposal. Additional advantages are that forests are often deficient in major nutrients found in sludges, and that forests are not major contributors to food chain crops for human consumption (Cole et al., 1983). Either liquid or solid (dewatered) sludges can be applied to forest stands of various ages or to harvested forest sites on which tree seedlings will soon be planted. Logic suggests that sludge application is best for sites with nutrient deficiencies or poor physical traits. There, the sludge is most likely to increase forest productivity and larger quantities of it can be applied than on sites of better quality. A considerable amount of research has been done in the USA since the early 1970s on application of liquid and dewatered sludges to forest land. The results show that proper application is environmentally safe and increases forest productivity. Volumes have been published on this topic (Cole et al., 1983, 1986; Bledsoe, 1981; Smith & Evans, 1977; Sopper & Kardos, 1973).

Land has been degraded by mining for metal ores, coal, kaolin, phosphate, bauxite, gravel, and sand. Serious degradation also has been caused by erosion that follows improper land-use practices. Of the several million hectares of degraded land in the USA, most was originally forest. Degraded land is characterized by the absence of vegetation and animal communities, loss of topsoil either by erosion or by deliberate removal, low organic matter content, and poor hydrologic properties. Any one of these characteristics can limit casual revegetation attempts. Degradable land has few of the attributes of the original forest soil. Such land will not be revegetated through normal plant succession in a reasonable time frame. Accelerated revegetation requires a new root growth medium, artificial introduction of adaptable plants, or alteration of soil physical, chemical, and microbiological conditions in which new plant communities can develop.

From 1930 through 1992, the mining industry extracted minerals from or deposited wastes on nearly 3.7 million hectares in 27 states in the USA. During this period, the mining industry reclaimed about 1.5 million hectares (Johnson & Paone, 1981; U.S. Dep. Interior, 1992). Additional millions of hectares are in borrow pits, where soil was taken for construction of roads, dams, airports, and other improvements.

During the past 20 yr, researchers have created a large body of knowledge on the feasibility of using municipal sludge for the revegetation of degraded land. Recently, Sopper (1992) published an excellent review on this subject. He discussed the effects of sludge on vegetation growth response; physical, chemical, and biological properties of the mined soil; soil percolation and groundwater quality; and animal health and nutrition. He concluded that municipal sludges, if applied properly according to present guidelines, can be used to revegetate mined lands in an environmentally safe manner with no major adverse effects on the vegetation, soil, groundwater quality, or animal and human health.

Since books have been published on the application of municipal sludges to forests and degraded lands in the USA, this chapter will cite only a few published studies as examples of this practice in the southeastern USA. Most of these studies were done at and funded by the Department of Energy (DOE), Savannah River Site (SRS), Aiken, South Carolina, in cooperation with the Savannah River Forest Station, United States Department of Agriculture-Forest Service (USDA-FS). Berry (1987) summarized results of many of these studies.

BORROW PIT APPLICATIONS

Many construction projects, such as dams, highways, and buildings require extensive earth fill to meet design criteria. When insufficient soil fill is available on the site it is necessary to "borrow" soil from another location. Generally, the entire A and B soil horizons have been removed from borrow pits. Compared with other disturbed areas such as surface mines, individual borrow pits may be relatively small, but collectively they represent a significant amount of surface area throughout the country. For example, borrow pits occupy about 1% of the DOE-SRS surface area of 81 000 ha.

Blauch (1978) discusses the problems of borrow-pit reclamation. Prescriptions for effective reclamation vary considerably with differences in soil type, slope, desired vegetation, climatic conditions, and other factors. In the past, most borrow pit revegetation consisted of little more than a single application of fertilizer, perhaps some scarification of the surface soil, and sowing grass seed or planting tree seedlings. It is doubtful, however, that such a minimal effort ever resulted in satisfactory reclamation of a borrow pit. On the SRS, located in the upper Coastal Plain of South Carolina, a great deal more effort appears to be needed because the exposed clay surfaces are highly compacted, eroded, impervious to root penetration and growth, and extremely low in available water, fertility, and organic matter.

Early borrow pit studies at the SRS showed that various grasses responded to N and P fertilization but not to K or lime applications (Granade, 1976). Planted slash and loblolly pine (*Pinus elliottii* Engelm. and *P. taeda* L.) seedlings responded to ammonium sulfate applications (McGregor & Goebel, 1968). However, the effects of these treatments lasted only a few growing seasons, then soil nutrients were depleted and plant growth slowed.

Three experiments were conducted on a large borrow pit at the SRS in 1976 to 1977. The borrow pit, which originally had Fuquay (loamy, siliceous, themic Arenic Plinthic, Kandiudults) and Wagram (loamy, siliceous, themic Arenic, Kandiudults) soils, was created in 1950 to 1952 by removing 2 to 6 m of soil leaving exposed a heavy clay. In 1953, reclamation was attempted by planting loblolly pine seedlings. In 1976, surviving trees were severely stunted, yellow, and their roots barely penetrated the soil surface. Many surviving trees were windthrown, but those still standing were only 2.5 to 5 m tall. Many standing trees had such limited root development and soil penetration that they could readily be pulled from the soil by hand. Although a thin layer of litter was present under some trees, no understory grasses or shrubs were growing to retard rapid surface runoff. The site was cleared of trees, subsoiled 0.9 m deep in opposing directions on 1.2-m centers, and double-disked. Municipal sludge used for the three studies had undergone secondary treatment with anaerobic digestion and sand-bed drying. It was obtained from treatment plants in Athens, Georgia, and contained about 2% N, 1% P, 0.5% K, 50% organic matter and less than 10 mg kg^{-1} Cd and 250 mg kg^{-1} Zn on a dry weight basis.

In the first study (Berry & Marx, 1980), sludge application was compared to fertilizer (560 kg ha^{-1} of 10-10-10) and lime (2242 kg ha^{-1} dolomitic limestone). Tree bark or bottom furnace ash treatments were applied to each of these. Sludge, bark and ash were applied at a rate of 125 m^3 ha^{-1}, or approximately 1.3 cm deep. For sludge, this rate was equal to 34 Mg ha^{-1} dry weight. After disking in amendments, the entire area was seeded to fescue (*Festuca arundinacea* Schreb.) in the fall of 1975 and planted the following spring with loblolly pine seedlings. Half of the seedlings had ectomycorrhizae formed naturally in the nursery, and half had *Pisolithus tinctorius* (Pers.) Coker & Couch ectomycorrhizae resulting from nursery soil inoculation (Marx et al., 1984).

After 3 yr, the effect of sludge overwhelmed effects of other amendments and specific ectomycorrhizae (Fig. 14-1). Seedlings without *P. tinctorius* ectomycorrhizae at planting formed this specific ectomycorrhiza from indigenous inoculum on site and thus precluded any ectomycorrhizal treatment effect. Seedling volumes (diameter squared times height, D^2H) were 28 times greater and grass biomass was five times greater on sludge plots than on nonsludge plots (Berry & Marx, 1980). At age five, the plots were thinned by removing every other tree. Fescue was no longer present in any plots by age five because canopy closure in sludge plots reduced light significantly and nonsludge plots were nutrient deficient. Trees in sludge plots af-

Fig. 14–1. Response of loblolly pine after 3 yr to amendments of municipal sewage sludge to a borrow pit in South Carolina. Stunted trees are growing in plots receiving a variety of other amendments and fertilizers but no sludge.

ter 10 yr were still dramatically larger in volume than trees on nontreated control and fertilized and limed plots (Berry, 1987). Compared to other plots, soils in sludge plots had more than two times as much N, more than 25 times as much P, and nearly three times more cation exchange capacity (CEC) and organic matter in the upper 20 cm of soil (Table 14-2). Increase in CEC was likely due to increased organic matter. Fertilizer and lime treatments were identical to control plots except for more Ca and Mg. We remeasured these trees at age 18 (October 1993), and found sludge effects persisting. Sludge increased tree volumes by more than 80-fold over those in control plots (Table 14-2). Trees in fertilized and limed plots were 10 times larger than control trees. Using standard site index curves, we estimate that trees on sludge plots will have a height for 60 ft (18.3 m) at age 25 (site index 60). On nonsludge plots, tree heights suggest that site index at age 25 will be 30 for the fertilized and limed plots and less than 20 for the nontreated control plots. In examining the age 3, 10, and 18 data, it becomes evident that the fertilizer and lime treatment effects have slowed considerably whereas sludge is still contributing to rapid growth after 18 yr.

    In the second loblolly pine study on this borrow pit, treatments were the same as the first except that bark and ash were not used and the nonamended control was not included. Other treatments were container-grown pine seedlings with *Pisolithus, Thelephora terrestris* (Ehrh.) Fr., or no ectomycorrhizae at planting (Ruehle, 1980). This study differed from the first in that seedlings were planted in the fall of 1977 rather than the spring of 1976. Amendments in this study were in soil without trees for two grow-

Table 14–2. Growth of loblolly pine after 18 yr and soil chemical properties after 10 yr on a borrow pit amended with sludge or fertilizer plus lime (Berry & Marx, 1980; Berry, 1978).

Treatment	Growth measurements after 18 yr*			
	Height	DBH	Tree volume	Site index†
	m	cm	$cm^3 \times 10^3$	
Control‡	4.1c	3.9c	4c	>20
Fertilizer & lime§	6.2b	7.5b	44b	30
Sludge¶	13.8a	15.4a	336a	<60

Treatment	Soil chemical properties (0–20 cm) after 10 yr*							
	N	P	K	Ca	Mg	pH	CEC	OM
	mg kg^{-1}					c mol kg^{-1}		g kg^{-1}
Control‡	174b	2b	30b	29c	17b	4.5ab	1.5b	3b
Fertilizer & lime§	142b	2b	35b	103a	59a	4.9a	1.9b	5b
Sludge¶	390a	55a	49a	76b	22b	4.3b	4.4a	13b

* Within columns, means sharing a common letter do not differ significantly at $P = 0.05$.
† Estimated height in feet at age 25 yr.
‡ Subsoiled one direction 2.5 m apart, disked and planted.
§ Subsoiled, 560 kg ha^{-1} of 10-10-10 fertilizer and 2242 kg ha^{-1} dolomitic limestone broadcast, disked and planted.
¶ Subsoiled, 34 Mg ha^{-1} (dry wt) dewatered sludge broadcast, disked and planted.

ing seasons and a winter (15 mo) compared to only the winter for the first study.

Two years after planting on sludge-amended plots, seedlings initially with *Pisolithus* ectomycorrhizae had greater height, root-collar diameter, and seedling volume (D^2H) than seedlings with no mycorrhizae or *Thelephora* ectomycorrhizae at planting (Fig. 14–2). *Pisolithus* seedlings had nearly four times more volume than other seedlings (Table 14–3). On fertilized plots, seedlings with *Pisolithus* ectomycorrhizae were larger than controls. There was no difference in survival among mycorrhizal treatments on the sludge plots.

After averaging the ectomycorrhizal treatments in the amendment treatments, the sludge effect on growth was over 20 times that of fertilizer and lime. The reason is that sludge plots had more than four times the N, nine times the P, and four times the organic matter found in the fertilizer and lime plots after 2 yr of tree growth and 3 yr of fescue growth. The poor growth performance of seedlings initially free of ectomycorrhizae shows again the importance of ectomycorrhizae on seedlings at planting, especially on stressed sites. The value of the stress adapted ectomycorrhizal fungus, *P. tinctorius*, in the sludge-amended soils also was demonstrated in this study. Apparently, the long time between adding amendments and seedling planting reduced the indigenous inoculum of *P. tinctorius* on the site so that ectomycorrhizal integrity was maintained on seedlings. The loss of integrity of the ectomycorrhizal treatments in the first study was attributed to indigenous *P. tinctorius* inoculum.

Fig. 14–2. Response of loblolly pine with specific ectomycorrhizae after 2 yr on a borrow pit amended with municipal sewage sludge in South Carolina. Seedlings had ectomycorrhizae formed by *Pisolithus tinctorius* (1), *Thelephora terrestris* (2), or no ectomycorrhizae (3) at planting.

Fig. 14-3. Response of sweetgum after 5 yr to amendments of municipal sewage sludge to a borrow pit in South Carolina. Stunted trees in foreground are growing in fertilizer and lime plots. Large trees in background are growing in sludge plots.

In the third study on this borrow pit, 0, 17, 34, or 68 Mg ha^{-1} of sludge were applied and incorporated as in the other studies. Plots were planted to sweetgum (*Liquidambar styraciflua* L.) in late winter of 1976.

During the first 5 yr (Kormanik & Schultz, 1985), sludge treatments significantly improved establishment of fescue and growth of sweetgum. However, all combinations of fescue with sludge significantly reduced 1st-yr survival of sweetgum, probably due to competition with fescue. Fifth-year height growth of sweetgum on the nonamended plots was only 0.7 m. On plots amended at even the lowest rate of sludge, height growth exceeded 2.75 m (Fig. 14-3). Heights (approximately 3.6 m) after 5 yr on plots amended at the highest sludge rate were equal to or greater than 5-yr heights reported for sweetgum on good quality reforestation sites (Kormanik, 1990). After 10 yr (Berry, 1987), treatment ranking remained the same but differences

Table 14-4. Growth of sweetgum and soil chemical properties after 10 yr on a borrow pit amended with different amounts of sewage sludge (Kormanik & Schultz, 1985; Berry, 1987).

Treatment	Growth measurements after 10 yr*		
	Height	Root collar diameter	Tree volume
	m	cm	$cm^3 \times 10^2$
Control†	0.63c	3.6b	2c
17 Mg/ha‡	4.11b	8.6a	36b
34 Mg/ha‡	5.49a	10.6a	71a
68 Mg/ha‡	5.57a	10.8a	74a

Treatment	Soil chemical properties after 10 yr*							
	N	P	K	Ca	Mg	pH	CEC	OM
	$mg\ kg^{-1}$						$c\ mol\ kg^{-1}$	$g\ kg^{-1}$
Control†	102b	2c	33b	65a	21a	4.6a	1.3c	3c
17 Mg/ha‡	368a	26b	47a	65a	17a	4.4b	5.6b	9b
34 Mg/ha‡	492a	43b	46a	58a	15a	4.2b	7.9a	13a
68 Mg/ha‡	596a	72a	52a	76a	21a	4.2b	7.9a	14a

* Within columns, means sharing a common letter do not differ significantly at $P = 0.05$.
† Subsoiled two directions 1.2 m apart, disked and planted.
‡ Subsoiled, dewatered sludge broadcast at 17, 34, or 68 Mg ha^{-1} (equivalent to 0.64-, 1.27-, or 2.54-cm deep) disked and planted.

in growth were greater (Table 14–4). The lowest amount of sludge increased seedling volumes by 18 times over that of controls. The two highest amounts of sludge increased volumes by over 35 times and did not differ from each other. Soil chemical properties after 10 yr showed the fertilizer value and soil improvement effects of sludge (Table 14–4). As in the other studies, sludge amendments improved amounts of N, P, K, and organic matter. Cation exchange capacity was increased by four times in the lowest sludge treatment and by more than six times in the two highest treatments, these increases were correlated with increases in organic matter. Sludge application increased N by nearly five times and P by 13 to 36 times, depending on the amount applied.

If the soil is compacted or has a shallow hardpan or other impervious subsoil layer, deep subsoiling is considered essential preparation for the establishment of trees. This practice is currently being used on reforestation sites and on degraded lands in the USA and other parts of the world (Berry, 1985). In 1979, a subsoiling and sewage sludge study with loblolly pine was installed on another borrow pit at the SRS. This site was originally overlain with Gunter sand (fine loamy, mixed thermic Aeric Ochruults), but removal of several meters of soil exposed a highly compacted deep clay. The stunted pines, planted 20 yr earlier, were removed with a bulldozer. Nine soil physical treatments were installed separately on plots amended with 17 Mg ha^{-1} sludge and 1120 kg ha^{-1} of fertilizer and 2242 kg ha^{-1} of lime. Physical treatments involved ripping soil to 45- or 90-cm depths, ripping on 1.1- or 2.2-m centers, or ripping in one direction or two opposing directions, and disking only. Half as much sludge and twice as much fertilizer and lime were applied as in the previous loblolly pine study. The source and chemical composition of the sludge were the same as the other borrow pit studies. Seedlings with abundant *P. tinctorius* ectomycorrhizae were planted on all plots in late winter 1979.

Root systems of several trees were excavated after 2 yr (Fig. 14–4). In all cases, root penetration was only as deep as the depth of tillage, 20 cm for disked plots and 45 or 90 cm for subsoiled plots (Berry, 1987). After 4 yr, trees on sludge plots grew an average of 37% more in height and 76% more in diameter breast height (DBH) than trees on fertilizer plots. Trees grew faster on fertilizer plots subsoiled 45 cm deep than on plots subsoiled 90 cm deep. Apparently, there was more leaching of nutrients and more available water in the deeper trenches. On sludge plots, trees grew faster when subsoiled 90 cm deep than when subsoiled 45 cm deep. Other variations in subsoiling, i.e., distance between furrows and whether in one direction or two, had little effect on growth of trees at this time. On sludge plots, all subsoiling treatments produced better growth than disking. Production of herbaceous biomass was significantly greater on sludge plots and on subsoiled fertilizer plots than on fertilizer plots that had only been disked.

At age seven (Berry, 1987), the sludge effect was still dominant. Regardless of soil physical treatments, trees on sludge plots had four times more volume than those on fertilizer and lime plots. Any type of subsoiling improved tree volumes by 30 to 50% compared to the disked treatment. Sludge plots at age seven had more than two times as much N, 10 times more P, twice the organic matter content, and three times the CEC as plots that were

Fig. 14–4. Above- and belowground response of loblolly pine after 2 yr to subsoiling in a borrow pit amended with municipal sewage sludge in South Carolina. Seedling on left is from a disced (20-cm-deep) plot. Seedlings in the middle and right are from plots subsoiled 45- and 90-cm deep, respectively.

fertilized (Table 14-5). Subsoiling did not affect soil chemical properties. By this time, essential nutrients from the fertilizer treatment were tied up in biomass and quantities remaining in the soil were not sufficient to maintain appreciable growth rates of trees. At age 14, we remeasured these trees. Tree volumes on sludge plots were five to six times greater than on fertilizer plots (Table 14-5). Depth of subsoiling also had significant effects on the sludge plots. Generally, sludge more than doubled the site index on this borrow pit. Not shown in Table 14-5 is windthrow. Few trees were windthrown or severely leaning on the subsoiled plots as compared to the disked plots. Improved tree anchorage due to deep root penetration following subsoiling is vital for long-term maintenance of trees on degraded lands such as borrow pits. A single heavy application of fertilizer and lime does little to maintain tree growth over the long term (i.e., 5-10 yr). Sludge, even at only 17 Mg ha^{-1}, can maintain growth for this period. However, twice this amount, as used in the earlier studies, assures greater tree productivity for a longer period due to greatly improved soil chemical properties.

Table 14-5. Growth of loblolly pine after 14 yr and soil chemical properties after 7 yr on a borrow pit amended with sludge or fertilizer plus lime and subsoiled to different depths (Berry, 1985, 1987).

Treatment	Growth measurements after 14 yr			
	Height	DBH	Tree volume	Site index‡
	m	cm	cm$^3 \times 10^3$	
	Disked			
Sludge‡	10.5a	12.9b	175b	60
Fertilizer & lime§	5.6b	7.1c	32c	28
	Subsoiled 45 cm deep¶			
Sludge‡	10.9a	13.6ab	206ab	62
Fertilizer & lime§	6.2b	7.6c	39c	28
	Subsoiled 90 cm deep††			
Sludge‡	11.8a	14.2a	242a	64
Fertilizer & lime§	6.1b	7.5c	38c	28

Treatment	Soil chemical properties (0-20 cm) after 7 yr*							
	N	P	K	Ca	Mg	pH	CEC	OM
	mg kg^{-1}						c mol kg^{-1}	g kg^{-1}
Sludge‡	219a	44a	32a	124a	22b	4.3a	6.2a	8a
Fertilizer & lime§	99b	4b	41a	127a	56a	4.8a	2.1b	4b

* Within columns, means sharing a common letter do not differ significantly at $P = 0.05$.
† Estimated height in feet at age 25 yr.
‡ Subsoiled, 17 Mg ha^{-1} (dry wt) dewatered sludge broadcast, disked and planted.
§ Subsoiled, 1121 kg ha^{-1} of 10-10-10 fertilizer and 2242 kg ha^{-1} dolomitic limestone broadcast, disked and planted.
¶ Subsoiled 45 cm deep, two directions on 1.3-cm centers.
†† Subsoiled 90 cm deep, two directions on 1.3-cm centers.

ERODED LANDS APPLICATION

The Tennessee Copper Basin is unique in the eastern USA in that the original mixed stands of oak and pine were eliminated on nearly 3000 ha and the surrounding 7000 ha were reduced to grasslands in the mid-to-late 1800s by air pollution resulting from primitive copper ore smelting. Trees were cut to produce fuel for heap roasting of iron pyrite ore. This process emitted SO_2 fumes, which killed remaining trees and natural regeneration from stump sprouts and seeds. Severe erosion of topsoil left deep gullies over thousands of hectares. Through the early 1970s, chronic air pollution continued to retard growth and reduce survival of trees planted by various agencies. Due to innovations in ore processing, SO_2 emissions rarely occur today. However, even with clean air, several other obstacles to reforestation exist. The surface soil is void of essential nutrients, organic matter, associated beneficial microorganisms, and compacted textural layers are present at shallow depths. Retention of soil water is low, and dry winds severely dessicate planted trees.

The Basin is in the transition zone between the Blue Ridge and Ridge and Valley physiographic provinces. It has a humid, mesothermal climate. Soils are principally Hapludults with some Palenaults. Since the mid-1970s, various techniques have been evaluated for reforesting the degraded Basin. Berry and Marx (1978) nearly doubled volumes of planted loblolly pine and Virginia pine (*P. virginiana* Mill.) after two growing seasons by planting seedlings with *P. tinctorius* ectomycorrhizae. Berry (1979) also found that a 90-g pellet of dried sewage sludge or a 9-g or 21-g commercial fertilizer tablet applied in the seedling closing-hole stimulated volume growth of loblolly pine by 87, 103, and 208%, respectively after 2 yr. None of these spot applications, however, had long-term effects on growth.

In the most significant study in the Basin, sludge was broadcast and incorporated into the soil (Berry, 1982). The study was installed on a severely eroded hilltop with less than 3% slope. Sludge from Athens, Georgia, comparable in chemical content to that described earlier, was broadcast at a rate of 34 Mg ha^{-1} before subsoiling 0.7 deep on 1.2-m centers and disking. For comparison, fertilizer (10-10-10 analysis) at 900 kg ha^{-1} and burnt lime (CaO) at 1400 kg ha^{-1} were applied and incorporated identically. The entire site was seeded to fescue and planted at high density (9000 seedlings ha^{-1}) to loblolly, Virginia, and shortleaf (*P. echinata* Mill.) pine seedlings in the spring of 1975. High-density planting of trees allows rapid domination of the prepared site, excludes undesirable vegetation by early canopy closure, obtains maximum utilization of soil amendments with rapid root closure, and controls erosion by reducing the velocity of rain through the closed canopy. Results after 4 yr (Berry, 1982) were as dramatic as those from the borrow pits. Sludge application increased tree volumes by more than four times on loblolly pine plots and more than three times on plots of the two other pine species. Distribution of aboveground biomass of the trees was affected by sludge at age five. McNab and Berry (1985) found that the proportion of total tree weight in wood was 8% higher on sludge plots than on others. The proportion of tree weight in foliage was lower on sludge plots. That difference suggests a greater photosynthetic efficiency of proportionally less foliage for trees growing in the sludge-amended soil or more allocation to roots in nonamended soils. Berry (1987) remeasured these trees and analyzed soil at age 10 and found the effect of the sludge still evident (Table 14-6). Tree volumes on sludge plots were still three to four times greater than on fertilizer plots for all pine species. Sludge plots contained twice as much N, three times as much organic matter and five times as much CEC. Soil P in the fertilizer plots after 10 yr was only 1 mg kg^{-1}, sludge plots had 18 mg kg^{-1}. There is little doubt that on such impoverished sites as the Copper Basin, sewage sludge helps to restore needed vegetation. Nevertheless, because of the public fear of sludge and its high transportation cost, sludge is not being used to reclaim the Copper Basin. Subsoiling on contour, application of inorganic fertilizer, and planting N-fixing trees are having some success in the Copper Basin.

Littleleaf disease is the most serious disease of shortleaf pine. The feeder root pathogen *Phytophthora cinnamomi* Rands, and other Pythiaceous fungi and nematodes are the biological causal agents (Otrosina & Marx, 1975). Severe erosion of surface soil also is a cause. This disease occurs on eroded sites on over two million hectares on the Piedmont Plateau in Alabama, Georgia, and South Carolina. Most littleleaf disease sites are void of top-

Table 14-6. Growth of pines and soil chemical properties after 10 yr on a severely eroded site in the Tennessee Copper Basin amended with sludge or fertilizer plus lime (Berry, 1982, 1987).

Treatment	Growth measurements*		
	Height	RCD	Tree volume
	m	cm	cm$^3 \times 10^2$
	Loblolly pine		
Sludge†	7.02a	13.4a	135a
Fertilizer & lime‡	4.50b	8.4b	40b
	Virginia pine		
Sludge†	6.08a	12.4a	103a
Fertilizer & lime‡	3.63b	7.2b	26b
	Shortleaf pine		
Sludge†	4.75a	7.7a	37a
Fertilizer & lime‡	3.25b	5.4b	14b

Treatment	Soil chemical properties (0–20 cm)*							
	N	P	K	Ca	Mg	pH	CEC	OM
	mg kg^{-1}						c mol kg^{-1}	g kg^{-1}
Sludge†	374a	18a	64a	61a	14a	4.3a	7.1a	10a
Fertilizer & lime‡	184b	1b	61a	82a	15a	4.7a	1.4b	.3b

* Within columns and species, means sharing a common letter do not differ significantly at $P = 0.05$.
† 34 Mg ha^{-1} (dry wt) dewatered sludge broadcast, subsoiled, disked, and planted.
‡ 900 kg ha^{-1} of 10-10-10 fertilizer and 1400 kg ha^{-1} CaO broadcast, subsoiled, disked, and planted.

soil, and exposed clays have poor internal drainage, low fertility, and low organic matter content. Berry (1977) subsoiled and applied solid sewage sludge to a high-hazard littleleaf disease site with the intent of eliminating soil edaphic factors contributing to the disease. Sludge from Athens, Georgia, was applied at 0, 17, 34, and 68 Mg ha^{-1}, 1 yr after subsoiling. One-half of the plots were subsoiled and all plots were disked. Midway through the first growing season after planting shortleaf and loblolly pine seedlings, growth of ragweed (*Ambrosia* spp.) and hairy crabgrass [*Digitaria sanquinalis* (L.) Scop.] was so great on the sludge plots that tree survival and productivity were minimal. Total weed biomass was five times greater on the high sludge plots than on controls. Effects of subsoiling could not be detected. This severely eroded littleleaf disease site was successfully revegetated by application of sludge, but the dominant plants were herbaceous plants or annuals rather than trees.

TREE NURSERY APPLICATION

Forest nurseries occupy approximately 3430 ha of land in the USA and produce two to three billion tree seedlings each year. Particularly in the South, tree nurseries must maintain high soil organic matter and nutrient content to grow high-quality seedlings (Davey & Krause, 1980). Sewage sludge is a good source of organic matter as well as nutrients. Many sludges, however, contain heavy metals and excessive amounts of salts that are potentially harmful to seedling growth (Bickelhaupt, 1980). Nevertheless, Berry (1981) obtained favorable results from a single application of 136 Mg ha^{-1} of dried sewage sludge in a Florida nursery. Slash pine (*P. elliottii* Englem. var. *elliottii*) seedlings were larger with sludge application than with the standard nursery applications of fertilizer. Screened compost (sewage sludge composted with wood chips) was used in Maryland for the production of high-quality hardwood seedlings (Gouin & Walker, 1977; Gouin et al., 1978). Bledsoe and Zasoski (1981) found that six conifers and two *Populus* species grew two to three times taller and produced two to five times more biomass in a soil/sludge mixture (3:1) than in soil alone.

Berry (1985) examined the nutritional and phytotoxic properties of several municipal sludges and their effects on ectomycorrhizal development and growth of loblolly pine in a microplot nursery test. He grew seedlings in fumigated soil amended with 34 or 68 Mg ha^{-1} of one of five sewage sludges: (i) old sludge from Athens, Georgia, (ii) fresh sludge from Athens, Georgia, and sludges of undetermined age from (iii) Aiken, South Carolina, (iv) Newberry, Florida, and (v) Chicago, Illinois. Newberry sludge and fresh Athens sludge supported growth comparable to the inorganic fertilizer control. Seedlings grown with Chicago sludge grew well at the 68 Mg ha^{-1} but were not as large as those grown with fertilizer. Seedlings grown with old Athens or Aiken sludge did not grow well and most of them were not of plantable size. *Pisolithus tinctorius* did not form ectomycorrhizae on seedlings grown with the sludges as well as on seedlings grown with fertilizer (in most cases <23% of short roots compared to 63% for controls). *Thelephora terrestris* formed ectomycorrhizae as readily on seedlings grown with the sludges as those grown with fertilizer when inoculation with the fungus was artificial. Natural ectomycorrhizae formed by *T. terrestris* from airborne spores, however, did not form as readily with some sludges as with fertilizer.

Somewhat surprising, in most cases seedlings with the highest concentration of heavy metals in tissues were not from sludge with the highest concentration of the specific heavy metals. An exception to this trend, however, was Cd. Its concentration was highest in seedlings grown with Chicago sludge, which had the highest concentration of Cd of all sludges tested. Zinc concentrations were high in seedlings grown in soil amended with Aiken sludge, which had the highest Zn concentration of any of the sludges. Chicago sludge, on the other hand, had virtually the same Zn content as Aiken sludge and induced growth of seedlings with significantly lower Zn concentrations in tissues.

During the one growing season of this study, Cd and Zn uptake by seedlings represented only about 1% of the amount present in the amended soil. Residual amounts remained high in soil. Berry (1985) concluded that municipal sewage sludges will support production of high-quality tree seedlings and could be used to supply organic matter and nutrients. He cautioned, however, that the lack of detrimental concentrations of certain heavy metals should be confirmed.

REGENERATION AND FOREST LAND APPLICATION

In the Pacific Northwest and North Central Regions of the USA, improvements in tree growth on forest sites from sludge application are well documented (Bledsoe, 1981). For southern tree species, however, there is little information about optimal rates of sludge application, stand ages when sludge application is beneficial, site characteristics for highest growth response, or possible interactions with indigenous pests. Three sludge studies at the SRS in South Carolina are noteworthy.

Stone and Powers (1989) amended a good-quality regeneration site (Typic Kandiudults, Orangeburg soil series) with anaerobically digested liquid municipal sewage sludge. Application rates of 0, 85 and 170 ML ha^{-1} provided total N at 0, 336, and 672 kg ha^{-1}. The low rate treatment contained 126, 67, 11, 60, 11 and 4500 kg ha^{-1} of NH_4-N, P, K, Ca, Mg, and organic matter, respectively; the high rate was twice these amounts. Loblolly pine seedlings of two genetic sources, one resistant to fusiform rust [*Cronartium quercuum* (Berk) Miyabe ex Shirai f. sp. *fusiforme*] disease, were planted in the winter of 1981. Results of combined data for both sources of loblolly pine after 6 yr were highly significant (Table 14-7). The sludge treatments increased height, diameter, and volume growth with no differences between the two sludge application rates. Associated with sludge treatments was increased competition from indigenous herbaceous vegetation which undoubtedly limited tree responses. Loblolly pine does not survive or grow well in

Table 14-7. Growth of loblolly pine after 6 yr on a good quality forest site amended with different amounts of liquid municipal sewage sludge (Stone & Powers, 1989).

Treatment	Height	DBH	Tree volume
	m	cm	$cm^3 \times 10^2$
Control	5.6a*	1.28b	9.5b
336 kg N ha^{-1}†	5.9a	1.38a	11.6a
672 kg N ha^{-1}‡	5.9a	1.40a	12.0a

* Within columns, means sharing a common letter do not differ significantly at $P = 0.05$.
† Anaerobically digested liquid municipal sewage sludge applied at approximately 85 ML ha^{-1}, disked and planted.
‡ Anaerobically digested liquid municipal sewage sludge applied at approximately 170 ML ha^{-1}, disked and planted.

its early years with dense competition from associated vegetation (Nelson et al., 1981; Knowe et al., 1985; Haywood & Melder, 1982). An unexpected effect of sludge treatment was observed on the incidence of fusiform rust disease. Treatments decreased rust infection on seedlings of both the nursery-run and the rust-resistant sources. Usually, cultural practices that stimulate seedling growth increase the incidence of rust disease on loblolly pine (Miller, 1977). Stone and Powers (1989) speculated that reduced disease associated with the sludge treatments may be due to changes in seedling physiology and susceptibility, or an interaction between increased herbaceous cover and reduced insect attack (Powers & Stone, 1988).

Kormanik and Berry (unpublished data) installed a solid sludge study with sweetgum on a forest site with Orangeburg soil. In July 1983, 34 Mg ha^{-1} of sludge from Athens, Georgia, were applied; 280 kg ha^{-1} diammonium phosphate were applied as the fertilizer treatment. In September, half of the amended and the control plots were subsoiled on 1.2-m centers, 0.76 m deep, and then all plots were disked. Sweetgum seedlings with or without endomycorrhizae were planted in the spring of 1984. Various herbicides were applied prior to planting sweetgum but were not effective in controlling competing vegetation. Plots were disked periodically for the first few years after planting to control weeds and herbaceous competition. Results after 2 yr were discussed by Berry (1987). Trees on both sludge and fertilizer plots were significantly larger than controls, but there were no differences in growth between the amendments. Trees on subsoiled plots also were significantly larger than those in disked plots when effects of soil amendments were combined. There was no effect of endomycorrhizal treatment. As in the other studies, soil chemical properties were significantly improved by applying sludge, even on this good-quality forest site (Table 14-8). Two years after application, there were no differences in soil analyses between the fertilizer and control treatments except for small differences in K. However, sludge plots had much higher soil concentrations of total N, available P, K and Ca and significantly higher CEC and organic matter. At age eight, the effect of sludge was still highly significant (Table 14-8), but the subsoiling effect was no longer evident. Sweetgum volume growth on sludge plots was twice that on fertilizer plots, and trees on control plots had less than half the volume of those

Table 14-8. Growth of sweetgum after 8 yr and soil chemical properties after 2 yr on a good quality forest site amended with sludge or fertilizer (Berry, 1987; Kormanik & Berry, unpublished data).

| Treatment | Growth measurements after 8 yr* | | |
	Height	DBH	Tree
	m	cm	$cm^3 \times 10^2$
Control	6.3c	7.4c	403c
Fertilizer‡	7.8b	10.0b	857b
Sludge§	9.0a	13.4a	1691a

| Treatment | Soil chemical properties (0–20 cm) after 2 yr* | | | | | | | |
	N	P	K	Ca	Mg	pH	CEC	OM
		mg kg^{-1}					c mol kg^{-1}	g kg^{-1}
Control	238b	31b	47b	111b	17a	5.0a	1.4b	16b
Fertilizer‡	248b	36b	42c	94b	15a	4.8a	1.3b	16b
Sludge§	605a	106a	57a	168a	18a	4.8a	2.2a	22a

* Within columns, means sharing a common letter do not differ significantly at $P = 0.05$.
† Data from subsoiled vs. disking treatments combined for all amendments.
‡ 280 kg ha^{-1} diammonium phosphate.
§ 34 Mg ha^{-1} (dry wt) dewatered sludge broadcast (2.5-cm-deep layer).

on fertilizer plots. Several different sweetgum biomass studies had been installed over the past 14 yr at the SRS on the better quality upland soils. In all cases, the best fertilizer practices have yielded 8.9 to 13.4 m^3 ha^{-1} yr^{-1} after 10 yr. This volume of biomass production compares favorably with that from intensive management of alder (*Alnus* spp.) and poplar (*Populus* spp.) over 12- to 15-yr rotations. It exceeds the 6.7 m^3 ha^{-1} yr^{-1} averaged obtained with loblolly pine on all site types on 25- to 30-yr rotations. With sludge in this study, however, yields of almost 27 m^3 ha^{-1} yr^{-1} were realized after only 8 yr. This is even more remarkable when one considers that sweetgum typically produces little volume for the first 2 to 3 yr after planting. Volume increases due to sludge in the last 5 to 6 yr of the study were tremendous.

The last sludge study we will discuss involved four ages of loblolly pine. The principal study objective was to evaluate alternative techniques for forest land disposal of sewage sludge. The goal was to increase biomass production in loblolly pine plantations without degrading the environment or wood quality. In 1980, single applications of 632 kg N ha^{-1} in solid aerobic sludge or 0, 402, or 804 kg N ha^{-1} in a liquid anaerobic sludge were made in plots in 1-, 3-, 8-, and 28-yr-old loblolly pine stands. The liquid sludge contained 7% total N, the solid sludge contained 1.3% total N. Liquid sludge came from Augusta, Georgia, and the solid sludge came from Aiken County, South Carolina. Prior to sludge applications, all physical and chemical characteristics of the trees and soils were determined. Over a 5-yr period, several subjects were studied.

For brevity, we cite the comprehensive report by Davis and Corey (1989), which describes in detail the results obtained by the various authors. After 4 yr, low rates of N increased diameter and basal area growth of the 8- and

28-yr-old trees. Higher rates did not result in greater increase. Solid sludge did not improve growth of the 28-yr-old trees. Solid and liquid sludge applied at plantation established (i.e., 1-yr-old) increased growth only modestly. Growth stimulation depended on control of insects and competing vegetation and incorporation of the sludge by disking. Incorporated solid sludge improved plantation establishment. Three-year-old trees did not respond to any sludge treatment, probably because competition from associated vegetation was intense. Liquid sludge reduced specific gravity of early wood in 8- and 28-yr-old trees but did not significantly change wood quality. Understory biomass production was increased by sludge applications in plantations of all ages. The herbaceous component showed the greatest response.

With a few exemptions, sludge treatments did not result in nitrate or heavy metal leaching into groundwater. Content of NO_3-N in soil water exceeded 60 mg L^{-1} at 1-m depth 4 mo after 800 kg N ha^{-1} were applied to plantations at age 0 and 28. Liquid sludge application of 400 kg N ha^{-1} or solid-sludge application of 630 kg N ha^{-1} did not increase NO_3-N above the 10 mg kg^{-1} standard for water quality in the 3-, 8-, and 28-yr-old stands. Nitrate-nitrogen was lower in stemflow, i.e., water flowing down stems, than in bulk precipitation or in throughfall. Ammonium-nitrogen was higher in stemflow of control plots than in any other samples except the high liquid treatment. In the 8- and 28-yr-old plots, both rates of liquid sludge increased litterfall weight 20 to 30% over control and solid–sludge treatments. Eighteen months after application, over 20% of the N from liquid sludge was in the forest floor of the 8-yr-old stand, 30% was in the forest floor of the 28-yr-old stand. In the 28-yr-old stand treated with solid–sludge, 67% of the N remained in the forest floor. The N increase in the forest floor of the 8-yr-old stand was caused by increased litterfall and understory development since there was little forest floor present at time of sludge application. The high retention percentage in the forest floor of the 28-yr-old stand is attributed to the presence of a well-established forest floor at time of application.

Eighteen months after treatment, the 8-yr-old stand had 40 to 70% of the Zn, Cu, Pb, Ni, Cd, and Cr from the liquid sludge in the forest floor. From 50 to 100% of these elements from liquid sludge were found in the forest floor of the 28-yr-old stand. The latter also contained 55 to 66% of these heavy metals from solid sludge.

Measurements of N availability in the 28-yr-old stand showed that concentrations in foliage, wood, fine roots, and needlefall all increased after sludge applications. Litterfall dry weight, total aboveground production, and N in litterfall were positively correlated with N availability. Nitrogen use efficiency decreased with sludge additions.

Sludge applications also affected soil biology. Soil mesofauna populations were reduced by liquid sludge in all stands, while solid sludge increased populations in the 28-yr-old stand. Low rates of sludge application increased cellulase activity while high rates decreased it.

Davis and Corey (1989) concluded that liquid sludge at rates of 400 kg N ha^{-1} or less is an effective silvicultural treatment to fertilize pulp and sawtimber stands of loblolly pine. Forest productivity was without significant environmental or wood-quality degradation. They suggested that pulpwood rotations could be shortened with low-level sludge applications at ages 8 to 10 and 12 to 14. Applications to young sawtimber (the 28-yr-old stand) could be made every 3 to 5 yr. Applications to stands less than 8 yr old (before crown closure) would dispose of sludge but would stimulate competing vegetation so much that rates of tree growth would decrease.

CONCLUSIONS

Application of liquid and solid municipal sewage sludge to forests and degraded lands has been studied extensively. The results clearly show that these properly applied and monitored treatments can physically, chemically, and biologically rehabilitate degraded lands. They also can increase productivity on regeneration sites and on established tree plantations. Depending on goals, competing vegetation may have to be controlled. Tremendous quantities of sludge could be recycled through the vast areas of forests and degraded land of the USA. Much of this land is close enough to municipalities for practical hauling of sludge. Recycling of the organic matter and the inorganic nutrients in sludge would eliminate a hugh waste disposal problem now faced by municipalities. Concerns over land application of sludges containing significant amounts of heavy metals undoubtedly will continue to be an issue. However, application of such sludges into degraded or forest soils greatly reduces the likelihood of these heavy metals entering the human and animal food chains. Application of these sludges to food crops or their disposal into landfills does not furnish these safeguards. Logic suggests that these nutrient-rich wastes should be applied to forests and degraded land. This approach offers environmentally sound recycling of the wastes and it enhances forest and soil productivity.

REFERENCES

Bastian, R.K., A. Montague, and T. Numbers. 1982. The potential for using municipal wastewater and sludge in land reclamation and biomass production as an I/A technology: An overview. p. 13–54. *In* W.E. Sopper et al. (ed.) Land reclamation and biomass production with municipal wastewater and sludge. Pennsylvania State Univ. Press, Univ. Park, PA.

Berry, C.R. 1977. Initial response of pine seedlings and weeds to dried sewage sludge in rehabilitation of an eroded forest site. USDA-FS Note SE-249. USDA-FS, Asheville, NC.

Berry, C.R. 1979. Slit application of fertilizer tablets and sewage sludge improve initial growth of loblolly pine seedlings in the Tennessee Copper Basin. Reclam. Rev. 2:33–38.

Berry, C.R. 1981. Sewage sludge effects soil properties and growth of slash pine seedlings in a Florida nursery. p. 46–51. *In* Proc. Southern Nursery Conf., Lake Barkeley, KY. 2–4 Sept. 1980. Tech. Publ. SE-TP-17. USDA-FS State & Private Forestry, Atlanta, GA.

Berry, C.R. 1982. Dried sewage sludge improves growth of pines in the Tennessee Copper Basin. Reclam. Reveg. Res. 1:195–201.

Berry, C.R. 1985. Subsoiling and sewage sludge aid loblolly pine establishment on adverse sites. Reclam. Reveg Res. 3:301–311.

Berry, C.R. 1987. Use of municipal sewage sludge for improvement of forest sites in the Southeast. USDA-FS Res. Pap. SE-266. USDA-FS, Asheville, NC.

Berry, C.R., and D.H. Marx. 1978. Effects of *Pisolithus tinctorius* ectomycorrhizae on growth of loblolly and Virginia pines in Tennessee Copper Basin. USDA-FS Res. Note SE-264. USDA-FS, Asheville, NC.

Berry, C.R., and D.H. Marx. 1980. Significance of various soil amendments to borrow pit reclamation with loblolly pine and fescue. Reclam. Reveg. Res. 3:87–94.

Bickelhaupt, D.H. 1980. Potential hazards of some organic wastes that may be used in nurseries. p. 166–173. *In* Proc. Workshop North Am. Forest Tree Nursery Soils. Syracuse, NY. 28 July–1 August. USDA-FS, Canadian For. Serv., and State Univ. of New York.

Blauch, E.W. 1978. Reclamation of lands disturbed by stone quarries, sand and gravel pits and borrow pits. p. 619–628. *In* R.W. Schaller, and P. Sutton (ed.) Reclamation of drastically disturbed lands. ASA, CSSA, and SSSA, Madison, WI.

Bledsoe, C.D., and R.J. Zasoski. 1981. Seedling physiology of eight tree species grown in sludge-amended soils. p. 93–100. *In* C.S. Bledsoe (ed.) Municipal sludge application to Pacific Northwest forest lands. Proc. Symp., Seattle, WA. 8–10 July 1980. College For. Resour., Univ. Washington, Seattle.

Bledsoe, C.S. (ed.) 1981. Municipal sludge application to Pacific Northwest forest lands. Proc. Symp., Seattle, WA. 8–10 July 1980. College For. Resour., Univ. Washington, Seattle.

Cole, D.W., C.L. Henry, W.L. Nutter (ed.) 1986. The forest alternative for treatment and utilization of municipal and industrial wastes. Proc. Symp., Seattle, WA. 25–28 June 1985. College For. Resour., Univ. Washington, Seattle

Cole, D.W., C.L. Henry, P. Schiess, and R.J. Zasoski. 1983. Forest systems. p. 125–143. *In* A.L. Page et al. (ed.) Proc. Workshop Utilization Wastewater and Sludge on land, Univ. of California, Riverside.

Davey, C.B., and H.H. Krause. 1980. Functions and maintenance of organic matter in forest nursery soils. p. 130–145. *In* Proc. Workshop, North Am. Forest Tree Nursery Soils. Syracuse, NY. 28 July–1 August. USDA-FS, Canadian For. Serv., and State Univ. of New York.

Davis, C.E., and J.C. Corey. 1989. Forest land application of sewage sludge on the Savannah River Plant (U.). DP-1763. Westinghouse Savannah River Co., Aiken, SC.

Gouin, F.R., C.B. Link, and J.F. Kundt. 1978. Forest seedlings thrive on composted sludge. Compost Sci. Land Util. 19:28–30.

Gouin, F.R., and J.M. Walker. 1977. Deciduous tree seedlings response to nursery soil amended with composted sewage sludge. Hort. Sci. 12:45–47.

Granade, G.V. 1976. Soil changes and plant response to lime and fertilizer on reseeded borrow pits. M.S. thesis. Univ. of Georgia, Athens.

Haywood, J.D., and T.W. Melder. 1982. How site treatments affect pine and competing plant cover. p. 224–230. *In* Proc. South. Weed Sci. Soc., New perspectives in weed science, Atlanta, GA. 19–21 January. WSSA, Champaign, IL.

Johnson, W., and J. Paone. 1981. Land utilization and reclamation in the mining industry, 1930–80. U.S. Dep. of Interior Inform. Circ. 8862. U.S. Gov. Print. Office, Washington, DC.

Knowe, S.A., L.R. Nelson, D.H. Gjerstad, B.R. Zutter, C.R. Glover, P.J. Minogue, and J.H. Dukes, Jr. 1985. Four-year growth and development of planted loblolly pine on sites with competition control. South. J. Appl. For. 9:11–15.

Kormanik, P.P. 1990. *Liquidambar styraciflua* L. p. 400–405. *In* R.M. Burns and B.H. Honkala (ed.) Silvics of North America. Vol. 2. USDA-FS Agric. Handb. 654. U.S. Govt. Print. Office, Washington, DC.

Kormanik, P.P., and R.C. Schultz. 1985. Significance of sewage sludge amendments to borrow pit reclamation with sweetgum and fescue. USDA-FS Res. Note SE-329. USDA-FS, Asheville, NC.

Marx, D.H., C.E. Cordell, D.S. Kenney, J.G. Mexal, J.D. Artman, J.W. Riffle, and R. Molina. 1984. Commercial vegetative inoculum of *Pisolithus tinctorius* and inoculation techniques for development of ectomycorrhizae on bare-root tree seedlings. For. Sci. Monogr. 25.

McGregor, W.H.D., and N.B. Goebel. 1968. Effectiveness of nitrogen fertilizers and mulch for the amelioration of severe planting sites. p. 65–72. *In* C.H. Youngberg and C.B. Davey (ed.) Proc. 3rd North Am. For. Soils Conf., Oregon State Univ. Press, Corvallis.

McNab, W.H., and C.R. Berry. 1985. Distribution of aboveground biomass in three pine species planted on a devastated site amended with sewage sludge or inorganic fertilizer. For. Sci. 31:373–382.

Miller, T. 1977. Fusiform rust management strategies in concept. p. 110–115. *In* R.J. Dinus and R.A. Schmidt (ed.) Proc. Symp. Manage. of Fusiform Rust in Southern Pines, Gainesville, FL. 7–8 Dec. 1976. USDA-FS, Atlanta, GA.

Nelson, L.R., R.C. Pedersen, L.L. Autry, S. Dudley, and J.D. Walstad. 1981. Impacts of herbaceous weeds in young loblolly pine plantations. South. J. Appl. For. 5:153-158.

Otrosina, W.J., and D.H. Marx. 1975. Populations of *Phytophthora cinnamomi* and *Pythium* spp. under shortleaf and loblolly pines in littleleaf disease sites. Phytopathology 65:1224-1229.

Powell, D.S., J.L. Faulkner, D.R. Darr, Z. Zhu, and D.W. MacCleery. 1993 Forest Resources of the United States, 1992. USDA-FS Rocky Mountain For. and Range Exp. Stn., Gen. Tech. Rep. RM-234.

Powers, H.R. Jr., and D.M. Stone. 1988. Control of tip moth by carbofuran reduces fusiform rust infection on loblolly pine. USDA-FS Res. Paper SE-270. USDA-FS, Asheville, NC.

Ruehle, J.L. 1980. Growth of containerized loblolly pine with specific ectomycorrhizae after 2 years on an amended borrow pit. Reclam. Rev. 3:95-101.

Smith, W.H., and J.O. Evans. 1977. Special opportunities and problems in using forest soils for organic waste application. p. 429-454. *In* J.F. Elliott and F.J. Stevenson (ed.) Soils for management of organic works and waste waters. ASA, CSSA, and SSSA, Madison, WI.

Sopper, W.E. 1992. Reclamation of mine land using municipal sludge. p. 351-431. R. Lal and B.A. Stewart (ed.) Advances in soil science. Vol 17. Springer-Verlag, New York.

Sopper, W.E., and L.T. Kardos. (ed.) 1973. Recycling treated municipal wastewater and sludge through forest cropland. Proc. Sympos., University Park, PA. 21-24 Aug. 1972. Pennsylvania State Univ. Press, University Park, PA.

Stone, D.M., and H.R. Powers. 1989. Sewage sludge increases early growth and decreases fusiform rust infection of nursery-run and rust resistant loblolly pine. South. J. Appl. For. 13:68-71.

U.S. Department of Interior. 1992. Surface coal mining reclamation: 15 years of progress. 1977-1992. Part I. U.S. Gov. Print. Office, Washington, DC.

NOTES

NOTES

NOTES

NOTES

NOTES